# Studies in Computational Intelligence

Volume 623

**Series editor**

Janusz Kacprzyk, Polish Academy of Sciences, Warsaw, Poland
e-mail: kacprzyk@ibspan.waw.pl

## *About this Series*

The series "Studies in Computational Intelligence" (SCI) publishes new developments and advances in the various areas of computational intelligence—quickly and with a high quality. The intent is to cover the theory, applications, and design methods of computational intelligence, as embedded in the fields of engineering, computer science, physics and life sciences, as well as the methodologies behind them. The series contains monographs, lecture notes and edited volumes in computational intelligence spanning the areas of neural networks, connectionist systems, genetic algorithms, evolutionary computation, artificial intelligence, cellular automata, self-organizing systems, soft computing, fuzzy systems, and hybrid intelligent systems. Of particular value to both the contributors and the readership are the short publication timeframe and the worldwide distribution, which enable both wide and rapid dissemination of research output.

More information about this series at http://www.springer.com/series/7092

Vassil Sgurev · Ronald R. Yager
Janusz Kacprzyk · Vladimir Jotsov
Editors

# Innovative Issues in Intelligent Systems

*Editors*
Vassil Sgurev
Institute of Information and Communication Technologies
Bulgarian Academy of Sciences
Sofia
Bulgaria

Janusz Kacprzyk
Systems Research Institute of the Polish Academy of Sciences
Intelligent Systems Laboratory
Warsaw
Poland

Ronald R. Yager
Iona College
Machine Intelligence Institute
New Rochelle, NY
USA

Vladimir Jotsov
University of Library Studies and Information Technologies
Sofia
Bulgaria

ISSN 1860-949X ISSN 1860-9503 (electronic)
Studies in Computational Intelligence
ISBN 978-3-319-27266-5 ISBN 978-3-319-27267-2 (eBook)
DOI 10.1007/978-3-319-27267-2

Library of Congress Control Number: 2015958079



Printed on acid-free paper

This Springer imprint is published by SpringerNature
The registered company is Springer International Publishing AG Switzerland

# Preface

In this book, ‘Innovative Issues in Intelligent Systems,’ a broad variety of different contemporary IT methods and applications is displayed. Every book chapter represents a detailed, specific, far-reaching, and original research in a respective scientific and practical field. However, all of the chapters share the common point of strong similarity in a sense of being innovative, applicable, and mutually compatible with each other. In other words, the methods from the different chapters can be viewed as bricks for building the next generation “thinking machines” as well as for other futuristic logical applications that are rapidly changing our world nowadays.

In chapter “Intelligent Systems in Industry,” a realistic overview of intelligent systems in industry has been presented by Arthur Kordon (USA). The author has an extended experience from applying these systems in a large global corporation. The main fields of research presented in this chapter are the neural networks, fuzzy logic, evolutionary computation, swarm intelligence, and intelligent agents. Here the main outcomes are in the areas of big data, data analytics, market analysis, product discovery, process monitoring and control, supply chain and their various financial and other applications. The main factors of success in key intelligent systems application areas have been broadly discussed.

Jawad Shafi, Plamen Angelov, and Muhammad Umair (UK, Spain, Pakistan) in their chapter “Prediction of the Attention Area in Ambient Intelligence Tasks” introduce the ANFIS aiming at prediction the human attention area on visual display with ordinary web camera and ambient intelligence applications. The proposed system should be able to make inferences in an interactive way with the user. Here the main objective is to adopt a suitable learning method for eye-gaze identification. For this purpose a hybrid learning algorithm has been applied. Simulated games are used for ANFIS training. The problem of prediction has been successfully solved. As a result, ANFIS appears as a better choice compared to the linear regression.

In the chapter entitled “Integration of Knowledge Components in Hybrid Intelligent Control Systems,” the authors Mincho Hadjiski and Venelina Boishina (Bulgaria) show some novel results for design of hybrid intelligent control systems

(HICS). This research is focused on the HICS design based on loosely coupled building blocks in order to be used for control in hierarchical and distributed structures. The integration problems are explored in HICS with a loose integration in the directions : ‘Consideration the HICS with more than two intelligent building blocks, with emphases on knowledge oriented elements—Ontology (O), Case–Based Reasoning (CBR), Rule-Based Reasoning (RBR)’; ‘Structural problems of HICS in dependence on control and behavioral goals, information and available knowledge’; and ‘Investigation of the building blocks blending methods in order to overcome their individual limitations and disadvantages.’ The results are considered as a promising way for creating large number of new hybrid intelligent control systems with increased solving capabilities and improved performance and robustness.

Georgi Dimirovski (MK, Turkey) in his chapter “Learning Intelligent Controls in High Speed Networks: Synergies of Computational Intelligence with Control and Q-Learning Theories,” has introduced efficient design solutions for learning intelligent flow controls in high-speed networks. Here a simple but efficient Q-learning model-independent flow controller is proposed. It possesses the ability to predict the network behavior thus improving the performance results.

In chapter “Logical Operations and Inference in the Complex s-Logic,” the author Vassil Sgurev (Bulgaria) introduces imaginary, *i*-logic, four-valued, and complex (summary) *s*-logic, six-valued, which to some extent ‘upgrade’ the classical real logic, here referred as *r*-logic. New logical structures are proposed in which through imaginary logical variables the classical propositional logic is extended. A possibility is provided for problems to be solved which are unsolvable in the framework of the classical propositional logic. This new class of logic may be used in the realization of real application models.

In chapter “Generalized Nets as a Tools for the Modelling of Data Mining Processes,” the author Krassimir Atanassov (Bulgaria) considers the possibilities of the apparatus of the generalized nets as tools for modeling of data mining processes. These nets can be considered as a major extension and a modification of the Petri nets that have clear advantages from a generalization and structural viewpoint. In this way, the described research on the generalized nets is a step forward compared to the currently existing similar approaches in the literature.

The next chapter “Induction of Modular Classification Rules by Information Entropy Based Rule Generation,” is written by Han Liu and Alexander Gegov (UK). Here the replicated sub-tree problem has been introduced and the well-known Prism algorithm has been further developed in order to avoid the existing limitations of the current Prism algorithm. For this reason, a new rule generation method IEBRG using the separate and conquer approach has been introduced. The experimental study in this chapter has shown that the IEBRG has the potential to avoid under-fitting of rule sets and to generate fewer, but more general rules as well as to perform relatively better in dealing with clashes.

Vladimir Jotsov (Bulgaria) in chapter, “Proposals for Knowledge Driven and Data Driven Applications in Security Systems,” has presented the research on data analytics for big data, multiagent, and other security applications. The main synthetic methods are named Puzzle, Funnel, Kaleidoscope, and Conflict. They evolve under the cover of the evolutionary metamethod Frontal.

Krassimir Atanassov (Bulgaria) and Janusz Kacprzyk (Poland) are the authors of the next chapter, "On Some Modal Type Intuitionistic Fuzzy Operators." In it, six new modal operators are introduced and some of their basic properties are discussed.

In chapter entitled "Uncertain Switched Fuzzy Systems: A Robust Output Feedback Control Design" the authors, Vesna Ojleska, Tatjana Kolemishevska-Gugulovska (MK), and Imre Rudas from Hungary, discuss the problem of robust output feedback control for a class of uncertain switched fuzzy time-delay systems using T-S fuzzy models. Controllers are designed by employing the parallel distributed compensation strategy.

In chapter entitled "Multistep Modeling for Approximation and Classification by Use of RBF Network Models," the author Gancho Vachkov (Bulgaria) describes the concept and the learning algorithms for creating two types of nonlinear radial basis function network (RBFN) models. Both of the models are able to gradually improve their performance by enlarging their complexity through the learning process. The Growing RBFN models are learned by adding a new RBF kernel at each optimization step thus increasing the structural complexity of the model and improving its accuracy. The Incremental RBFN model is a composite model that consists of several sub-models, trained sequentially. The first one is a simple linear model, while all others are small nonlinear RBFN models that in connection with all previously created sub-models gradually decrease the approximation error. Particle swarm optimization algorithm is used for tuning all the parameters of the models. Applications in the area of nonlinear approximation and classification are also given and discussed in the chapter.

Vassil Sgurev<br>
Ronald R. Yager<br>
Janusz Kacprzyk<br>
Vladimir Jotsov

# Contents

# Intelligent Systems in Industry

## A Realistic Overview

Arthur Kordon

**Abstract** The objective of this paper is to give a realistic overview of the current state of the art of intelligent systems in industry based on the experience from applying these systems in a large global corporation. It includes a short analysis of the differences between academic and industrial research, examples of the key implementation areas of intelligent systems in manufacturing and business, a discussion about the main factors for success and failure of industrial intelligent systems, and an estimate of the projected industrial needs that may drive future applications of intelligent systems.

## 1 Introduction

Intelligent systems, based on neural networks, fuzzy logic, evolutionary computation, swarm intelligence, and intelligent agents, show a growing number of applications, mostly in large global corporations, such as General Electric [4], Ford [10], The Dow Chemical Company [15], etc. Some areas with successful application record in manufacturing are inferential sensors, intelligent control, and complex supply-chain scheduling. Recently, there is a trend of growing implementations of intelligent systems in the business application areas of data mining and advanced analytics [24]. Despite these success stories, however, intelligent systems are virtually unknown to industry at large. Promoting this emerging technology for real-world applications on a broader scale to middle-size and small businesses is a serious challenge, having in mind the high competition from other established modeling techniques, based on first-principles, statistics, and classical optimization. In order to make this step more effective, a realistic analysis of the current state of the art of intelligent systems industrial applications is needed, focusing on their

A. Kordon (✉)
Advanced Analytics Group, The Dow Chemical Company,
2301 N Brazosport Blvd, Freeport, TX 77566, US
e-mail: arthur@kordon-consulting.com



V. Sgurev et al. (eds.), *Innovative Issues in Intelligent Systems*,
Studies in Computational Intelligence 623, DOI 10.1007/978-3-319-27267-2_1

advantages, disadvantages, and potential for future growth. The objective of this paper is to give a realistic overview of the current intelligent systems applications based on the experience of large manufacturing corporations.

The paper is organized as follows. First, the differences between academic and industrial research, which play a major role in understanding the principles of transforming mature academic ideas into industrial applications are clarified. Second, the key intelligent systems application areas in manufacturing and business are briefly described, followed by a summary of the key factors for success and issues in applying intelligent systems. Third, some projected industrial needs related to intelligent systems are briefly discussed in the last section of the paper.

## 2 Differences Between Academic and Industrial Research

The first step in improving the efficiency in promoting a mature research idea into practice is a good understanding of the current mode of operation of industrial R&D. To begin with the drastic changes in doing applied science in the last 10 years. Once upon a time, there were industrial laboratories, such as Bell Lab, with broad scientific focus and purely academic mode of operation. Not anymore. Globalization and the push of shareholders for maximal profit has significantly transformed applied research in the direction of perpetual cuts and short-term focus. The crusade for cost reduction imposed a new strategy for applying emerging technologies, such as intelligent systems, which assumes minimal exploratory efforts. As a result, the new methodology for applied research must significantly reduce the risk and the time for introducing new technologies. At the basis of the proposed methodology, fully described in [15] are the following key components: value creation, optimal integration of different modeling methods, and a proper balance among applied methods, needed infrastructure, and people.

### *2.1 Value Creation as the Key Industrial Research Objective*

At the basis of the different mode of operation between academic and industrial research is the factor of value. The key objective of academic research is to create new knowledge at any level of Nature (from quantum to the Universe) and the definition of success is the quality and quantity of publications. In contrast, the key objective of industrial research is to create value by exploring and implementing knowledge from almost any level of Nature. The definition of success is the increased profit and improved competitive position in the market.

Since the value factor in industry is a question of survival, it has a dominant role and dictates different ways of doing research than in the academic world. University professors can satisfy their curiosity at any level and depth of the knowledge ladder even without funding. Industrial researchers don't have this luxury and must concentrate on those levels of the knowledge ladder where the profit is maximal. As a result, assessment of value creation in almost any phases of industrial research is a must. Unfortunately, this important fact is practically ignored in the literature. Some guidance how to estimate the value of a research method is given in [15].

## 2.2 Integrate and Conquer

Since the value creation capability is the key driving force in real-world applications, the strategy for applying intelligent systems is based on factors like minimizing modeling cost and maximizing model performance under a broad range of operating conditions. An obvious result of this strategy is the increased efforts in robust empirical model building, which is very often at an economic optimum. Unfortunately, the robustness of the empirical solutions, i.e. their capability to operate reliably during minor process changes, is difficult to accomplish with one modeling method only. Very often the nasty reality of real-world problems requires the joint work of several modeling techniques. In order to meet this need, it is necessary to develop a consistent methodology that effectively combines different modeling approaches to deliver high quality models with minimal efforts and maintenance. It is based on the almost infinite number of ways to explore the synergetic benefits between the modeling methods. The result is an integrated methodology which significantly improves the performance and compensates the disadvantages of the corresponding individual methods [14].

The good news about building an integrated system is that the ways to explore the synergistic capabilities of its components are practically unlimited. A clear example is when a neural network or a fuzzy system model optimizes its structure using a genetic algorithm. Exploring the integration potential of intelligent systems gives tremendous application opportunities at a relatively low cost and only a slightly complicated model development process. The bad news is that it is practically unrealistic to devote theoretical efforts to analyze all of possible combinations. As a result, most of the designed and used integrated systems lack a solid theoretical basis. However, the experience from several big multinational companies, such as General Electric [4], Ford [10], The Dow Chemical Company [15], demonstrate the broad application of integrated systems in different areas of manufacturing, new product design, and financial operations.

From application point of view, three layers of integration are important. The first layer includes the synergy between the intelligent systems methods themselves, the second layer integrates intelligent systems methods with first-principles models, and the third layer explores the bridges between intelligent systems and statistics. Examples of these layers are given in [15].

The motto for successful intelligent systems applications is clear: Integrate the modeling methods and conquer the real world. From our experience it is the winning strategy which opens the door of industry to intelligent systems.

## 2.3 The Troika

The winning application strategy is also based on three key components—people, scientific methods, and infrastructure. We call it the Troika of successful applied research. The first component (People) represents the most important factor—the people involved in the whole implementation cycle, such as researchers, managers, programmers, and different types of final users. The second component (Methods), where most of the current attention in literature is focused, includes the theoretical basis of intelligent systems. The third component (Infrastructure) represents the necessary infrastructure for implementing the developed intelligent systems solution. It includes the needed hardware, software, and all organizational work processes for development, deployment, and support. Promoting this broader view of intelligent systems and especially clarifying the critical role of the human component for the success of practical applications is the leading philosophy in applied research.

## 2.4 What is a Successful Intelligent System Application?

From the practical point of view, the most valuable feature of human intelligence is the ability to make correct predictions about the future based on all available data from the changing environment. We define applied intelligent system as a system of methods and infrastructure that enhance human intelligence by learning and discovering new patterns, relationships, and structures in complex dynamic environments for solving practical problems.

The definition of success for applied intelligent systems, according to the definition, is enhanced human intelligence as a result of the continuous adaptation and automatic discovery of new features in the applied system. The knowledge gained may deliver multiple profit-chains from new product discovery to almost perfect optimization of very complex industrial processes under significant dynamic changes.

It is a common perception that human intelligence is the driving force for economic development, productivity gains, and, at the end, increasing living standards. From that perspective, enhancing human intelligence by any means increases the chances for more effective value creation. Of special importance are the recent challenges from the data avalanche in the Internet and the dynamic global market operating in high fluctuations of energy prices and potential political instabilities. The fast emerging area of Big Data, based on extracting patterns and business insight from smart phones, social networks, such as Facebook and Tweeter, RFID tags, pictures, videos, etc., has captured the attention of industry [23]. As a result, human intelligence had to learn how to handle an enormous amount of information and make the right decisions with a high level of complexity in real time. A significant factor is the growing role of data analysis, pattern recognition, and learning. Intelligent systems can play a critical role in this process.

# 3 Key Intelligent Systems Application Areas

Intelligent systems have been applied and created value in many areas (see a comprehensive overview in [15]. In this section we'll focus on the key applications in manufacturing and business.

## *3.1 When Is an Intelligent System the Right Solution?*

Some answers to this question that may open or close the door to potential intelligent systems applications are discussed briefly below:

### 3.1.1 Competitive Advantage

The most important criterion for applying intelligent systems is the competitive advantage it gives to the user. It will be extremely difficult to sustain long-term success if there is no clear vision of the sources of competitive advantage in implementing this emerging technology in specific business areas.

### 3.1.2 Complex Problems

Often practitioners look at intelligent systems after exhausting the capabilities of other methods. The key reason is the high level of complexity of the problem, especially related to nonlinear interactions. This category also includes problems with very high dimensionality and any type of social systems modeling. The methods that contribute mostly to complexity modeling are intelligent agents, neural networks, support vector machines, and evolutionary computation.

### 3.1.3 High Level of Uncertainty

Another issue that could be resolved by using intelligent systems is the imprecise nature of available expertise combined with inaccurate data. Fuzzy systems, especially in combination with neural networks and intelligent agents can reduce this type of uncertainty and build reliable models. Another form of uncertainty is the minimal a priori assumptions required for model development. In contrast to first-principles and statistical models, neural networks, support vector machines, or evolutionary computation do not need strict assumptions for building models. In addition, machine learning allows reducing uncertainty through continuously learning about the changes in the environment and updating model parameters accordingly.

### 3.1.4 Novelty Generation

Intelligent systems are very effective for businesses where innovation is critical. Intelligent agents can capture emergent behavior from local interactions, which could be identified as novelty. Evolutionary computation automatically generates novel structures, such as electric circuits, optical lenses, and control systems. Neural networks, support vector machines, and evolutionary computation capture unknown dependencies between variables that could be used for process monitoring and new product design.

### 3.1.5 Business Support

Demonstrating the technical advantages of applying intelligent systems to a specific business is one side of the equation. On the other side is the clear commitment of the business to allocate the necessary resources for the required period of time. The support must be formally communicated to the organization or the key stakeholders in a short document, like a statement of direction. It is preferable to clarify the application funding in advance as well.

## *3.2 Key Application Areas in Manufacturing*

Advanced manufacturing is based on good planning, comprehensive measurements, process optimization and control, and effective operating discipline. This is the application area where all of the intelligent systems methods have demonstrated impact. For example, intelligent agents can be used for planning. Neural networks, genetic programming, and support vector machines are used for inferential sensors and pattern recognition. Evolutionary computation and swarm intelligence are widely implemented for process optimization. The list of industrial neural networks-based and fuzzy logic-based control systems is long and includes chemical reactors, refineries, cement plants, etc. Effective operating discipline systems can be built, combining pattern recognition capabilities of neural networks with fuzzy logic representations of operators' knowledge and decision-making abilities of intelligent agents. Some examples are given below.

### 3.2.1 New Product Discovery

A typical list of new product invention tasks is: identification of customer and business requirements for new products; product characterization; new product optimal design; and design validation on a pilot scale. Two intelligent systems approaches—evolutionary computation and support vector machines—are especially appropriate in solving problems related to discovery of new products.

Evolutionary computation generates automatically or interactively novelty on almost any new idea that can be represented by structural relationships and supported by data. Support vector machines are capable of delivering models from very few data records, which is the norm in new product development.

An example of applying genetic programming on a typical industrial problem of modeling a set of structure–property relationships for specified materials is shown in [15]. A large number of variables like molecular weight, molecular weight distribution, particle size, the level and type of crystallinity, etc. affect these properties. These factors all interact with each other in different ways and magnitude to give the specific performance attributes. By traditional experimental design there would be a huge number of experiments to perform. Of even more difficulty, many of these variables cannot be controlled systematically and independently of each other. As a result, validating the first-principles models is time-consuming and developing black-box models is unreliable due to insufficient data. As an alternative an "accelerated" fundamental model building methodology that uses proto-models, generated by genetic programming is used. It reduces significantly the hypothesis search space and the development time in comparison to the classical first-principles model building process (10 hours vs. 3 months in this specific case study).

Another fruitful direction in applying intelligent systems for new products is building hybrid fundamental and empirical models. The empirical models are either conventional linear models or nonlinear models derived using genetic programming. This modeling technique is useful in multi-scale modeling for constructing compact representations of experimental data or data generated using complex fundamental models. These representations, which can have tunable fundamental and empirical characteristics, allow efficiently leveraging information between multiple scales and modeling platforms. As a result, the total cost for model development is significantly reduced. An application of the hybrid approach to model vapor-liquid equilibrium and polymer viscosity is shown in [26].

### 3.2.2 Process Monitoring

It is assumed that process monitoring gives sufficient information to estimate the current process state and make correct decisions for future actions, including process control. Classical process monitoring was based mostly on hardware sensors and first-principles models. The modern process monitoring, however, includes also empirical model-based estimates of critical process parameters as well as process trend analysis. In this way the decision-making process is much more comprehensive and has a predictive component linked to process trends.

Several intelligent systems approaches contribute to enhanced process observability. The key application is inferential sensors with thousands of applications based on neural networks, and, recently, on symbolic regression via genetic programming. Inferential sensors are discussed in this paper. Neural networks and support vector machines are the basic techniques for trend analysis and related fault-detection and process health monitoring systems. These systems create significant value by early detection of potential faults and they reduce the losses from low quality and process shutdowns.

### 3.2.3 Process Control

It is a well-known fact that the majority of the control loops in industry use the classical Proportional-Integral-Derivative (PID) control algorithm. Optimal tuning of these controllers on complex plants is not trivial, though, and requires high-level expertise. The losses due to non-optimal control are substantial and probably in the order of at least hundreds of millions of dollars. Intelligent systems can reduce these losses through several approaches. For example, genetic algorithms is one of the most appropriate methods for performing automatic tuning of control parameters to their optimal values.

A growing part of the remaining industrial control loops are based on nonlinear optimal control methods. Three intelligent systems technologies—neural networks, evolutionary computation, and swarm intelligence—have the potential to implement nonlinear control. Neural networks are at the core of advanced control systems. Hundreds of industrial applications, mostly in the petrochemical and chemical industries, have been reported by the key vendors—Pavilion Technologies and Aspen Technologies [15].

### 3.2.4 Supply Chain

The main activity with greatest impact in the supply chain is optimal scheduling. It includes actions like vehicle loading and routing, carrier selection, replenishment time minimization, management of work-in-progress inventory, etc. Two intelligent systems methods—evolutionary computation and swarm intelligence—broaden the optimal scheduling capabilities in these problems with high dimensionality and noisy search landscapes with multiple optima. For example, Southwest Airlines saved $10 million/year from a Cargo Routing Optimization solution designed using evolutionary computation. Procter & Gamble saved over $300 million each year thanks to a Supply Network Optimization system based on evolutionary computation and swarm intelligence [15]. Another important supply chain application, based on intelligent systems, is optimal scheduling of oil rigs at the right places at the right times, which maximizes production and reserves while minimizing cost.

## *3.3 Key Application Areas in Business*

In contrast to manufacturing, business-related applications of intelligent systems have to overcome significant challenges, related to data, the nature of derived models, and the fast and complex dynamics of social and economic processes. In most of the cases business data are very messy, with different nature—numerical, categorical, textual, unstructured, and with relatively short history within a fixed business structure. In addition, often the data sources are spread across different data bases that create issues in data integration and harmonization. However, the biggest challenge in business-related applications is that humans are in the loop either hidden in the data due to past economic and social events or explicitly in the decision-making process with delivered solutions. This introduces a different meaning of causality of the derived relationships than the dependences extracted from manufacturing data based on physical or chemical processes. The last key challenge in applying intelligent systems in business is the complex nature of social and economic processes, especially due to the recent trends of globalization, highly de-centralized communications, growing influence of social networks, and so on.

Relative to manufacturing, business-related intelligent systems applications in industry are still in their early phase. The key application areas are discussed briefly below.

### 3.3.1 Forecasting

Recently, the demand of using economic forecasts in business decision-making grew rapidly after the global recession of 2008–2009 [24]. In addition to classical time-series statistical approaches, there is a growing interest towards development and implementation of nonlinear models in forecasting, some of them based on intelligent systems methods [6]. Established application areas for nonlinear models are macroeconomics and finance. For example, most of real business cycle models are highly nonlinear. The application list includes also bond pricing models, product diffusion processes, and almost all other continuous time finance models. The key idea behind the nonlinear nature of economic time series is that major economic phenomena, such as output growth in a market economy is represented by the presence of two or more regimes (e.g. recessions and expansions), as is the case of financial variables (with periods of high and low volatility). Other types of nonlinearity might include the possibility that the effects of shocks accumulate until a process reach a catastrophic event (as is the case of the financial crisis in 2008).

Neural networks became the foster child of empirical modeling in almost any activity in the financial industry. The list of users includes almost any big banks, such as Chase, Bank One, Bank of New York, PNC Bank, etc. and insurance giants like State Farm Insurance, Merrill Lynch, CIGNA, etc. The age of the Internet allows accumulation of huge volumes of consumer and trading data that paves the way for neural network models to forecast almost everything in the financial world, from interest rates to the movement of individual stocks.

The other intelligent systems method that is appropriate for nonlinear forecasting is GP. Numerous studies have applied GP to time series forecasting with favorable results (see many references in the survey paper of [2]. The main innovation of GP is that it automatically selects and self-adjusts functional form and the time period on which its forecasts are based. GP has been used for US demand of natural gas forecast, daily exchange rate forecast, real estate prices forecast of residential single family homes in south California, oil revenue in Kuwait forecast, stock trading rules generation, etc.

### 3.3.2 Market Analysis

Analysis of current and future markets is one of the key application areas where intelligent systems can make a difference and demonstrate value potential. Typical activities include, but are not limited to, customer relation management, customer loyalty analysis, and customer behavior pattern detection. Several intelligent systems methods, such as neural networks, evolutionary computation, and fuzzy systems give unique capabilities for market analysis. For example, self-organizing maps can automatically detect clusters of new customers. Evolutionary computation also generates predictive models for customer relation management. Fuzzy logic allows quantification of customer behavior, which is critical for all data analysis numerical methods. Some specific examples of the impact of using intelligent systems in market analysis include: a 30 % system-wide inventory reduction at Procter & Gamble as a result of customer attention analysis and a system for customer behavior prediction with a loyalty program and marketing promotion optimization [15].

Intelligent agents can also be used in market analysis by simulating customers' responses to new products and price fluctuations.

### 3.3.3 Financial Modeling

Financial modeling includes computation of corporate finance problems, business planning, financial securities portfolio solutions, option pricing, and various economic scenarios. The appropriate intelligent systems techniques for this application area include neural networks, evolutionary computation, and intelligent agents. Neural networks and genetic programming allow the use of nonlinear forecasting models for various financial indicators derived from available data of past performance, while intelligent agents can generate emergent behavior based on economic agents' responses. For example, State Street Global Advisors is using GP for derivation of a stock selection models for low active risk investment style. Historical simulation results indicate that portfolios based on GP models outperform the benchmark and portfolios based on traditional models. In addition, GP models are more robust in accommodating various market regimes and have more consistent performance than the traditional models [3]. An additional benefit of evolutionary computing in financial modeling is the broad area of optimizing different portfolios.

### 3.3.4 Fraud Detection

Fraud is a significant problem in many industries, such as banking, insurance, telecommunication, and public service. Detecting and preventing fraud is difficult, because fraudsters develop new schemes all the time, and the schemes grow more and more sophisticated to elude easy detection. Intelligent systems, based on neural networks, support vector machines, and artificial immune systems are used to resolve this problem. For example, the London-based HSBC, one of the world's largest financial organizations, uses neural network solutions (the software package Fair Isaac's Fraud Predictor) to protect more than 450 million active payment card accounts worldwide. The empirical model calculates a fraud score based on the transaction data at the cardholder level as well as the transaction and fraud history at the merchant level. The combined model significantly improves the percentage of fraud detection [15].

### 3.3.5 Business Decision-Making

According to Davenport, 40 % of major business decisions are based not on facts, but on the manager guts [8]. Intelligent systems technologies can significantly reduce this dangerous subjectivity in business decision-making. A typical example is the decision-making chain in improving customer relationships. Most of the businesses need to segment their customers and identify their best ones. They want to understand customer behaviors, predict their customers' needs, and offer fitting products and promotions. They price products for maximum profitability at levels they know their customers will pay. Finally, they identify the customers at greatest risk of attrition, and intervene to try to keep them. Examples for using decision trees, cluster analysis, neural networks, etc. in business decision-making in companies, such as Avis Europe, GE Capital, CreditCorp, and so on are given in [8].

## 4 Factors for Successful Industrial Applications of Intelligent Systems

The focus in this section is on discussing the critical role of competitive advantage of intelligent systems, which is the economic driver for their applications, and the key factors for a successful industrial application. An example of a successful application for inferential sensors is given at the end of the section.

### 4.1 Intelligent Systems Competitive Advantage

Competitive advantage of a research approach, such as intelligent systems, is defined as technical superiority that cannot be reproduced by other technologies and can be translated with minimal efforts into a position of competitive advantage in the marketplace [19]. The definition includes three components. The first component requires clarifying the technical superiority (for example, better predictive accuracy) over alternative methods. The second component assumes minimal total cost of ownership and requires assessment of the potential implementation efforts of the approach and the competitive technologies. The third component is based on the hypothesis that the technical gains can improve the business performance and contribute to economic competitive advantage.

One of the most important steps in defining competitive advantages of intelligent systems is identifying the key technological rivals in the race, such as first-principles modeling, statistics, heuristics, and classical optimization. One of the comparative analysis issues is that the specific intelligent systems technologies have a very broad range of features. For example, the advantages of fuzzy systems are significantly different from the benefits of swarm intelligence. In such high-level analysis, however, we compare the competitors with the wide spectrum of features that all intelligent systems technologies, such as fuzzy systems, neural networks, support vector machines, evolutionary computation, swarm intelligence, and intelligent agents, offer for practical applications. It is possible to implement the same methodology for each specific intelligent systems technique at a more detailed level. A detailed comparative analysis is given in [15].

One of the most important competitive advantages of intelligent systems technology is their impressive potential for growth. Usually this potential is linked to the efficiency of intellectual labor. Since the key role of intelligent systems is enhancing human intelligence, the potential influence to growth is big. In addition, intelligent systems can contribute to various aspects of growth. As was discussed earlier in the previous section, new product discovery, which is critical for growth, could be significantly accelerated using intelligent systems. Another form of growth by geographical expansion is decisive for successful globalization. Through its unique capabilities of market analysis and supply-chain optimization, intelligent systems could significantly reduce the cost of global operations. Intelligent systems also have technical competitive advantage in different areas of manufacturing which can deliver productivity growth. Last, but not least, intelligent systems, based on computational intelligence, are one of the fastest-growing areas in delivering intellectual capital by filing patents. The growth of computational intelligence-related patents in the period between 1996 and 2010 is clearly seen in Fig. 1.

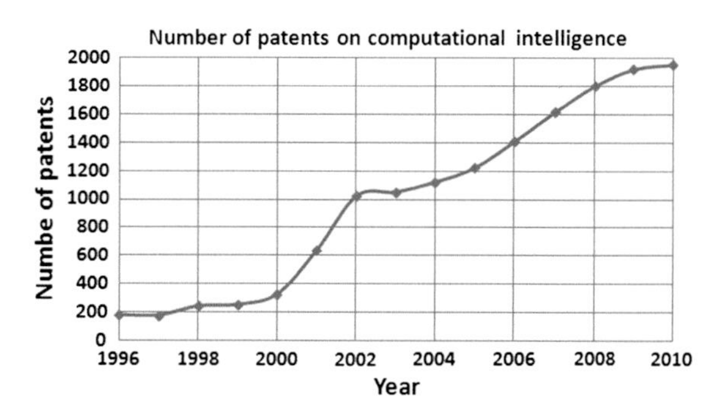


**Fig. 1** Number of published patents on computational intelligence between 1996 and 2010

## *4.2 Factors for Success*

### 4.2.1 Identified Business Needs

The driving force for using intelligent systems is some business needs that are difficult to solve using existing methods. As a first step, it is necessary to identify and prioritize the generic needs. Examples of such generic needs are: improved process monitoring systems, process control based on inferential sensors, and developing robust empirical models for new products. In many cases, the existing solutions for each of these needs, based either on first-principles models or on statistical models, have exhausted their potential. Most intelligent systems technologies offer capabilities, such as low-cost data-driven models or pattern-recognition algorithms, which may deliver the desired solutions.

### 4.2.2 Data Quality

Since most intelligent systems methods are data-driven, data quality becomes a critical factor for success. Firstly, data availability must be checked very carefully. It is possible that historical records are too short to capture seasonal effects or trends. Secondly, the ranges of the most important factors in the data have to be as broad as possible to represent the nonlinear behavior. Data-driven models developed on narrow data ranges have low robustness and require frequent readjustment. Thirdly, the frequency of data collection must be adequate to the nature of modeling. For example, dynamic modeling requires more frequent collection and data sampling. Steady-state models, on the other hand, assumes slow data collection frequency that filters the dynamic effects. Fourthly, the noise level has to be at acceptable limits to avoid the classical Garbage-In-Garbage-Out (GIGO) effect. In the case when some of these requirements are not met, it is recommended to create an adequate data collection infrastructure and to begin the application only after collecting the right data. Making a compromise with data quality is one of the most frequent mistakes in applying intelligent systems. The perception that poor data can be compensated for with the "magic" and sophistication of advanced modeling methods are one of the leading causes of an application fiasco.

### 4.2.3 Available Expertise

The access to and the quality of domain knowledge is of great importance for applying intelligent systems. However, allocating the necessary resources and winning their support for the application is not trivial. Most subject-matter experts are excited to participate in projects with advanced technologies, especially if their role is well recognized. It is possible, however, that some domain gurus feel challenged by the potential smart computerized "competitor" and may not cooperate enthusiastically. Having the key experts "on board" the application efforts is time-consuming and may require management intervention as well.

### 4.2.4 Risk-Taking Culture

It is difficult to apply intelligent systems in organizations with a low level of taking risks. Unfortunately, it is not trivial to assess in advance the potential for this problem. Some patterns for detecting technological conservatism can be defined as: no investment in high-tech in the last 5 years, no work process to bring in and support innovation and process improvements, and no incentives for taking risk.

## *4.3 Example of a Successful Application: Inferential Sensors*

Some critical parameters in chemical processes are not measured on-line (composition, molecular distribution, density, viscosity, etc.) and their values are captured either by lab samples or off-line analysis. However, for process monitoring and quality supervision, the response time of these relatively low-frequency (several hours or even days) measurements are very slow and may cause loss of production due to poor quality control. One of the approaches to address this issue is through development and installation of expensive hardware on-line analyzers. Another solution is by using soft or inferential sensors that infer the critical parameters from other easy-to-measure variables like temperatures, pressures, and flows (see a comprehensive overview of the technology in [9]).

### 4.3.1 Identified Business Need

The sources of economic benefits from inferential sensors are as follows:

- Soft sensors allow tighter control of the most critical parameters for final product quality and, as a result, the product consistency is significantly improved;
- On-line estimates of critical parameters reduce process upsets through early detection of problems;
- Inferential sensors improve working conditions by reducing or eliminating laboratory measurements in a dangerous environment;
- Very often soft sensors are at the economic optimum. Their development and maintenance cost is lower in comparison with the alternative solutions of expensive hardware sensors or more expensive first principles models;
- One side effect of the implementation of inferential sensors is the optimization of the use of expensive hardware i.e. they reduce capital investments;
- Soft sensors can be used not only for parameter estimation but also for running "What-If" scenarios in production planning.

These economic benefits have been realized rapidly by the industry and from the early 1990s a spectacular record of successful applications is reported by the vendors and in the literature (for details, see [12]). For several years, a number of well-established vendors, such as Pavilion Technologies, Aspen Technology, Siemens, Honeywell, Yokogawa, etc., have implemented thousands of inferential sensors in almost any industry. The benefit from improved quality and reduced process upsets is estimated in hundreds of millions of dollars, but the potential market is much bigger.

### 4.3.2 Inferential Sensors Based on Symbolic Regression

Different inference mechanisms, such as modeling based on first-principles, statistics, Kalman filters, neural networks, support vector machines, and so on, can be used for soft sensor development. However, our attention will be focused on one modeling approach that is increasingly being used in industry—genetic programming. GP is of special interest to soft sensor development due to its capability for symbolic regression [20]. GP-generated symbolic regression is a result of simulation of the natural evolution of numerous potential mathematical expressions. The final results is a list of "the best and the brightest" analytical forms according to the selecting objective function. Of special importance to industry are the following unique features of GP: no a priori modeling assumptions, derivative-free optimization, and natural selection of the most important process inputs, and parsimonious analytical functions as a final result.

The last feature has double benefit. On one side, a simple inferential sensor often has a potential for better generalization capability, increased robustness, and needs less frequent re-training. On the other side, process engineers and developers prefer to use non black-box empirical models and are much more open to take the risk to implement inferential sensors based on functional relationships. Recently, a Pareto-front version of genetic programming (based on multi-objective optimization between performance and complexity) significantly improves the quality of the derived models by selecting only functions with good performance and optimal complexity (for details about the Pareto-front genetic programming technology, see [28]). By using Pareto-front genetic programming the complexity of the generated analytical expressions (represented by the number of nodes) is explicitly taken into account in model development. As a result, only the most parsimonious models with equal accuracy are selected. An example of Pareto-front GP is given in Fig. 2.

In Fig. 2 each point corresponds to a certain model with the $x$-coordinate referring to model complexity and the $y$-coordinate is the model error. The points marked with circles form the Pareto front of the given set of models. Models at the Pareto front are non-dominated by any other model in both criteria simultaneously. The Pareto front itself is divided into three areas (see Fig. 2). The first area contains the simple under-fit models that occupy the upper left part of the Pareto front. The second area of complex over-fit models lie on the bottom right section of the Pareto front. The third and most interesting area of best-fit models is around the tipping point of the Pareto front where the biggest gain in model accuracy for the corresponding model complexity is. Usually this "sweet zone" is limited to several models, which significantly reduced the time for model selection and development. An additional advantage of Pareto-front GP is the capability to design an inferential sensor based on an ensemble of models with different inputs, which increases the robustness towards inputs failures in on-line operation.

An example of an inferential sensor for propylene prediction based on an ensemble of four different models is given in [11]. The models were developed from an initial large manufacturing data set of 23 potential input variables and 6900 data points. The size of the data set was reduced by variable selection to seven significant inputs and the models were generated by 20 independent GP runs. As a result of the model selection, a list of 12 models on the Pareto front was proposed for

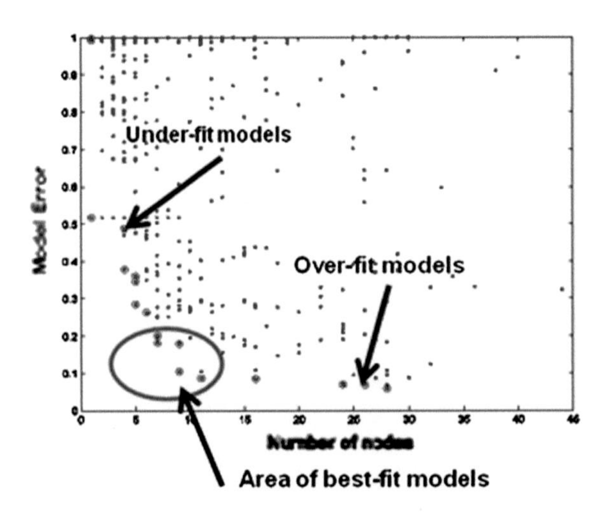


**Fig. 2** Results from a Pareto-front GP model generation

further evaluation by process engineers. All twelve models have high performance ($R^2$ of 0.97–0.98) and low complexity. After evaluating their extrapolation capabilities with "What-If" scenarios, diversity of model inputs, and by physical considerations, an ensemble of four models was selected for on-line implementation. Two of the models are shown below:

$$GP\_Model1 = A + B\left(\frac{Tray64\_T^4 Vapor^3}{Rflx\_flow^2}\right)$$

$$GP\_Model2 = C + D\left(\frac{Feed^3 \sqrt{Tray46\_T - Tray56\_T}}{Vapor^2 Rflx\_flow^4}\right)$$

where *A, B, C*, and *D* are fitting parameters, and all model inputs in the equations are continuous process measurements, such as temperatures and flows.

The models are simple and interpretable by process engineers. The difference in model inputs increases the robustness of the estimation scheme in case of possible input sensor failure. The inferential sensor has been in operation in Dow Chemical since May 2004.

### 4.3.3 Competitive Advantages of Symbolic Regression-Based Inferential Sensors

The dominant method for inferential sensor until recently is neural networks. However, the proposed symbolic regression-based inferential sensors have the following competitive advantages that are at the basis of the successful implementation strategy:

*Inferential sensors are based on analytical expressions*
In contrast to soft sensors, which are based on black-box models (neural networks), symbolic regression-based inferential sensors are represented as explicit functional relationships between the cheap measurements (inputs) and the expensive measurements (outputs).

*Inferential sensors models have optimal trade-off between accuracy and complexity*
Model complexity is critical for empirical model performance in changing operating conditions as is the case with on-line operation of inferential sensors. By using Pareto-front GP the complexity of the generated analytical expressions is explicitly taken into account in model development. As a result, only the most parsimonious models with equal accuracy at the tipping point of the Pareto front are selected. All other solutions have either worse accuracy (higher prediction error) or their complexity is much higher (i.e., their robustness is much worse).

*Robustness towards process changes*
One of the key drawbacks of neural networks is their high sensitivity to process changes that result in frequent model re-development. Symbolic regression-based inferential sensors resolve this issue by generating functional models with optimal

balance between accuracy and complexity. These types of models can tolerate minor process changes, such as measurement drifts, raw materials composition variation, small process parameters fluctuations, etc. As a result, the amount of model re-tuning is significantly reduced.

*Possible model interpretation*
The explicit functional form allows possible physical interpretation of the derived empirical relationship, which can build significant credibility and allow more reliable extrapolation in the new operating conditions. An additional advantage of the explicit relationships that helps model interpretation is the analysis of the response surface of the inferred variable while changing the process parameters. Only models with responses according the expected behavior based on process physics are accepted by the user.

*No need for specialized software for on-line implementation*
One of the disadvantages of neural network-based soft sensors is that they require a specialized run-time software license for on-line implementation. Since these capabilities are not built-in the current process monitoring and control systems, it means even adding additional servers on which to install the specialized software. In addition to the hardware cost, yearly licenses fees and upgrades are also required. In contrast, an explicit equation can be implemented in almost any software environment and needs neither additional software nor hardware.

### 4.3.4 Leveraging Symbolic Regression-Based Inferential Sensors

The benefits of the proposed symbolic regression-based inferential sensors have been gradually leveraged in The Dow Chemical Company since the first application in 1997. The following leveraging steps have been implemented:

*Pilot application*
The objective of the first pilot symbolic regression-based inferential sensor was an early detection of complex alarms in a chemical reactor [16]. In most cases, late detection of the alarms causes unit shutdown with significant losses. The alternative solution of developing hardware sensors was very costly and required several months of experimental work. An attempt to build a neural network-based inferential sensor was unsuccessful, due to the frequent changes in operating condition.

Twenty-five potential inputs (hourly averaged reactor temperatures, flows, and pressures) were selected for model development. The output was a critical parameter in the chemical reactor, measured by lab analysis of a grab sample every 8 h. The selected model was an analytical function of two temperatures in the reactor and the predicted output which was implemented directly in the process monitoring system.

*Scale-up pilot application on the same type of reactors*
The robust performance for the first 6 months gave the process operation the confidence to ask for leveraging the solution to all three similar chemical reactors in

the unit. No new modeling efforts were necessary for model scale-up and the only procedure to fulfill this task was to fit the parameters of the GP-generated function to the data set from the other two reactors. Since the fall of 1998 the symbolic regression-based inferential sensors have been in operation without need for retraining. Their robust long-term performance convinced the process operation to reduce the lab sampling frequency from once a shift to once a day since July 1999. The maintenance cost after the implementation is minimal and covers only the efforts to monitor the performance of the three soft sensors (see details in [16]).

*Validating technology on different processes*
The next step in applying the proposed symbolic regression-based inferential sensors was to implement them in variety of processes. Some examples of these applications are given below.

The technology was successfully applied for interface level estimation in a chemical manufacturing process with multiple product types [13]. The engineers in the plant were very interested in keeping the level under control during the product-type transition. The loss of control usually leads to a process upset. The developed inferential sensor, based on a simple equation of several process flows has been on-line since October 2000. A unique feature of this particular inferential sensor is that the output data used for development are not quantitative measurements but qualitative estimates of the level determined manually by process operators using a scale from one to five.

Another different type of process where symbolic regression-based inferential sensors were applied was biomass estimation. Usually the biomass concentrations are determined off-line by lab analysis every 2–4 h. This low measurement frequency, however, can lead to poor control, low quality, and production losses, i.e. on-line estimates are needed. Several neural network-based soft sensors have been implemented since the early 1990s. Unfortunately, due to the batch-to-batch variations, it is difficult for a single neural network-based inferential sensor to guarantee robust predictions in the whole spectrum of potential operating conditions. As an alternative, an ensemble of GP-generated predictors was developed and tested in a real fermentation process [18]. Data from eight batches was used for model development (training data) and the test data included three batches. Seven process parameters like pressure, agitation, oxygen uptake rate (OUR), carbon dioxide evolution rate (CER), etc. were used as inputs to the model. The output was the measured Optical Density (OD) which is proportional to the biomass. The implemented inferential sensor was based on an ensemble of five models with an average $R^2$ performance above 0.94. The selected model was scaled up to 20 fermenters.

An application of this technology for propylene estimation in a distillation column has been described at the beginning of this section. A different class of processes—for emission estimation has been explored as well. Details for applying the symbolic regression-based inferential sensors for NOx emissions monitoring can be found in [30]. Many other applications are described in internal reports. The total number of applications has grown to more than 100.

### 4.3.5 Technology Improvement

The growing business support of the proposed symbolic regression-based inferential sensors based on the successful record of applications in various chemical and bioprocesses opened the door for internal R&D support for technology improvement. As a result, the following contributions have been made:

- a novel methodology for development of soft sensors based on integration of genetic programming, analytical neural networks, and support vector machines has been developed and refined [14, 17]
- a Pareto-front GP approach has been developed and explored [28]
- a methodology for development of inferential sensors based on an ensemble of symbolic regression models has been explored [11]
- variable selection, based on Pareto-front GP has been explored [29]
- internal MATLAB toolboxes for Pareto-front GP, Support Vector Machines, and Analytical Neural Networks have been developed.

### 4.3.6 Evolving Inferential Sensors

One of the biggest issues that even the proposed symbolic regression-based inferential sensors cannot resolve is handling the high extrapolation levels in future operation modes. A performance analysis of the implemented solutions show that the on-line operating conditions for at least 30 % of the applied inferential sensors exceed the initial off-line model development ranges on average >20 % outside the off-line model development range. This extrapolation level is very high and is a challenge for any of the used empirical modeling technique, including symbolic-regression-based inferential sensors. Unfortunately, the high extrapolation level requires model re-design, including derivation of entirely new structure. In this case, modeling expertise is needed and, as a result, maintenance cost is increased.

A potential solution is the growing research area of evolving intelligent systems with their potential to create low-cost self-healing systems with minimum maintenance. An example of such a system is the evolving fuzzy inferential sensor *eSensor* with embedded evolving Takagi–Sugeno models that brings the ability to self-develop, self-calibrate, and self-maintain [1]. The learning method of evolving Takagi–Sugeno is based on two stages that are performed during a single time interval: (i) automatically separating the available data through evolving clustering and defining the fuzzy rules; and (ii) recursive linear estimation of the parameters of the consequents of the defined rules.

The proposed novel approach for inferential sensing has been tested on four problems from the chemical industry (prediction of the properties of three compositions and propylene in a simulated online mode). The four test cases include a number of challenges, such as operating regime change, noise in the data, and a large number of initial variables. These problems cover a wide range of real issues in industry when an inferential sensor is to be developed and applied. In all of these

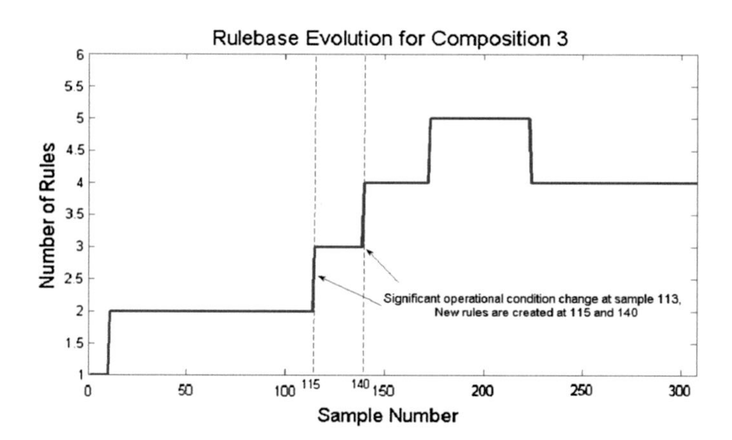


**Fig. 3** Evolution of the fuzzy rule base of *eSensor*

*FINAL RULE-BASE for COMPOSITION 2:*

$R_1$: ***IF*** *($x_1$ is around 183.85)* ***AND*** *($x_2$ is around 170.31)* ***THEN*** $(\bar{y} = 0.84 - 0.96\bar{x}_1 + 0.61\bar{x}_2)$

$R_2$: ***IF*** *($x_1$ is around 178.09)* ***AND*** *($x_2$ is around 166.84)* ***THEN*** $(\bar{y} = 0.87 - 0.98\bar{x}_1 + 0.54\bar{x}_2)$

$R_3$: ***IF*** *($x_1$ is around 172.70)* ***AND*** *($x_2$ is around 166.01)* ***THEN*** $(\bar{y} = 0.87 - 1.02\bar{x}_1 + 0.64\bar{x}_2)$

**Fig. 4** Example of a final rule base of *eSensor*

tests, the proposed new *eSensor* proved to be capable of being an advanced replacement for the existing less-flexible solutions (see details in [1]).

The operation sequence of the *eSensor* is as follows. First, it starts to learn and generate its fuzzy rule-base from the first data sample it reads. Second, *eSensor* evolves the fuzzy rule-base structure on a sample-by-sample basis (see an example of the evolution of the number of rules for Composition 3 in Fig. 3). Third, it adjusts the parameters of each rule in the rule-base online. In this way, the evolving fuzzy inferential sensor continuously adapts its structure and self-calibrates.

Another advantage of *eSensor* is the simplicity and interpretability of the evolving models. An example of a Takagi–Sugeno model at the end of the simulated period for the test case of Composition 2 is shown in Fig. 4. It could be easily interpreted by plant engineers and process operators.

## 5 Issues in Applying Intelligent Systems

Applying any emerging technology, such as intelligent systems, is not trivial and requires some level of risk-taking even when the competitive advantage is clear. In the case of intelligent systems the process is even harder due to the different nature of the comprising methods, the lack of marketing, affordable professional tools and application methodology. Another important factor slowing down intelligent systems applications is the wrong perception of the technology. To many potential users it looks like it's either too expensive or it is a rocket science. The pendulum of expectations also swings from one extreme of anticipating a silver bullet to all problems to the other extreme of awaiting the next technology fiasco.

The topic of this section is to focus on the most important application issues of intelligent systems. Their understanding and resolution is critical for the success of industrial applications.

## *5.1 Wrong Expectations*

Probably the most difficult issue of applied intelligent system is to help the final user in defining the proper expectations from the technology. Very often the dangerous combination of lack of knowledge, technology hype, and negative reception from some applied research communities creates incorrect anticipation about the real capabilities of intelligent systems. The two extremes of wrong expectations either by exaggeration or by underestimation of the intelligent systems capabilities cause almost equal damage to promotion of the technology in industry.

### 5.1.1 Magic Bullet

The expectation for technical magic is based on the unique capabilities of applied intelligent systems to handle uncertainty, complexity, and to generate novelty. The impressive features of the broad diversity of methods, such as fuzzy logic, machine learning, evolutionary computation, and swarm intelligence, contribute to such a Harry Potter-like image even when most of the users do not understand the principles behind them. Another factor adding to the silver bullet perception of intelligent systems is the technology hype from the vendors, the media, and some high-ranking managers.

As a result, potential users look at applied intelligent systems as the last hope to resolve very complex and difficult problems. Often, they begin looking at the technology after several failed attempts of using other methods. In some cases, however, the problems are ill-defined and not supported by data and expertise. In order to avoid the magic bullet trap, it is strongly recommended to identify the requirements, communicate the limitations of the appropriate methods, and to define realistic expectations in the very early phase of potential intelligent systems applications.

### 5.1.2 GIGO 2.0

The worst-case scenario of the magic bullet image is the GIGO 2.0 effect. In contrast to the classical meaning of GIGO 1.0 (Garbage-In-Garbage-Out), which represents the ignorant expectations of a potential solution, GIGO 2.0 embodies the next level of arrogant expectations defined as Garbage-In-Gold-Out. In essence, this is the false belief that low-quality data can be compensated for with sophisticated data analysis. Unfortunately, intelligent systems with its diverse capabilities to analyze data are one of the top-ranking technologies that create GIGO 2.0 arrogant expectations. It is observed that the bigger the disarray with data the higher the hope of exotic unknown technologies to clean up the mess. Usually this behavior is initiated by top management who are unaware of the nasty reality of the mess.

It is strongly recommended to protect potential intelligent systems applications from the negative consequences of the GIGO 2.0 effect. The best winning strategy is to define the requirements and the expectations in advance and to communicate clearly to the user the limitations of the methods. Better to reject an impossible implementation than to poison the soil for many feasible intelligent systems applications in the future.

### 5.1.3 Skepticism

In contrast to the magic bullet optimistic euphoria, disbelief and lack of trust in the capabilities of applied intelligent systems is the other extreme of wrong expectation, but in this case on the negative side. Usually skepticism is the initial response of the final users of the technology on the business side. Several factors contribute to this behavior, such as lack of awareness of the technical capabilities and the application potential of intelligent systems, lessons from other over-hyped technology fiascos in the past, and caution from ambitious R&D initiatives pushed by management.

Skepticism is a normal attitude if risk is not rewarded. Introducing emerging technologies, like intelligent systems, requires a risk-taking culture from all participants in this difficult process. The recommended strategy for success and reducing skepticism is to offer incentives to the developers and the users of the technology.

## *5.2 Marketing Efforts Needed*

The most critical issue of applied intelligent systems from a practical point of view is the lack of professional marketing of this emerging technology. As a result, we have the paradox when the technology is introduced to potential users and markets through the "back door" by individual efforts of R&D enthusiasts. The need to open the field to industry through the "front door" with professional marketing is discussed briefly in this section. The key issues that need to be resolved are addressed below (a detail overview of this important topic is given [15]).

### 5.2.1 Product not Clearly Defined

The first challenge in professional marketing of applied intelligent systems is the nontrivial definition of the final product. There are several sources of confusion, such as the wide diversity of methods and the broad application areas, to name a few. The deliverables are also extremely different. Here are some obvious examples of the main types of products, derived from applied intelligent systems: predictive models, problem classifiers, complex optimizers, system simulators, and search engines. They could be the basis of a broad definition of the expected products from this emerging technology. For marketing purposes it could be advertised that applied intelligent systems enhances the productivity of human intelligence through complex predictive models, problem classifiers, optimizers, system simulators, and search engines.

### 5.2.2 No Advertisement

A significant part of the marketing efforts is advertising the technology to a very broad nontechnical audience of potential users. The advertisement objective is to define a clear message for representing the technology and capturing attention. The key principle is focusing on the unique deliverables which give competitive advantage. Examples of advertising applied intelligent systems to nontechnical and technical audiences are given in [15].

## 5.3 *Looks Too Academic*

Another issue of applying intelligent systems is its academic image in industry. Part of the problem is the lack of popular references that would help the potential users to understand the diverse intelligent systems approaches. The dynamic growth of the technology makes it difficult to track the state of the art even for researchers, not to mention practitioners. These key factors are discussed in this section.

### 5.3.1 Difficult to Understand

One of the reasons for the academic image is the lack of understanding of the technology outside several closed research communities. A number of issues contribute to this situation. First, undergraduate and graduate courses on intelligent systems are offered in very few universities. The technology is virtually unknown to most students in related technical disciplines and this significantly narrows the application opportunities. It is extremely difficult for potential users to find popular explanation of the key intelligent systems approaches without scientific jargon and heavy math. It has to be taken into account that some of the methods, especially support vector machines, are not easy to translate into plain English. The other option of understanding and learning the capabilities of intelligent systems by playing with popular and user-friendly software is also virtually non-existent.

### 5.3.2 Diverse Approaches

An additional factor contributing to the academic image of intelligent systems is the wide diversity of the comprising approaches. They differ in scientific principles, mathematical basis, and user interaction. It is extremely challenging even for a person with a sound technical and mathematical background to be easily introduced to all key approaches. The confusion is also increased by the still divisive environment among the research communities developing the different methods. Usually each community tries to glorify the role of the specific approach as the ultimate technology, very often at the expense of the others. The spirit of competition prevails over the efforts of pursuing the synergetic benefits. There are very few popular references and practical guidelines on integration of these approaches, which is critical for the success of real-world applications.

### 5.3.3 It's Not Yet Ready for Industry

A negative consequence from the fast scientific progress of intelligent systems is the perception of technology and incompleteness it creates among practitioners. In principle, it is much more difficult to convince a potential user to apply a technology that is still in the high-risk dynamic development phase. In addition, management is reluctant to give support for a technology which may require continuous internal R&D development.

The limited knowledge about industrial success stories based on intelligent systems and the lack of affordable professional software contribute to the image of technology immaturity as well.

## 5.4 Maintenance and Support

Unfortunately, this is the most underestimated and ignored step in applying intelligent systems. The different intelligent systems methods require different maintenance efforts. Usually neural network-based models are more sensitive to process changes and have higher frequency of model adjustment. Part of the neural network maintenance procedure also needs historical data collection that captures the process changes. It is also recommended to add out-of-range indicators for the most statistically significant model inputs to avoid predictions in the case of extrapolation.

Model performance tracking is critical for model maintenance and support. That is why model robustness is so important and becomes a significant economic factor.

It is recommended that some built-in model self-assessment criteria are applied to warn the user if the performance is unacceptable. If the performance deterioration becomes a trend, model readjustment or complete redesign is needed. An alternative solution is to explore the capabilities of evolving intelligent systems [1].

# 6 Projected Industrial Needs for Intelligent Systems

The future directions in applying intelligent systems in industry depend on the expected demand. The projected industrial needs themselves are based on the key trends of increased globalization, progressively growing computational power, and expanding wireless communication and social networks. It is believed that these industrial requirements could be fulfilled with solutions generated by applied intelligent systems. The selected needs are discussed shortly below.

## *6.1 Big Data*

Big data, driven by the current business environment, is characterized by any or all of three "V" words: Volume, Variety and Velocity of the data (see details in [23]). The key issue in dealing with big data, from analysis point of view, is to identify unstructured data and ideally to generate structured data as a source of different modeling approaches. Many machine learning algorithms for clustering and classification, especially those capable of parallelization, can be used for this task. One of the key application areas is sentiment analysis from the textual information spread in many potential sources, which delivers quantitative leading indicators for critical parameters in a business. Typical example is customer loyalty estimation based on comments in blogs, Facebook, Tweeter, LinkedIn, etc. Looking ahead, it is anticipated that big data will begin to include more audio and video information analysis. The role of natural language processing, natural visual interpretation and visual machine learning will grow.

## *6.2 Advanced Analytics*

No doubt that advanced analytics is the fastest growing area in business-related applications. More than 80 % of Forbs 500 companies are using advanced analytics and the growth has increased after the recession in 2008 [8]. Advanced analytics provides algorithms for complex analysis of either structured or unstructured data. It includes sophisticated statistical models, machine learning, neural networks, text analytics, and other advanced data mining techniques. Among its many use cases, it can be deployed to find patterns in data, prediction, optimization, forecasting, and for complex event processing/analysis. Examples include predicting churn, identifying fraud, market basket analysis, or understanding website behavior. Advanced analytics does not include database query and reporting. The current state of the art of advanced analytics is given in [8].

## 6.3 Predictive Marketing

Successful acceptance of any new product or service by the market is critical for industry. Until now most modeling efforts have been focused on product discovery and manufacturing. Usually the product discovery process is based on new composition or technical features. Often, however, products with expected attractive new features are not accepted by potential customers. As a result, big losses are recorded and the rationale of the product discovery and the credibility of the related technical modeling efforts are questioned.

One possible solution to this generic issue in industry is to improve the accuracy of market predictions with better modeling. Recently, there has been a lot of activities in using the enormous amount of data on the Internet for customer characterization and modeling by intelligent agents [25]. Intelligent systems may play a significant role in this growing new type of modeling. The key breakthrough can be achieved by modeling customer perceptions. The subject of marketing is the customer. In predicting her/his response to a new product, the perception of the product is at the center point of the decision-making process. Unfortunately, perception modeling is still an open area of research. However, this gap may create a good opportunity to be filled by combining the capabilities of agent-based systems with the new intelligent systems approach—computing with words [31].

## 6.4 High-Throughput Innovation

Recently, high-throughput combinatorial chemistry is one of the leading approaches for generating innovative products in the chemical and pharmaceutical industries [5]. It is based on intensive design of experiments through a combinatorial sequence of potential chemical compositions and catalysts on small-scale reactors. The idea is fast experiment combined with data analysis to discover and patent new materials. The bottleneck, however, is the speed and quality of model generation. It is much slower than the speed of experimental data generation. Applied intelligent systems could increase the efficiency of high-throughput innovation by adding additional modeling methods that could generate robust empirical dependencies from small data sets and capture the accumulated knowledge during the experimental work.

By combining the capabilities of statistics, evolutionary computation and swarm intelligence, it is possible to develop high-throughput systems with self-directing experiments, which could minimize the new product discovery time. The biggest benefit, however, will be in integrating the high-throughput system with a system for predictive marketing. In this way, a critical feedback of potential market response to the newly discovered material will be given from agent-based simulations of targeted customers [22]. As a result, the risk involved with investment in the new product could be significantly reduced.

## 6.5 Manufacturing at Economic Optimum

Some of the best manufacturing processes operate under well-tuned PID controllers or model predictive control systems. It is assumed that the set points of the controllers are calculated based on either optimal or sub-optimal conditions. However, in the majority of cases the objective functions includes technical criteria and are not directly related to economic profit. Even when economics is explicitly used in optimization, the resulting optimal control is not adequate to fast changes and swings in raw material prices. The high dynamics and sensitivity to local events of the global economy require process control methods that continuously fly over the moving economic optimum. It is also desirable to include a predictive component based on economic forecasts [24].

One potential approach that could address the need for continuously tracking the economic optimum and could revolutionize process control is swarm-based control. It combines the design and controller development functions into a single coherent step through the use of evolutionary reinforcement learning. In several case studies, the designed swarm of neuro-controllers finds and continuously tracks the economic optimum, while avoiding the unstable regions. The profit gain in case of a bioreactor control is >30 % relative to classical optimal control [7].

## 6.6 Emerging Simplicity

An analysis of survivability of different approaches and modeling techniques in industry shows that the simpler the solution the longer it is used and the lower the need to be replaced with anything else. A typical case is the longevity of the PID controllers which for more than 60 years have been the backbone of manufacturing control systems. According to a recent survey, PID is used in the majority of practical control systems, ranging from consumer electronics such as cameras to industrial processes such as chemical processes [21]. One of the reasons is their simple structure that appeals to the generic knowledge of process engineers and operators. The defined tuning rules are also simple and easy to explain.

Intelligent systems technologies can deliver simple solutions when multiobjective optimization with complexity as an explicit criterion is used. An example of this approach with many successful applications in the chemical industry is using symbolic regression generated by Pareto front genetic programming. Most of the implemented empirical models are very simple, as was demonstrated and were easily accepted by process engineers. Their maintenance efforts were low and the performance over changing process conditions was acceptable.

The demand for developing capabilities to simplify potential solutions will grow in the big data environment. Some ideas for application areas are given in [27].

### *6.7 Handling the Curse of Decentralization*

The expected growth of wireless communication technology will allow mass-scale introduction of "smart" components of many industrial entities, such as sensors, controllers, packages, parts, products, etc. at relatively low cost. This tendency will give a technical opportunity for the design of totally decentralized systems with built-in self-organization capabilities. On the one hand, this could lead to the design of entirely new flexible industrial intelligent systems capable of continuous structural adaptation and fast response to the changing business environment. On the other hand, the design principles and reliable operation of a totally distributed system of thousands, even millions, of communicating entities of a diverse nature is still a technical dream. Most existing industrial systems avoid the curse of decentralization through their hierarchical organization. However, this imposes significant structural constraints and assumes that the designed hierarchical structure is at least rational, if not optimal. The problem is the static nature of the designed structure, which is in striking contrast to the expected increased dynamics in the industry of the future. Once set, the hierarchical structure of industrial systems is changed either very slowly or not changed at all, since significant capital investment is needed. As a result, the system does not operate effectively or optimally in dynamic conditions.

The appearance of totally decentralized intelligent systems of free communicating sensors, controllers, manufacturing units, equipment, etc., can lead to the emergence of optimal solutions in real time, and effective response to changes in the business environment is needed. Intelligent systems technologies could be the basis in the design of such systems and intelligent agents could be the key carrier of the intelligent component of each distributed entity.

## 7 Conclusions

Intelligent systems have been successfully applied in manufacturing in areas, such as new product discovery, process monitoring and control, and supply chain. Recently, there is a growing trend in business-related applications in areas, such as forecasting, market analysis, financial modeling and complex business decision-making. The key factors that contributed to successful applications of intelligent systems are: defining their competitive advantage relative to the other methods, identified business needs, data quality, available expertise, and risk-taking culture. An example of a very successful intelligent systems application is symbolic regression-based inferential sensors. The key issues that need to be resolved in applying intelligent systems are: wrong expectation, lack of marketing, their academic image, and complicated maintenance and support. The future intelligent systems applications might be driven by the projected industrial needs in the areas, such as big data, advanced analytics, predictive marketing, high-throughput innovation, manufacturing at economic optimum, emerging simplicity, and handling complex decentralized systems.

## References


1. Angelov, P., Kordon, A.: Evolving inferential sensors in the chemical industry. In: Angelov, P., Filev, D., Kasabov, N. (eds.) Evolving Intelligent Systems: Methodology and Applications, pp. 313–336. Wiley, New York (2010)
2. Brabazon, A., O'Neil, A., Dempsey, I.: An introduction to evolutionary computation in finance. IEEE Comput. Intell. Mag. **3**, 42–55 (2008)
3. Becker, Y., Fei, P., Lester, A.: Stock selection: an innovative application of genetic programming methodology. In: Riolo, R., Soule, T., Worzel, B. (eds.) Genetic Programming Theory and Practice IV, pp. 315–335. Springer, Berlin (2007)
4. Bonissone, P., Chen, Y., Goebel, K., Khedkar, P.: Hybrid soft computing systems: industrial and commercial applications. Proc. IEEE **87**(9), 1641–1667 (1999)
5. Cawse, J. (ed.): Experimental Design for Combinatorial and High-Throughput Materials Development. Wiley, New York (2003)
6. Clements, M., Franses, P., Swanson, N.: Forecasting economic and financial time-series with non-linear models. Int. J. Forecast. **20**, 169–183 (2004)
7. Conradie, A., Aldrich, C.: Development of neurocontrollers with evolutionary reinforcement learning. Comput. Chem. Eng. **30**(1), 1–17 (2006)
8. Davenport, T., Harris, J., Morrison, R.: Analytics at Work: Smarter Decisions Better Results. Harvard Business Press (2010)
9. Fortuna, L., Graziani, S., Rizzo, A., Xibilia, M.: Soft Sensors for Monitoring and Control of Industrial Processes. Springer, Berlin (2007)
10. Gusikhin, O., Rychtyckyj, N., Filev, D.: Intelligent systems in the automotive industry: applications and trends. Knowl. Inf. Syst. **12**(2), 147–168 (2007)
11. Jordaan, E., Kordon, A., Smits, G., Chiang, L.: Robust inferential sensors based on ensemble of predictors generated by genetic programming. In: Proceedings of PPSN 2004, pp. 522–531. Springer, Berlin (2004)
12. Kadlec, P., Gabrys, B., Strandt, S.: Data-driven soft sensors in the process industry. Comput. Chem. Eng. **33**, 795–814 (2009)
13. Kalos, A., Kordon, A., Smits, G., Werkmeister, S.: Hybrid model development methodology for industrial soft sensors. In: Proceedings of the IEEE ACC 2003, Denver, CO, pp. 5417–5422 (2003)
14. Kordon, A.: Hybrid intelligent systems for industrial data analysis. Int. J. Intell. Syst. **19**, 367–383 (2004)
15. Kordon, A.: Applying Computational Intelligence: How to Create Value. Springer, Berlin (2010)
16. Kordon, A., Smits, G.: Soft sensor development using genetic programming. In: Proceedings of GECCO 2001, San Francisco, pp. 1346–1351 (2001)

17. Kordon, A., Smits, G., Jordaan, E., Rightor, E.: Robust soft sensors based on integration of genetic programming, analytical neural networks, and support vector machines. In: Proceedings of WCCI 2002, Honolulu, pp. 896–901 (2002)
18. Kordon, A., Jordaan, E., Chew, L., Smits, G., Bruck, T., Haney, K., Jenings, A.: Biomass inferential sensor based on ensemble of models generated by genetic programming. In: Proceedings of GECCO 2004, Seattle, WA, pp. 1078–1089 (2004)
19. Kordon, A., Jordaan, E., Castillo, F., Kalos, A., Smits, G., Kotanchek, M.: Competitive advantages of evolutionary computation for industrial applications. In: Proceedings of CEC 2005, Edinburgh, UK, pp. 166–173 (2005)
20. Koza, J.: Genetic Programming: On the Programming of Computers by Means of Natural Selection. MIT Press, Cambridge (1992)
21. Li, Y., Ang, K., Chong, G.: Patents, software, and hardware for PID control. IEEE Control Syst. Mag. **26**(1), 42–54 (2006)
22. Luck, M., McBurney, P., Shehory, O., Willmott, S.: Agent Technology Roadmap. AgentLink III (2005)
23. Minelli, M., Chambers, M., Dhiraj, A.: Big Data, Big Analytics. Wiley, New York (2013)
24. Rey, T., Kordon, A., Wells, C.: Applied Data Mining for Forecasting Using SAS. SAS Press (2012)
25. Schwartz, D.: Concurrent marketing analysis: a multi-agent model for product, price, place, and promotion. Mark. Intell. Plann. **18**(1), 24–29 (2000)
26. Seavy, K., Jones, A., Kordon, A.: Hybrid genetic programming—first-principles approach to process and product modeling. Ind. Eng. Chem. Res. **49**(5), 2273–2285 (2010)
27. Siegel, A., Etzkorn, I.: Simple: Conquering the Crisis of Complexity. Twelve, New York (2013)
28. Smits, G., Kotachenek, M.: Pareto-front exploitation symbolic regression. In: O'Reiley, U.M., Yu, T., Riolo, R., Worzel, B. (eds.) Genetic Programming Theory and Practice II, pp. 283–300. Springer, New York (2004)
29. Smits, G., Kordon, A., Jordaan, E., Vladislavleva, C., Kotanchek, M.: Variable selection in industrial data sets using pareto genetic programming. In: Yu, T., Riolo, R., Worzel, B. (eds.) Genetic Programming Theory and Practice III, pp. 79–92. Springer, New York (2006)
30. Stefanov, Z., Chiang, L., Kordon, A.: Successful industrial application of robust inferential sensors for $NO_x$ emissions monitoring. In: Proceedings of AIChE (2008)
31. Zadeh, L.: Fuzzy logic = computing with words. IEEE Trans. Fuzzy Syst. **90**, 103–111 (1996)

# Prediction of the Attention Area in Ambient Intelligence Tasks

**Jawad Shafi, Plamen Angelov and Muhammad Umair**

**Abstract** With recent advances in Ambient Intelligence (AmI), it is becoming possible to provide support to a human in an AmI environment. This paper presents an Adaptive Neuro-Fuzzy Inference System (ANFIS) model based scheme, named as prediction of the attention area using ANFIS (PAA_ANFIS), which predicts the human attention area on visual display with ordinary web camera. The PAA_ANFIS model was designed using trial and error based on various experiments in simulated gaming environment. This study was conducted to illustrate that ANFIS is effective with hybrid learning, for the prediction of eye-gaze area in the environment. PAA_ANFIS results show that ANFIS has been successfully implemented for predicting within different learning context scenarios in a simulated environment. The performance of the PAA_ANFIS model was evaluated using standard error measurements techniques. The Matlab® simulation results indicate that the performance of the ANFIS approach is valuable, accurate and easy to implement. The PAA_ANFIS results are based on analysis of different model settings in our environment. To further validate the PAA_ANFIS, forecasting results are then compared with linear regression. The comparative results show the superiority and higher accuracy achieved by applying the ANFIS, which is equipped with the capability of generating linear relationship and the fuzzy inference system in input-output data. However, it should be noted that an increase in the number of membership functions (MF) will increase the system response time.

J. Shafi (✉) · P. Angelov
Data Science Group, School of Computer and Communication,
Lancaster Univeristy, Lancaster, UK
e-mail: j.shafi@lancaster.ac.uk; jawadshafi@ciitlahore.edu.pk

P. Angelov
e-mail: p.angelov@lancaster.ac.uk

P. Angelov
Chair of Excellence, Carlos III University, Madrid, Spain

J. Shafi · M. Umair
Department of Computer Sciences, COMSATS Institute of Information Technology,
Lahore, Pakistan
e-mail: mumair@ciitlahore.edu.pk



V. Sgurev et al. (eds.), *Innovative Issues in Intelligent Systems*,
Studies in Computational Intelligence 623, DOI 10.1007/978-3-319-27267-2_2

# 1 Introduction

The Ambient Intelligence (AmI) is aiming to provide pervasive support to humans. The visual attention is one of the primary tools to figure out the user attention area on a display [10, 17, 34]. It is a key challenge to figure out the appropriate representation and awareness of a human's attention states, some of research work relevant to human attention are discussed in [1, 2, 9, 30]. "Mindreading" may address different types of human states [14] e.g. intention, attention, emotions. More dedicated services can be provided in AmI environment, if the user attention area can be determined. Some research work have been already done but that is purely based on using special type of tracking devices [11, 26]. The uses of such devices in an application make it relatively expensive. This paper adopts non-linear regression prediction technique, using web camera coordinates with a combination of hand controlling coordinates. This makes it cost effective, reusable, robust and with better accuracy to figure out visual attention area on visual display.

To provide personalized services in an AmI environment, the system has to be able to make inferences based on the interaction with the user. To achieve this we apply computational intelligence (CI). CI are commonly used for learner modelling because of the complexity of relationships, which are difficult to represent. One research [36] listed the main issues related to the use of machine learning techniques for modelling purposes as: the need for large data sets, the need for labelled data and computational complexity.

This paper present methods for identification of the visual attention area on a display. The main objective is to adopt a suitable learning method for eye-gaze identification. We used Adaptive Neuro-Fuzzy Inference System (ANFIS) based on the acquired data from the environment in an application. Section 2 provides a brief review of the work which is based on ANFIS. Section 3 presents the structure of the proposed ANFIS based method. Section 4 presents the results and discussion of the proposed system. Finally, Sect. 5 presents concluding comments about future developments related to this work.

# 2 Related Work

There are several studies in the literature which deal with the detection of visual attention. The results of [33] are based on a neural network for classification of posture and a Hidden Markov Model for recognizing the state of interest. Another approach in [16] is based on the creation of computing and communication systems that can detect and reason about the human attention by fusing the information received from multiple sources. Another study based on neuro-fuzzy approach is reported in [7] to infer the attention level of a user in front of a monitor using a simple camera but this is based on combining eye gaze and head pose information, in a non-intrusive environment. These models use two inputs together with other biometrics to infer the user's attention.

There are other works in literature around this issue but they are all based on head pose estimation and also use more than one camera or extra equipment [25, 28]. In [11, 26] the authors suggested the gaze detection of visual attention but, as mentioned above, that is based on the use of an eye tracking device.

The Adaptive-Neuro Fuzzy Inference System is a hybrid system that combines the potential benefits of both, the artificial neural network (ANN) [15] and fuzzy logic (FL) [23] methods. ANFIS is being employed in numerous modelling and forecasting problems [21] and in different domains, e.g. chemical and biological engineering [6], renewable energy [27], energy economics [29], stock market [12], cancellation of EMG signals [35], time series prediction [37], speech recognition and signal processing [13] and industrial management [5]. To the best of our knowledge no body have used ANFIS to predict the user attention area in AmI so far.

## 3 Traditional Architecture

One of the widely used hybrid intelligent systems is the neuro-fuzzy combination. These systems take advantage of human-like ambiguity, interpretability, transparency, and ability to model non-linear vagueness on data in an environment using fuzzy logic [23]; also it has a flexible structure and superior capability of a neural network [15]. Modern neuro-fuzzy systems are multilayer feed-forward neural networks. In these systems the fuzzy rules are trained by the learning algorithm implemented and applied on a neural network [8]. Fuzzy logic does not incorporate any learning mechanism, while neural networks appeared as a black box approach; do not have explicit knowledge representation. A typical neuro-fuzzy system such as the Adaptive neuro-fuzzy inference system (ANFIS) introduced by Jang in 1993 [18] integrates both neural networks and fuzzy logic principles; it has potential to capture the benefits of both in a single framework. Its inference system corresponds to a set of fuzzy IF–THEN rules that have learning capability to approximate non-linear functions [3]. Hence, ANFIS is considered to be a universal estimator [21]. There are few characteristics that enable ANFIS to achieve a success [19]:

- It makes the complex system behaviour more refined using fuzzy IF-THEN rules.
- It is easy to implement.
- It enables fast and accurate learning.
- It is easy to incorporate both linguistic and numeric knowledge for problem solving.

### 3.1 Takagi Sugeno Fuzzy Inference System (TS-FIS)

TS models are a powerful practical engineering tool for modeling and control of complex systems [4]. The output of a TS model can be linear or a constant [22]. As ANFIS is based on Takagi–Sugeno fuzzy inference system [18], so the rules are of the following type:

$$\Re_j : IF(x_m is \aleph_{j1}) AND..AND(x_m is \aleph_{jp}) THEN(x_g = a_{j0} + a_{j1}x_1 + \ldots + a_{jn}x_n);$$
$$j = \{1, R\}$$

where $\Re_j$ denotes the *j*th fuzzy rule; $R$ is the number of fuzzy rules; $x_m$ is the input variable, $\aleph_{jp}$ denotes the antecedent fuzzy sets, $p = \{1, n\}$; $x_g$ is the output of the *j*th linear subsystem; $a_{jl}$ are its parameters, $l = \{0, n\}$. In ANFIS, the output linear membership functions of the consequent part of TS-FIS are automatically adjusted. TS-FIS consists of inputs, output(s), set of predefined rules and a defuzzification method. Aggregation is employed to unify the outputs of all the rules resulting into a single fuzzy set. Hence, the final output of the system is the weighted average of all the rule partial outputs.

### 3.2 Artificial Neural Network (ANN)

ANN is a structure of interconnected neurons arranged in a systematic manner to perform some computing task. It is widely used due to its capability of learning using the training data. Its architecture usually consists of input hidden and output layers. Each layer is composed of several processing neurons. Input layer includes the inputs, data values obtained from training. Hidden layer processes the inputs into the hidden layer. The training algorithm which is widely used in ANN is called error back-propagation algorithm [24].

### 3.3 Back-Propagation Algorithm

Back-propagation is a machine learning feed forward, multilayer network parameter optimization based algorithm for supervised mode of learning [24]. The number of input, hidden, output layers and neurons in each layer depend on the application requirements. The main objective of back-propagation learning algorithm is to adjust the values of weights in the training data set in such a manner so as to get the same value as the correct output value of the network using the validation data set. This process is shown in Fig. 1.

In the forward pass, input weights are injected to subsequent layer. The activation function is implemented to generate the weights for the next layer.

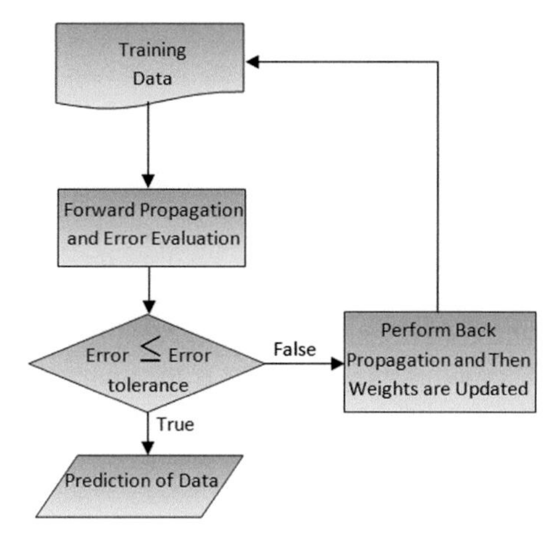


**Fig. 1** Back propagation in ANFIS

Finally, the output layer is ready to generate some output value. The generated and original values of the output are utilized to derive the error which is propagated further back to the input layer. The process will continue until the error is less than a pre-defined error tolerance and the network is ready to be used.

## 3.4 ANFIS Architecture

It maps the first order Takagi-Sugeno Fuzzy Inference System in multilayer feed-forward adaptive neural network to enhance performance with the above mentioned attractive features. The first order Takagi-Sugeno fuzzy model's inference mechanism and defuzzification process is shown in Fig. 2, [18]. It has five-layer architecture.

Fuzzy If-Then rules are used to describe in ANFIS [18] a system for example by two rules:

$R_1$ = IF $x$ is $C_1$ and $y$ is $D_1$, then $f1 = p_1x + q_1y + r_1$
$R_2$ = IF $x$ is $C_2$ and $y$ is $D_2$, then $f2 = p_2x + q_2y + r_2$

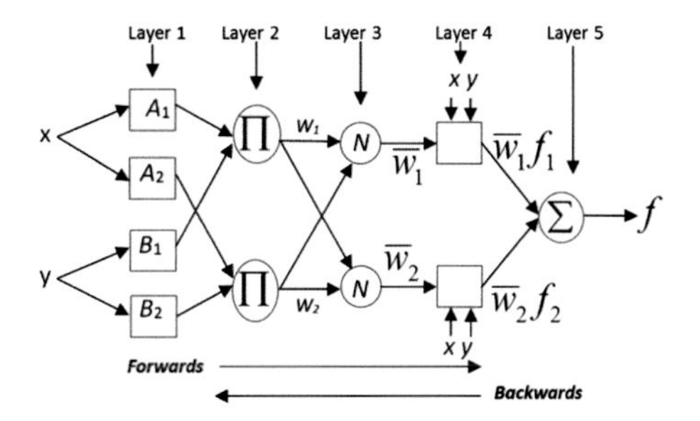


**Fig. 2** ANFIS 5 layer architecture, with forwards and backwards propagation

where $x$ and $y$ are the inputs, $C_i$ and $D_i$ are the fuzzy sets $f_i$ are the outputs within the fuzzy region specified by the fuzzy rule, and $p_i$, $q_i$, and $r_i$ are the design parameters of the output/consequent part that are determined during the training process. Each layer has a specific functionality.

Layer 1, consist of node function with adaptive parameters as follows:

$$o_j^1 = \mu_{Cj}(x), \quad j = 1, 2 \tag{1}$$

where $x$ is input value, $o_j^1$ is membership value of fuzzy variable $\mu_{Cj}$, $Cj$ shows constant coefficients and $j$ is the number of rules. Output of this layer is a fuzzy membership grade of the input, which can be described with Gaussian membership function as $\mu_{Cj}(x)$:

$$\mu_{Cj}(x) = \exp\left[-\left(\frac{x - c_j}{a_j}\right)^2\right], \tag{2}$$

where $a_j$ and $c_j$ are the parameters of the membership function.

In layer 2, fuzzy AND operators are used, to fuzzify the inputs. The output of this layer is represented as:

$$o_j^2 = w_i = \prod_{j=1,2} \mu_{Cj} = \mu_{Cj}(x) * \mu_{Dj}(y), \tag{3}$$

Layer 3, consists of fixed nodes, which are used to calculate the normalized firing strength. The output of this layer can be represented as:

$$o_j^3 = \bar{w}_j = \frac{w_j}{w_1 + w_2}, \quad j = 1, 2 \tag{4}$$

In layer 4, the nodes are adaptive. The output of each node in this layer is simply the product of the normalized firing strength and for a first order Takagi-Sugeno model. The output of this layer can be described as:

$$o_j^4 = \bar{w}_j f_j = \bar{w}_j(p_j x + q_j y + r_j), \quad j = 1, 2, \tag{5}$$

where $\bar{w}$ is the output of the previous layer, and $p_i$, $q_i$, and $r_i$ are the consequent parameters. In the last layer of ANFIS, it performs the summation of all incoming signals from the previous layer 4 as:

$$o_j^5 = \sum\nolimits_j \bar{w} f_j = \frac{\sum_j w_j f_j}{\sum_j w_j}. \tag{6}$$

### 3.5 *Hybrid Learning Algorithm (HLA)*

This algorithm is a combination of the gradient descent and the least squares methods which are used to minimize the error in the learning stage. The HLA consist of two passes known as forward and backward pass. In the forward pass, the info flows forward until $o_j^4$ and the consequent parameters are determined by the least square approach. In the backward pass, the error signals propagate backward and the premise parameters are updated using gradient descendent approach. This process is shown in Fig. 3. Hence, HL approach is much faster by reducing the search space dimensions of the [18].

$$f = \frac{w_1}{w_1 + w_2} f_1 + \frac{w_1}{w_1 + w_2} f_2 \tag{7}$$

$$= \bar{w}_1 f_1 + \bar{w}_2 f_2 \tag{8}$$

$$= (\bar{w}_1 x)p_1 + (\bar{w}_1 y)q_1 + (\bar{w}_1)r_1 + (\bar{w}_2 x)p_2 + (\bar{w}_2 y)q_2 + (\bar{w}_2)r_2 \tag{9}$$

where $p_1$, $q_1$, $r_1$, $p_2$, $q_2$, and $r_2$ are the linear consequent parameters.

For the optimal values of these parameters, the least square method is used. When the premise parameters are not fixed, the search space becomes larger and the convergence of the training becomes slower. In the forward pass least square method is used to optimize the consequent parameters. In the backward pass, the error signals propagate backward and premise parameters are updated with the help of gradient descent used to optimize the premise parameters method by keeping consequent parameter fixed. Parameters updating rule is given as [18]:

$$a_{jl}(t+1) = a_{jl}(t) - \varpi . \frac{\partial E}{\partial a_{jl}} \tag{10}$$

where $a_j$ are the adaptive parameters. $\varpi$ is the learning rate for parameter $a_{jl}$, gradient is obtained by using the chain rule as [31]:

$$\frac{\partial E}{\partial a_{jl}} = \sum_{j=1}^{l+1} \frac{\partial E_j}{\partial \mu_{A_{jl}}} \cdot \frac{\partial \mu_{A_{jl}}}{\partial a_{jl}} \tag{11}$$

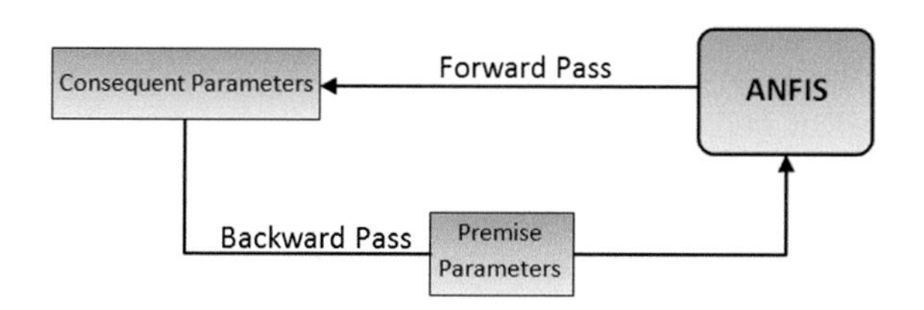


**Fig. 3** Forward and backward pass in the hybrid algorithm

The consequent parameters which are found during the forward pass, are used to calculate the output of ANFIS. The output error is used to adapt the premise parameters by means of a standard back propagation algorithm. It has been proven that this hybrid algorithm is highly efficient in training the ANFIS systems [19].

### 3.6 ANFIS for Visual Attention Area Modelling

Neuro-Fuzzy technique has been used for intelligent modelling and control of uncertain systems. It is based on the input/output data pairs collected by the system. ANFIS is selected to predict the value of eye-gaze coordinates on the visual display. We need to apply several steps to train the model using ANFIS. These steps include: to define input/output values, to define fuzzy sets for the input values; to define fuzzy rules; and to create and train the neural network.

To implement and test the ANFIS, a development tool is required. Matlab TS fuzzy logic toolbox (FLT) [20] was selected as the development tool. This tool provides an environment to build and evaluate fuzzy systems using a graphical user interface (GUI). It consists of a FIS editor, the rule editor, a membership function editor, the fuzzy inference viewer, and the output surface viewer.

The FIS editor displays general information about a fuzzy inference system. The membership function editor is the tool that displays and edits the membership functions associated with all input and output variables. The rule editor allows the user to construct the rule statements. The rule viewer allows users to interpret the entire fuzzy inference process at once. The ANFIS editor GUI menu bar can be used to load a FIS training initialization and save the trained FIS.

### 3.7 Learning in ANFIS Using a Simulated Game

The basic purpose of our environment is to predict the value of the eye-gaze using the ordinary web camera. Using the mouse coordinates we want to construct a system that learns from $x$ and $y$ coordinates. Based on this learning it will adapt predicted eye-gaze coordinates of the visual attention area in the environment. This model of learning is a three-step process: first, we gather the data which will be used for learning, then create the learning model based on the data which we gather in first step, and finally adapt the learning model and deploy it in the environment.

Selection of the input have high influence on the prediction model. For this reason, it is of great importance to collect significant inputs and then design the system based on them. We need to adopt such a technique from which we can get maximum advantage. A suitable technique, that is based on the assumption that the ANFIS model is with the smallest Root Mean Square Error (RMSE) using a small number of data [32]. In our model only two inputs $(x_g, y_g)$ are selected for the learning purpose.

## 3.8 ANFIS Training in the Environment

This process starts after receiving the training input/output data set. Two vectors are used to train the ANFIS, as training data is a set of input and output vectors. Training data set is used to find the parameters for the membership function. Also, in this process a threshold value for the error between the actual and desired output is defined. If this error is greater than the threshold value then the parameters are updated using the gradient decent method. It continues to change the premise parameters until the error value becomes less then the threshold value. The checking data set is then used to compare the model with the data [18].

ANFIS training rules use hybrid learning with combination of least square and gradient descent methods to determine the consequent parameters. The aim of using ANFIS is to identify the visual attention area on display. The further goal is to achieve the best performance possible; this is totally dependent on how we create a suitable training data-set to train the Neuro-Fuzzy system. Considering this reason, we need to gather training data from different users with different age and from male and female and on different display sizes and with different camera resolutions.

## 3.9 ANFIS Training in Matlab

We have used ANFIS Matlab toolkit [20] for this purpose. The first step is to prepare the training data in Matlab. ANFIS training uses the 'anfis.m' function. Evaluation of the system compared to the desired output is conducted using the 'evalfis.m' function. The data-set must be in a matrix form before we input it to the 'anfis.m' function. Where each column is used to represent the number of input and rows represent data-sets. The matrix contains as many input columns as needed but the last column must be the output.

As in traditional approaches you need to decide or ask expert's to create the membership function, or use command 'fismat.m', which provide help in the creation of the initial set of membership functions. After the creation of membership functions, the system training begins. We input the training data we defined using the command 'anfis.m'.

After the training process, the final training membership functions and training error are produced, Fig. 4. The checking data-set can be used in conjunction with the training data-set to improve accuracy. We can use as much data to train the system as needed, even ANFIS can be used with single training data-set but input of checking data increases the chance that the system quality will increase. Figure 4 illustrates the trend of errors using hybrid learning algorithms of ANFIS. These depict that error monotonically decreases with an increase in number of epochs.

As soon as the system training is complete, the process of evaluation of the system begins. We can enter the input data sets into our trained fuzzy system, but this

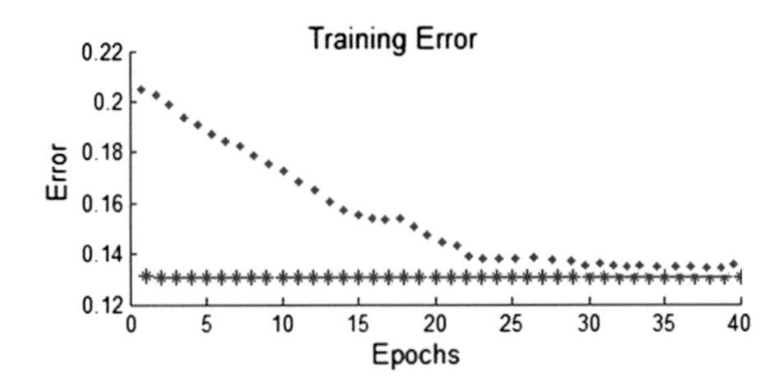


**Fig. 4** Trend of errors with ANFIS

data-set do not include the output column in the matrix. The function 'evalfis.m' represents the output of the ANFIS system in our developed environment. There are different measurement parameters to test the effectiveness and efficiency of the system; few of them are Root Mean Square

Error (RMSE), Mean Average Error (MAE), Mean Absolute Percentage Error (MAPE), Normalized Mean Square Error (NMSE), coefficient of determination ($R^2$) and Coefficient of Correlation (R) between the desired learner contexts and the learning content format as the system input/output. We will discuss them in latter section. Once the ANFIS is trained, we can test the system against different sets of data values to check its functionality.

The system training methodology is summarized in Fig. 5. For each time instant we have two separate input parameters which control the eye-gaze $x$ and mouse $x$ coordinates ($x_g$, $x_m$) and one output coordinate $\tilde{x}_g$. At the same time, other two input parameters were control the eye-gaze $y$ and mouse $y$ coordinates ($y_g$, $y_m$) and one other output coordinate $\tilde{y}_g$.

Further, 'genfis1.m' function is used to generate an initial single output FIS matrix from training data. First, we need to initiate our model with default values for membership function numbers "13*13" and further we need to define the type of membership functions (these may be Gaussian curve, Generalized bell-shaped and Triangular-shaped; with the following commands 'gaussmf.m', 'gbellmf.m' and 'trimf.m'). There are few others as well but in our model we are just using these three.

## 3.10 ANFIS Structure and Takagi–Sugeno fuzzy inference

The conditions that determine the learning content format depend completely on experts. Each of the two input conditions is represented by the following term sets. Display is divided in four grids, Each of $x_g$, $y_g$, $x_m$ and $y_m$ are shown in Fig. 6 representing the following linguistic terms for defining the membership functions, upper-left-corner (ULC), Lower-Left-Corner (LLC), Upper-Right-Corner (URC), Lower-Right-Corner (LRC), Upper-Mid_Centre (UMC), Lower-Mid_Centre (LMC), Centre (C), Right-Centre (RC), Left-Centre (LC), Grid-1-Center (G1C), Grid-2-Center (G2C), Grid-3-Center (G3C) and Grid-4-Center (G4C).

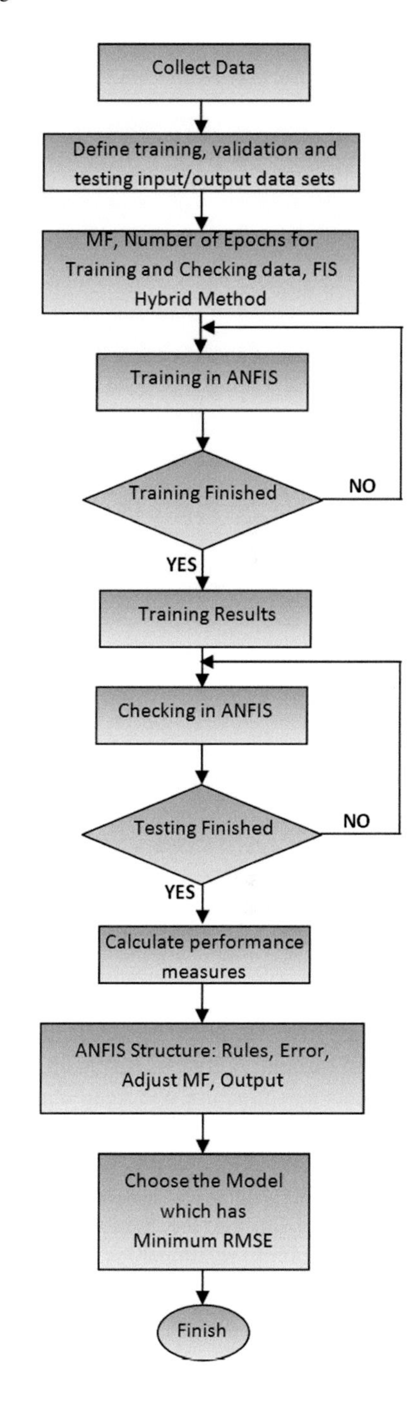


**Fig. 5** Flowchart of the PAA_ANFIS

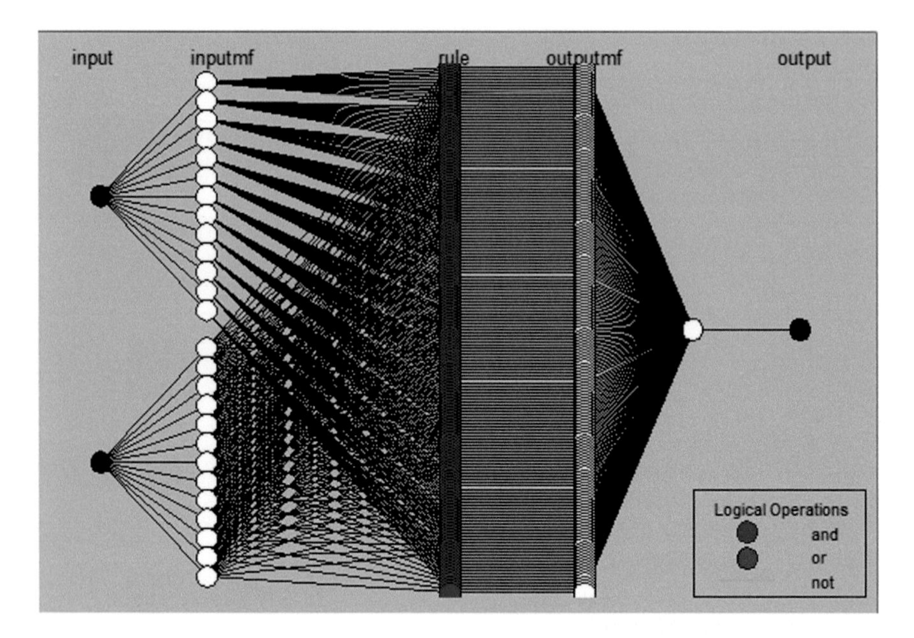


**Fig. 6** ANFIS structure with two inputs, one hundred sixty nine rules and one output

In our case, the number of membership functions for each input is 13, so the total numbers of membership functions used is 52. The generated TS-FIS structure contains 169 fuzzy rules. Each rule has one output membership function which is of type "linear". ANFIS has a multi-input single output (MISO) structure. Membership function parameters adjustment is done through the use of gradient based method, which reflects how well different set of training data are modelled; as this data represent different conditions in its parameters.

Several experiments have been conducted to assess whether the proposed model with combination of ANFIS has produced acceptable result. ANFIS network training involves mapping inputs through input membership functions and mapping output through output membership functions. The parameters associated with each membership function will keep changing throughout the learning process, but are fixed afterwards.

The process of training, in ANFIS begins by specifying the number of sets of each input variable and shape of their membership functions. Then, all the training data passes through the Neural Network (NN), to adjust the input parameters to find the input/output relationships. Another reason for passing the data in NN is that the error can be minimized.

In this paper, an ANFIS model based on both Neural Network (NN) and Takagi–Sugeno (TS) fuzzy inference system, has been developed to adapt learning content formats for our gaming environment. The experiments were divided into two ANFIS structures to demonstrate learning activities in different contexts. For this purpose, computer gaming environment using Matlab statistical validation indexes were used for determining the performance of the best proposed model. Once the input/output have been recorded it is necessary to validate the quality of the model results after training and before deploying it into the environment.

We trained the system with a variety of different settings such as dataset sample, epoch number, membership function type and number, and number of inputs to achieve the best performance. These results are discussed in the last section

### *3.11 Proposed Model for Eye-Gaze Prediction*

The steps involved in the implementation of the proposed model using ANFIS include:

Step1: Collecting data from the environment.
Step2: Determining training, validation and testing input- output sets.
Step3: Generating initial first order TSK fuzzy inference system.
Step4: Extracting rules.
Step5: Tuning antecedent and consequent parameters by ANFIS hybrid learning algorithm. After setting training parameters, learning algorithm will be continued to reach the specified error goal or maximum number of epochs.
Step6: Calculating performance measures.
Step7: Selecting the model with the minimum validation error.
Step8: Running the model.

By using TS-FIS and neural methodology in form of HLA, we generated the optimal ANFIS structure. This procedure establishes the first order TSK fuzzy inference system that models the data behaviour in a better way. The flowchart of our proposed model is shown in Fig. 5.

## 4 Results and Discussions

Many experiments were conducted, and data were divided into two separate sets; one for the training and other for the checking. The training data set was used to train the ANFIS, whereas the checking data set was used to verify the accuracy and the effectiveness of the trained ANFIS model, so it can be adopted in the environment. In order to find out the optimal solution, that can address the requirements of our problem of identifying the visual attention area with the help of eye-gaze. We need to consider the numbers of important factors which can affect the system efficiency before we deploy them in the environment.

Some of the aspects are taken into account in relation to ANFIS system training such as: number of membership functions, type of membership functions, training data samples and epoch number which generated different MAE and MASE.

### 4.1 Performance Measures

We applied the commonly used performance measures in forecasting problems to compare and evaluate the accuracy of our model. These performance measures are Root Mean Square Error (RMSE), Mean Average Error (MAE) and Coefficient of Determination ($R^2$) between the desired learner contexts and the learning content. Which are calculated by

$$MAE = \frac{1}{N}\sum_{q=1}^{Q}\left|y_q - \tilde{y}_q\right|, \tag{12}$$

RMSE is defined as the square root of the mean squared error as:

$$RMSE = \sqrt{\frac{1}{N}\sum_{q=1}^{Q}(y_q - \tilde{y}_q)^2} \tag{13}$$

The difference of the RMSE between observed and predicted values was computed for each trial with different epoch numbers, and the best structure was determined by the lowest value of the RMSE. Coefficient of Determination ($R^2$) is used, to measure of how well future outcomes are likely to be predicted by the framework. And can be given as:

$$R^2 = 1 - \frac{\sum_{q=1}^{Q}(y_q - \tilde{y}_q)^2}{\sum_{q=1}^{Q}(y_q - \bar{y}_q)^2} \tag{14}$$

where N is the number of testing patterns, $\tilde{y}_q$ is the predicted values, and $y_q$ is original value, $\bar{y}_q$ is the average of all original values to Q testing patterns. The value of $R^2$ lies in between 0 and 1.0. It is used to describe how well a regression line fits a set of predicted data. The value of $R^2$ near 1.0 indicates that a regression line fits the predicted data well, while value of $R^2$ closer to 0 indicates a regression line does not fit the predicted data very well. The model was trained for 200 epochs and it was observed that the most of the learning was completed in the first 90 epochs as the root mean square error (RMSE) settles down to almost 0.01785 at 130 th epoch. Figure 7 shows the training RMSE curve for the ANFIS model. After training the ANFIS, it is found that the shape of membership functions (Fig. 8) is slightly modified. This is because of the close agreement between the knowledge provided by the expert and input/output data pairs. Figures 9 and 10, compares the accuracy of eye-gaze and ANFIS prediction values.

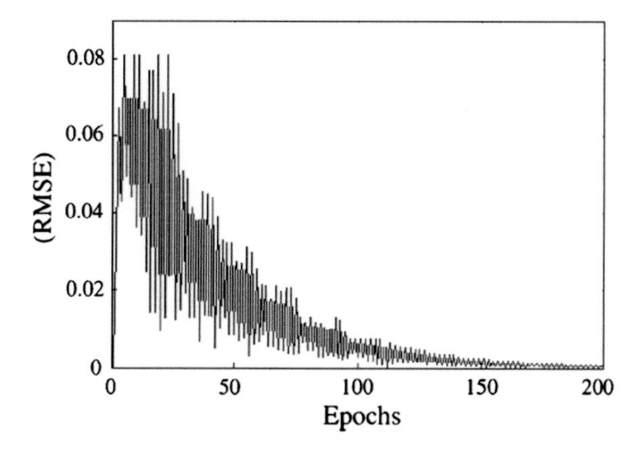


**Fig. 7** Training trend of errors with ANFIS

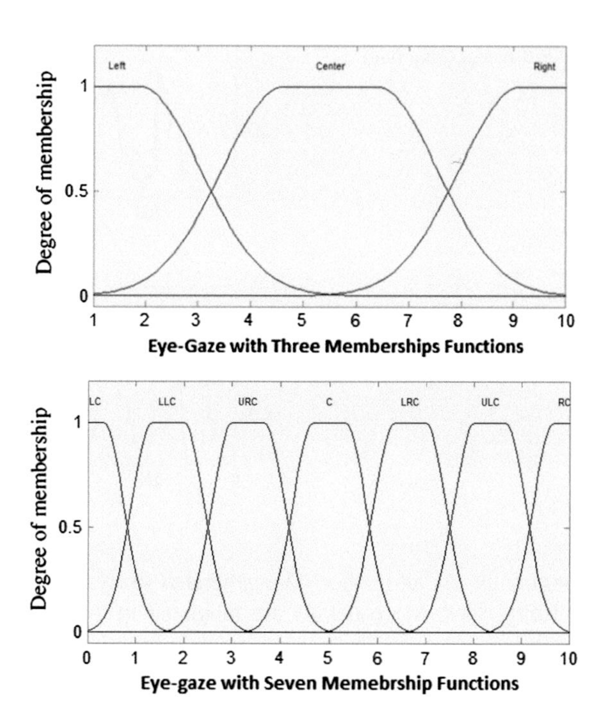


**Fig. 8** Membership functions of eye-gaze with three MF's (*Left*) and membership functions of eye-gaze with seven MF's (*Right*)

## 5 Numerical Results

The first system was structured by selecting two inputs and feeding them into the network given the name as function number (3*3). Two input parameters $x_m$ and $x_g$ were selected. There were three membership functions represented as "Centre, Right, Left". First, we tested our model with default values for membership function number (3*3); membership function types "Bell-shaped, Gaussian, Triangular" are used. These defaults provide membership functions on each of the

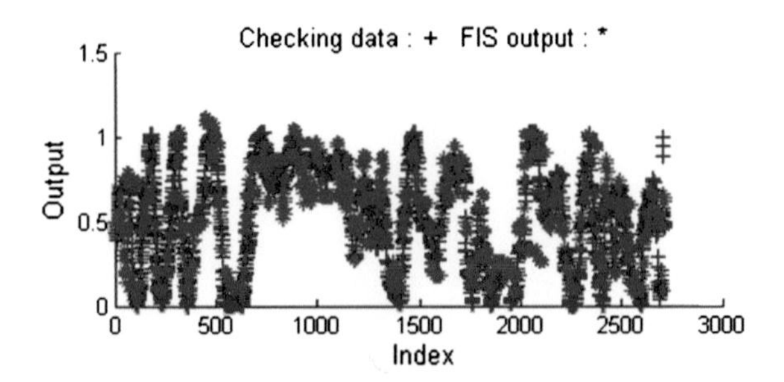


**Fig. 9** Modelling performance of estimated (*red*) versus real values (*blue*) using ANFIS in simulated environment

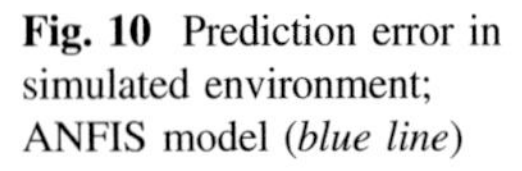
**Fig. 10** Prediction error in simulated environment; ANFIS model (*blue line*)

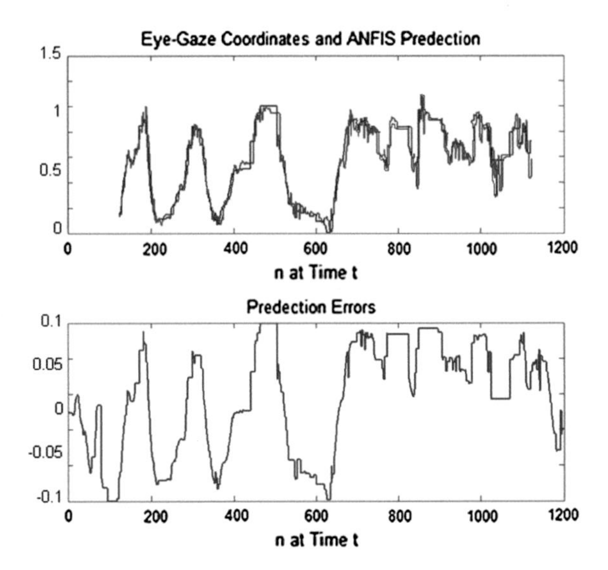


two inputs, six altogether. The generated fuzzy inference system structure contains 9 fuzzy rules. More details are tabulated in Table 1 with different measurement techniques.

Next, we use membership function number (13*13). The first system was structured by selecting two inputs and feeding them into the network. Two input parameters $x_m$ and $x_g$ were selected. The ANFIS has thirteen membership functions represented as "ULC, LLC, URC, LRC, UMC, LMC, RC, LC, G1C, G2C, G3C, G4C, C". Now we have used/consulted a human expert who has selected/suggested these membership functions. Again membership function types "Bell-shaped, Gaussian, Triangular" are used for number (13*13). These human experts provide membership functions on each of the two inputs, twenty-six altogether. The generated fuzzy inference system structure contains 169 fuzzy rules. More details are tabulated in Table 2 with different measurement techniques.

**Table 1** The scenario$_1$ (membership functions number "3*3")

| ANFIS parameter | Scenario$_{(1)}$ | | Scenario$_{(1)}$ | | Scenario$_{(1)}$ | |
|---|---|---|---|---|---|---|
| Number of inputs | 2 | | | | | |
| Membership function type | Bell-Shaped | | Gaussian | | Triangular | |
| Number of membership function | 3*3 (Right, Centre, Left) | | | | | |
| Training data set | 20 | 28 | 20 | 28 | 20 | 28 |
| Checking data set | 40 | 50 | 40 | 50 | 40 | 50 |
| Epoch number | 10 | 17 | 10 | 17 | 10 | 17 |
| Number of nodes | 35 | | 35 | | 35 | |
| Number of linear parameters | 27 | | 27 | | 27 | |
| Number of non-linear parameters | 18 | | 12 | | 18 | |
| Total number of parameters | 45 | | 39 | | 45 | |
| Number of fuzzy rules | 9 | | | | | |
| MAE | 0.03923 | 0.04175 | 0.01843 | 0.02888 | 0.02510 | 0.03479 |
| RMSE | 0.19807 | 0.20432 | 0.13576 | 0.16994 | 0.15843 | 0.18652 |

**Table 2** The scenario$_2$ (membership functions number "13*13")

| ANFIS parameter | Scenario$_{(2)}$ | | Scenario$_{(2)}$ | | Scenario$_{(2)}$ | |
|---|---|---|---|---|---|---|
| Number of inputs | 2 | | | | | |
| Membership function type | Bell-Shaped | | Gaussian | | Triangular | |
| Number of membership function | 13*13(ULC,LLC,URC,LRC,UMC,C,LMC,RC,LC,G1C, G2C,G3C,G4C) | | | | | |
| Training data set | 10 | 30 | 10 | 30 | 10 | 30 |
| Checking data set | 20 | 60 | 20 | 60 | 20 | 60 |
| Epoch number | 8 | 15 | 8 | 15 | 8 | 15 |
| Number of nodes | 195 | | 195 | | 195 | |
| Number of linear parameters | 305 | | 305 | | 305 | |
| Number of non-linear parameters | 28 | | 18 | | 28 | |
| Total number of parameters | 333 | | 323 | | 333 | |
| Number of fuzzy rules | 169 | | | | | |
| MAE | 0.01547 | 0.05432 | 0.01325 | 0.02234 | 0.02364 | 0.04587 |
| RMSE | 0.12441 | 0.23307 | 0.11511 | 0.14947 | 0.15375 | 0.21417 |

Next, we use membership function number (5*5). The first system was structured by selecting two inputs and feeding them into the network. Two input parameters $x_m$ and $x_g$ were controlled. Has thirteen membership functions represented as "ULC, LLC, URC, LRC, C". Again, membership function types "Bell-shaped, Gaussian, Triangular" are used for number (5*5). These human experts provide membership functions on each of the two inputs, ten altogether. The generated fuzzy inference

**Table 3** The scenario$_3$ (membership functions number “5*5”)

| ANFIS parameter | Scenario$_{(3)}$ | | Scenario$_{(3)}$ | | Scenario$_{(3)}$ | |
|---|---|---|---|---|---|---|
| Number of inputs | 2 | | | | | |
| Membership function type | Bell-Shaped | | Gaussian | | Triangular | |
| Number of membership function | 5*5 (ULC, LLC, URC, LRC, Centre) | | | | | |
| Training data set | 60 | 80 | 60 | 80 | 60 | 80 |
| Checking data set | 120 | 160 | 120 | 160 | 120 | 160 |
| Epoch number | 10 | 20 | 10 | 20 | 10 | 20 |
| Number of nodes | 265 | | 265 | | 265 | |
| Number of linear parameters | 440 | | 440 | | 440 | |
| Number of non-linear parameters | 20 | | 15 | | 20 | |
| Total number of parameters | 460 | | 455 | | 460 | |
| Number of fuzzy rules | 25 | | | | | |
| MAE | 0.04923 | 0.06875 | 0.02884 | 0.03888 | 0.03254 | 0.04847 |
| RMSE | 0.22188 | 0.26220 | 0.16982 | 0.19718 | 0.18042 | 0.22016 |

system structure contains 25 fuzzy rules. More details are tabulated in Table 3 with different measurement techniques.

Next, we use membership function number (7*7). The first system was structured by selecting two inputs and feeding them into the network. Two input parameters $x_m$ and $x_g$ were controlled. Has seven membership functions represented as “ULC, LLC, URC, LRC, RC, LC, C”. Again membership function types “Bell-shaped, Gaussian, Triangular” are used for number (7*7). A human expert provides membership functions on each of the two inputs, fourteen altogether. The generated fuzzy inference system structure contains 49 fuzzy rules. More details are tabulated in Table 4 with different measurement techniques.

The ANFIS models are evaluated based on their performance in training and checking sets as shown in Tables 1, 2, 3 and 4. The ANFIS models have shown significant performance variations against the evaluation criteria in terms of data sample, number and type of membership functions. It appears that the ANFIS models are more accurate and consistent in different subsets, where average values of RMSE and MAE are 0.20896 and 0.04367 respectively.

The experimental results shown in the tables above demonstrate that the fuzzy rule base selected by the human expert produces consistent results for the test data used i.e. 10 and 18 in Scenario$_{(1)}$,20 and 62 in Scenario$_{(2)}$, 60 and 80 in Scenario$_{(3)}$ and, finally, 60 and 100 in Scenario$_{(4)}$. However, not all situations are covered by expert’s fuzzy rules and some missing rules are detected by ANFIS. Tables 1, 2, 3 and 4 show that all situations for all input attributes are covered by the sets of 9, 169, 25, and 49 rules; however some of the rules have been found to produce illogical decisions because the training sample size was not large enough to cover all possible cases.

**Table 4** The scenario$_4$ (membership functions number "7*7")

| ANFIS parameter | Scenario$_{(4)}$ | | Scenario$_{(4)}$ | | Scenario$_{(4)}$ | |
|---|---|---|---|---|---|---|
| Number of inputs | 2 | | | | | |
| Membership function type | Bell-Shaped | | Gaussian | | Triangular | |
| Number of membership function | 7*7 (ULC,LLC,URC,LRC,RC,LC,C) | | | | | |
| Training data set | 60 | 100 | 60 | 100 | 60 | 100 |
| Checking data set | 100 | 210 | 100 | 210 | 100 | 210 |
| Epoch number | 15 | 30 | 15 | 30 | 15 | 30 |
| Number of nodes | 397 | | 397 | | 397 | |
| Number of linear parameters | 900 | | 900 | | 900 | |
| Number of non-linear parameters | 45 | | 30 | | 45 | |
| Total number of parameters | 945 | | 930 | | 945 | |
| Number of fuzzy rules | 49 | | | | | |
| MAE | 0.05923 | 0.08675 | 0.04843 | 0.07852 | 0.05963 | 0.08561 |
| RMSE | 0.24337 | 0.29453 | 0.22007 | 0.28021 | 0.24419 | 0.29259 |

The models are evaluated based on their performance measures (RMSE and MAE) in training and checking data-sets. The results shown in Table 1, 2, 3 and 4 reflect that ANFIS model have shown significant performance variations against the evaluation criteria in terms of data sample, number of membership functions, and type of membership functions. Scenario$_{(2)}$, which consists of two inputs and membership function(3*3), has minimum RMSE,MAE and the highest correlation. It appears that the ANFIS models are accurate and consistent in different subsets as the lowest value of RMSE is 0.11511 Scenario$_{(2)}$ and the highest value of the RMSE is 0.29453 Scenario$_{(4)}$.

Both the type and number of MF are important in building the ANFIS architecture. The required number of membership functions is determined through trial and error based on error values. It indicates that each ANFIS model is very sensitive to the type and number of MF. Increasing the number of MF's per input does not necessarily increase the model performance, but usually leads to model over fitting. The RMSE values were used to determine the best MF and epoch number in order to select the best fit model. From these results, the Gaussian membership function with epoch numbers of 8 and 15 was found to be the best fit model with the lowest RMSE value and with epoch 15 and 30 (Scenario$_{(2)}$ and Scenario$_{(4)}$).

Tables 1, 2, 3 and 4 show different approximations regarding different epoch numbers and membership functions (3*3, 13*13, 7*7, and 5*5). It is shown that the number of training samples has influenced ANFIS behaviour by adding new possibilities to produce more acceptable results. For Fuzzy Logics it is of great importance to train the system with different training data that tries to covers different possible data samples. Training is also important to build ANFIS architecture; as they include epoch numbers, the error, the initial step size, and decrease or increase rate of the step size.

In Scenario$_{(1)}$, Scenario$_{(2)}$, Scenario$_{(3)}$ and Scenario$_{(4)}$ different epoch numbers are used. One important point is that increasing the epoch number for a training data set does not necessarily improve the system performance significantly but it helps to overcome the problem of over-fitting. The trend of predicted data using ANFIS is illustrated in Figs. 11 and 12 along $x$ and $y$ axis respectively. Figure 11 illustrates that predicted data using ANFIS is converging more to the regression line. Approximately 91.23 % of the predicted data fits along the regression line for $x$-axis, while in case of Fig. 12, 89.01 % predicted data is converging to the regression line for $y$-axis.

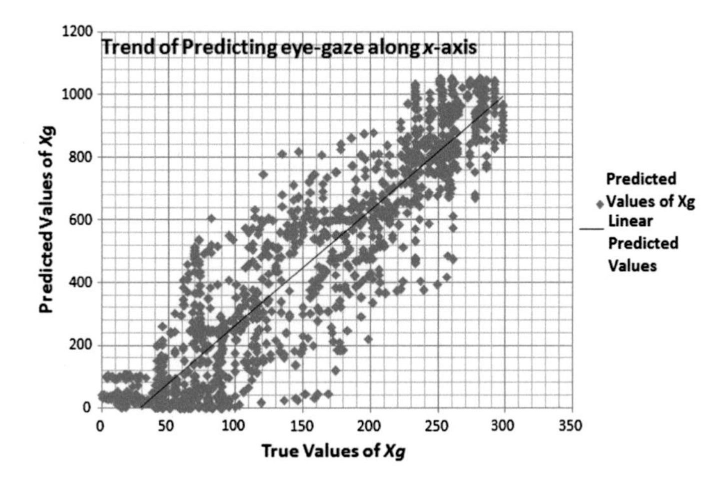


**Fig. 11** Trend of predicted $x_g$ using PAA_ANFIS

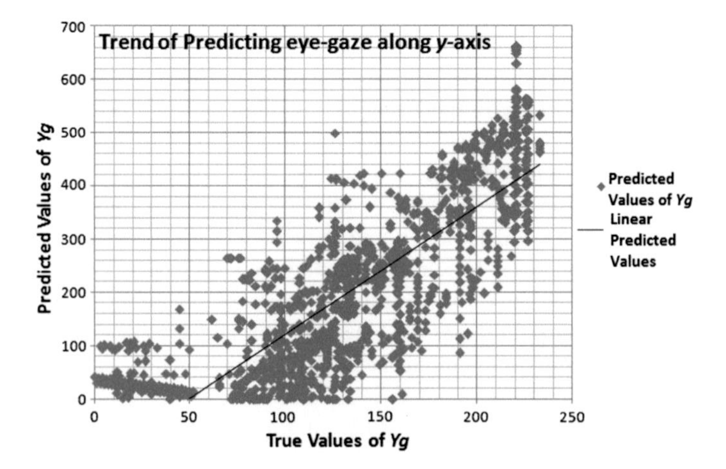


**Fig. 12** Trend of predicted $y_g$ using PAA_ANFIS

## 5.1 Comparative Performance Analysis

The performance analysis and a comprehensive difference of ANFIS and linear regression (LR) are presented in Table 5, on the basis of trend of generated data, fitness of test and predicted data, trend of error and execution time. The results obtained have been compared using 150 training data and 125 testing data pairs.

It is observed that ANFIS result is highly correlated with least RMSE as comparison with LR. Decreasing rate of error is higher because of its hybrid learning algorithm. Mean execution time is also computed which is competitively low. Figures 13 and 14 depict the fitness of generated data over test data. The CORR and RMSE between predicted and test data are shown in Table 6.

Results in the tables are sufficient proof, that ANFIS model performance is, in general, accurate and acceptable, where some rules are covered by human expert training data-sets and the missing rules are extracted by ANFIS. Furthermore ANFIS has capability of generating linear relationship in input-output data. The results of

**Table 5** The comprehensive analysis of ANFIS and LR for eye-gaze prediction

| Factors | ANFIS | LR |
|---|---|---|
| Capturing of fuzziness | Using FIS | Not available |
| Correlation in test and generated data | High | Less |
| Coefficient of determination | High | Less |
| Error rate | Low | High |
| Execution time | Less | High |

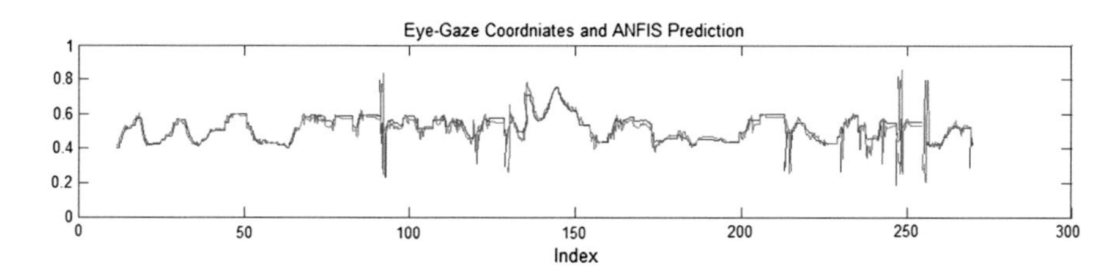


**Fig. 13** Trend of prediction of eye-gaze coordinates (*green line*); ANFIS model (*blue line*)

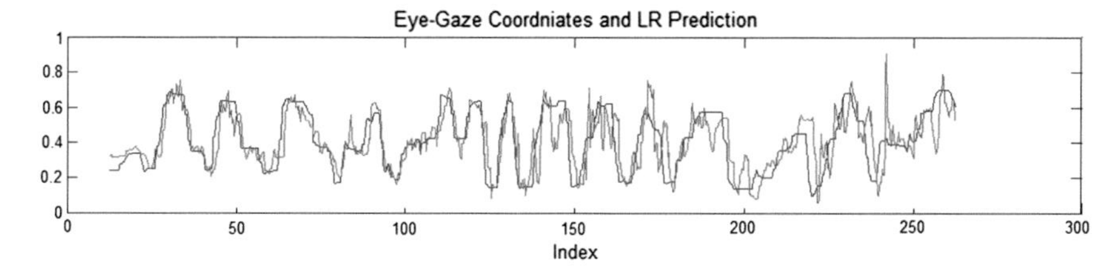


**Fig. 14** Trend of prediction of eye-gaze coordinates (*green line*); LR model (*blue line*)

**Table 6** Comparative analysis of ANFIS and LR for eye-gaze prediction

| | CORR | RMSE |
|---|---|---|
| ANFIS | 0.9134 | 0.1038 |
| LR | 0.8269 | 0.4873 |

the use of ANFIS models demonstrate that ANFIS can be successfully applied to predict the eye-gaze visual attention area in the environment.

## 6 Conclusions

In this research, we used ANFIS model to predict the user attention area using an ordinary web-camera. We implemented the PAA_ANFIS in developed game to investigate its accuracy. The performance of ANFIS was evaluated using standard error measurements techniques which revealed the optimal setting necessary for better predictability. PAA_ANFIS adopted a hybrid approach that combined the Fuzzy Inference System with the Neural Network in determining a complete fuzzy rule system. Combining a fuzzy inference system into an ANFIS Structure, we take advantage of neural methodology to train a fuzzy system that increases its capability for dealing with nonlinearity and imprecision in addition to its easier implementation.

The PAA_ANFIS approach has successfully solved the problem of prediction. The utilized performance measures confirmed higher forecasting accuracy of our model in comparison to Linear Regression. Hence ANFIS appears as a better choice then linear regression.

Future research on this scenario needs to study new techniques which are evolving and online. These techniques have advantage over ANFIS, that we do not need to ask expert to define membership function. As these memberships functions are even artificial which cannot reflect the real world scenario. So we expect that using an evolving technique will increase the feasibility of developing more effective and predictive systems.

**Acknowledgments** We gratefully acknowledge Prof. Dr. Plamen Angelov (leads the intelligent System Research Group at School of Computing and Communications Lancaster University) for providing valuable comments and suggestion during span of this research work, without which we have been unable to complete this research.

## References

1. Aarts, E., Harwig, R., Schuurmans, M.: Ambient intelligence. In: Denning, P. (ed.) The Invisible Future, pp. 235–250. McGraw Hill, New York (2001)
2. Aarts, E., Collier, R., van Loenen, E., de Ruyter, B. (eds.): Ambient intelligence. In: Proceedings of the First European Symposium, EUSAI2003. Lecture Notes in Computer Science, vol. 2875, pp. 432. Springer (2003)

3. Abraham, A.: Adaptation of fuzzy inference system using neural learning, fuzzy system engineering: theory and practice. In: Nedjah, N., et al. (eds.), ch. 3, pp. 53–83. Springer, Berlin, Germany (2005)
4. Angelov, P., Filev, D. (2004) An approach to on-line identification of takagi-sugeno fuzzy models. IEEE Trans. Syst. Man Cybern. Part B—Cybern **34**(1), 484–498. ISSN 1094-6977
5. Anifowose, F.A., Labadin, J., Abdulraheem, A.: Prediction of petroleum reservoir properties using different versions of adaptive neuro-fuzzy inference system hybrid models. Int. J. Comput. Inf. Syst. Ind. Manage. Appl. **5**, 413–426 (2013)
6. Areerachakul, S.: Comparison of ANFIS and ANN for estimation of biochemical oxygen demand parameter in surface water. Int. J. Chem. Biol. Eng. **6**, 286–290 (2012)
7. Asteriadis, S., Karpouzis, K., Kollias, S.: A neuro-fuzzy approach to user attention recognition. ICANN'08: Proceedings of the 18th international conference on Artificial Neural Networks, Part I, pp. 927–936. Springer, Berlin, Heidelberg (2008)
8. Berenji, H.R., Khedkar, P.: Learning and tuning fuzzy logic controller through reinforcements. IEEE Neural Netw. **3**(5), 724–740 (1992)
9. Bosse, T., Hoogendoorn, M., Klein, M., Treur, J.: A component-based ambient agent model for assessment of driving behaviour. In: Sandnes, F.E. et al. (eds.) Proceedings of the 5th International Conference on Ubiquitous Intelligence and Computing, UIC'08. Lecture Notes in Computer Science, vol. 5061, pp. 229–243. Springer (2008)
10. Bosse, T., Maanen, P., van, P., Treur, J.: Simulation and formal analysis of visual attention. Web Intell. Agent Syst. J. **7**, 89–105 (2009)
11. Bosse, T., Hoogendoorn, M., Memon, Z.A., Treur, J., Umair, M.: An adaptive human-aware software agent supporting attention-demanding tasks. In: Yang, J.-J., Yokoo, M., Ito, T., Jin, Z., Scerri, P. (eds.) Principles of practice in multi-agent systems, pp. 292–307. Springer, Berlin, Heidelberg, Standard: ISSN: 0302-9743 ISBN: 978-3-642-11160-0 (2009b)
12. Boyacioglu, M.A., Derya, A.: An adaptive network-based fuzzy inference system (ANFIS) for the prediction of stock market return: the case of the Istanbul Stock Exchange, expert systems with applications, vol. 37, pp. 7908–7912. Elsevier (2010)
13. Elwakdy, A.M., Elsehely, B.E.: Speech recognition using a wavelet transform to establish fuzzy inference system through subtractive clustering and neural network (ANFIS). Int. J. Circuits Syst. Signal Process. **2**(4) (2008)
14. Gärdenfors, P.: How Homo Became Sapiens: On The Evolution Of Thinking. Oxford University Press, Oxford (2003)
15. Haykin, S.: Neural Networks: A Comprehensive foundation second edition". Pearson Prentice Hall, Delhi India (2005)
16. Horvitz, E., Kadie, C., Paek, T., Hovel, D.: Models of attention in computing and communication: from principles to applications. Commun. ACM **46**(3), 52–59 (2003)
17. Itti, L., Koch, C.: Computational modeling of visual attention. Nat. Rev. Neurosci. **2**(3), 194–203 (2001)
18. Jang, J.: ANFIS: adaptive-network-based fuzzy inference system. IEEE Trans. Syst. Man Cybernet. **23**, 665–685 (1993)
19. Jang, J.S.R., Sun, C.T.: Neuro-fuzzy modeling and control. Proc. IEEE **83**(3), 378–406 (1995)
20. Jang, J.S., Gulley, N.: Fuzzy Logic Toolbox User's Guide, the Math Works Inc (1995)
21. Jang, J.S., Sun, C.T., Mizutani, E.: Neuro-fuzzy and Soft Computing: A Computational Approach to Learning and Machine Intelligence. Prentice-Hall, New Jersey (1997)
22. Jassbi, J.J.: A Comparison of Mamdani and Sugneo Fuzzy Inference Systems for a Space Fault Detection Application, IEEE, pp.1–8 (2008)
23. Klir, G.J.: Fuzzy Sets and Fuzzy Logic, PHI publications, ISBN:81-203-1136-1 (1995)
24. Mashrei, M.A.: Neural network and adaptive neuro-fuzzy inference system applied to civil engineering problems. In: Fuzzy Inference System - Theory and Applications, ISBN 978-953-51-0525-1, Hard cover, 504 pp. InTech (2012)
25. Mao, Y., Suen, C.Y., Sun, C., Feng, C.: Pose estimation based on two images from different views. In: Eighth IEEE Workshop on Applications of Computer Vision (WACV), Washington, DC, USA, IEEE Computer Society 9 (2007)

26. Memon, Z.A., Oorburg, R., Treur, J., Umair, M., de Vos, M.: A Software environment for a human-aware ambient agent supporting attention-demanding tasks. In: Tenth International Conference on Computational Science, ICCS'10, Elsevier, pp. 2033–2042, vol. 1 (2010)
27. Mellit, A., Soteris, A.K.: ANFIS-based modelling for photovoltaic power supply system: A case study. Renew. Energy **36**, 250–258 (2011)
28. Meyer, A., BÄohme, M., Martinetz, T., Barth, E.: A single-camera remote eye tracker. In: Lecture Notes in Artificial Intelligence, pp. 208–211. Springer (2006)
29. Moreno, J.: Hydraulic plant generation forecasting in Colombian power market using ANFIS. Energy Econ. **31**, 450–455 (2009)
30. Riva, G., Vatalaro, F., Davide, F., Alcañiz, M. (eds.) Ambient Intelligence. IOS Press (2005)
31. Rumelhart, D.E., Hinton, G.E., Williams, R.J.: "Learning internal representations by error propagation. In: Rumelhart, D.E., James, L., McClelland (eds.) Parallel Distributed Processing: Explorations in the Microstructure of Cognition, vol. 1, ch. 8, pp. 318–362. MIT Press, Cambridge (1986)
32. Shing, J., Jang, R.: Input selection for ANFIS learning. In: Proceedings of IEEE Fifth Int'l Conference Fuzzy Systems (1996)
33. Tarzia, S.P., Dick, R.P., Dinda, P.A., Memik, G.: Sonar based measurement of user presence and attention. In: Ubicomp'09: Proceedings of the 11th international conference on Ubiquitous computing, pp. 89–92. ACM, New York (2009)
34. Turatto, M., Galfano, G.: Color, form and luminance capture attention in visual search. Vision. Res. **40**, 1639–1643 (2000)
35. Vijila, C.K.S., Kumar, C.E.S.: Interference cancellation in EMG signal using ANFIS. Int. J. Recent Trends Eng. **2**(5), 244–248 (2009)
36. Webb, G.I., Pazzani, M.J., Billsus, D.: Machine learning for user modeling. User Model. User-Adap. Inter. **11**(1/2), 19–29 (2001)
37. Zounemat-Kermani, M., Teshnehlab, M.: Using adaptive neuro-fuzzy inference system for hydrological time series prediction. Appl. Soft Comput. **8**, 928–936 (2008)

# Integration of Knowledge Components in Hybrid Intelligent Control Systems

**M.B. Hadjiski and V.G. Boishina**

**Abstract** In the presented investigation some problems of Intelligent Building Blocks (BBs) integration are considered with emphasis on the Hybrid Intelligent Control System (HICS) design. The main target is to guaranty relevant operability of the complex system into each possible operation situation in autonomous mode. A design method is adopted using Case-Base Reasoning (CBR) approach. Each case has by—tuple description, involving arbitrary situation and corresponding control system design (strategy, structure, building blocks, algorithms and tuning parameters). The design procedure aims to incorporate all available information for the current situation—internal state (environment and equipment conditions), economic circumstances (throughput, quality, objective function), limitations (resources, ecological), environmental interconnections (disturbances, variability). A reconfiguration approach is accepted using situation-based control combined with a sequential improvement. The properties of IBB integration are studied giving prominence to knowledge oriented of them—ontology, case- and rule-based reasoning, intelligent agents and multi-agent systems (MAS). The loose integration is accepted in this investigation. A number of simulation results are presented and discussed for variety of Hybrid Intelligent Control Systems (HICS), which confirm the effectiveness of the proposed approach for complex, non-linear, faulty systems with large variability of feasible situations.

**Keywords** Control · Design · Hybridization · Integration · Intelligence · Knowledge · Reconfiguration · Situation

M.B. Hadjiski
Institute of Information Technologies, Bulgarian Academy of Sciences (BAS) and University of Chemical Technology and Metallurgy,
8 Kliment Ohridski Blvd, 1756 Sofia, Bulgaria
e-mail: hadjiski@uctm.edu

V.G. Boishina (✉)
University of Chemical Technology and Metallurgy,
8 Kliment Ohridski Blvd, 1756 Sofia, Bulgaria
e-mail: venelina_b@yahoo.com



V. Sgurev et al. (eds.), *Innovative Issues in Intelligent Systems*,
Studies in Computational Intelligence 623, DOI 10.1007/978-3-319-27267-2_3

# 1 Introduction

Traditionally the notion "Hybrid Systems (HS)" have been addresses to these structures where continuous and discreet elements work together in order to solve control tasks for tracking and disturbance rejection [1, 2]. In the last 10 years the spread of the same notion was radically changed due to the efforts and results of many investigators from different fields of interests [3–5]. According to the up-to-date understanding the Hybrid Systems (HS) could integrate a great variety of functional elements for data and signal acquisition, processing and saving, reasoning, decision making, learning, reactive and proactive action, knowledge representation, retaining, HMI and etc. Including components from the computational intelligence like Neural Networks (NN), Fuzzy Systems (FS), Genetic Algorithms (GAs), bio-inspired optimization methods (e.g. Ant Colony Optimization—ACO), knowledge-based elements such as Ontology (O), Rule-Based Reasoning (RBR), Case-Based Reasoning (CBR) as well as autonomous Agents (A) these integrated systems are now usually defined as Hybrid Intelligent Control Systems (HICS) [3, 4, 6].

The presented paper is an attempt to summarize the experience of the authors in design, analysis and application of HICS [7–12, 24]. The organization of this paper is as follows. In the next section the related works are considered. The main problems to be discussed as well the accepted generalized hybrid structures are presented in Sects. 3 and 4. The formal description and building blocks (BB) peculiarities of Hybrid Intelligent Control System (HICS) are summarized in Sect. 5. The design principles and reconfiguration-based control of HICS are under consideration in Sect. 6. A number of studied hybrid intelligent control structures are analyzed and compared in Sect. 7. The main simulation results are systematized in Sect. 8. Some conclusions are derived in the last section.

# 2 Related Works

Recently a trend toward dominating emphasis on development of new promising schemes for Intelligent Systems via integration of separate components could be observed [3–5]. The building blocks—NN, FL, GAs, O, RBR, CBR, MAS have reached already high maturity in functional and computational aspects [5, 13–15]. The results received show that the combination of the well known intelligent components in the complex system is more effective than the creation of new methods and techniques. In a number of investigations [6, 8, 9, 16, 17] is shown that HIS could reach considerably better performance in comparison with the systems, containing isolated BB. Via hybridization the weaknesses and limitations inherent to the separate intelligent building blocks could be overcome to gain better functionality paying by some increased complexity [3, 11, 12, 18, 19, 32].

There are a lot of HIS design, but the analysis of really coupled HIS [3–6] show that it looks as two main approaches are formed—with Tightly (T) and Loosely (L) integration [5, 12, 16, 23]. Into the T-approach functional fusion is accomplished by mutual algorithmic penetration of building blocks [6, 17]. The design in this case is both structural and parametrical. Unfortunately the T-integration is successful only for relatively small systems with 2–3 intelligent components. The main reached results are in integration between NN, FL, Gas, BBs in various combination and used techniques [5, 6, 17]. Only in few works there are coupled O and CBR [18, 20], FL and O [9, 21], CBR and RBR [19, 22] building blocks. Limited numbers of investigation are addressed to implementation of MAS in HICS. The analysis of existing results show the necessity of extensive elaboration of new methods for real fusion, relevant approaches for incorporating domain knowledge, software realization (e.g. common language interfaces). The T-integration appears to have preference for the systems with limited dimensionality, more compactness and homogeneity, with an automatic realization of accepted control actions.

The interest toward the theoretical aspects of the loose integration (L) is considerable during the last years [5, 13, 16, 23, 24]. The emphasis here is directed on building blocks functionality choice, interconnections, cooperative optimization and synchronization. The separate building blocks save relative independence but they are strongly coordinated and parametrically optimized to realize the common goal. The HICS could contain legacy control components and to remain open for evolving [3, 10, 11]. The L-integration seems to be more efficient for complex uncertain systems with a large dimensionality and higher level of abstraction, where some kind of domain knowledge could be used.

A wide variety of HICS application have been reported resent years: industrial plants and systems [3–5, 32], power plants [10, 12], ecological systems [17], transport and robotics [3, 5, 32]. In some investigations have been shown that the HICS give promising results in fulfilling specific tasks like data analysis [5], nonlinear control [6], adaptive control [12, 16], diagnosis [8, 10, 11, 25] and machine learning [16, 22].

## 3 Problem Formulation

This paper is focused on the HICS design based on loosely coupled building blocks to control in hierarchical and distributed structures. In contrast of previous works where the components are prevailing from the area of computational intelligence—NN, FL and GAs, we incorporate knowledge-based elements as Ontologies (O), Case-Based Reasoning (CBR) [13], Rule-Based Reasoning (RBR) [27]. As a main technique for realization of HICS we accept Multi-Agent Systems (MAS) containing reactive and proactive intelligent agents [15]. The Ant Colony Optimization (ACO) [14] was found as relevant optimization technique for all considered HICS systems both in structural and parameterization aspects. The loose coupling poses a lot of questions into the different design stages as a relevant procedure for specifying the components for knowledge representation and reasoning, queering, blending for the candidate solutions, learning and/or adaptation approaches.

Other dimension along which have been directed our integration tasks have been related with the synchronization of different intelligent building blocks, orchestrating in MAS. The independent problem represents the granularity optimization of the building blocks. From a practical point of view it is important to find realistic and effective way to integrate new HICS elements with the existing legacy components and conventional algorithms using JAVA as a basic language platform. In our research an attempt is made to consider the integration procedures of building blocks with their software realization via Multi-Agent System as united problem. Thus in this investigation the HICS could be considered as computational intelligence-, knowledge-, rule- and agent-based control systems.

In this investigation we have much broader scope in BBs using, than conventional or advanced control algorithms where are mainly control BBs have been encapsulated. We emphasis on knowledge oriented systems but using also components from computational intelligence in order to expand the functional capabilities of the control system, tacking into account business requirements, operational conditions and equipment safety. The system must be multiface, able to overcome all emerging situations, with a possibility to refine its behavior via learning and adaptation, to incorporate in a flexible and efficient way the existing legacy part. In our attempt to accomplish this scope at least in part we accept the holistic approach stressing on the importance of proper choice of the structure, BBs and their algorithms and tuning parameters.

The situation based control the plant is considered as laying in a sequences of situations $H_i$ (Fig. 1):

$$H_m \to H_{m+1} \to H_{m+2} \to \cdots \to H_{m+n} \tag{1}$$

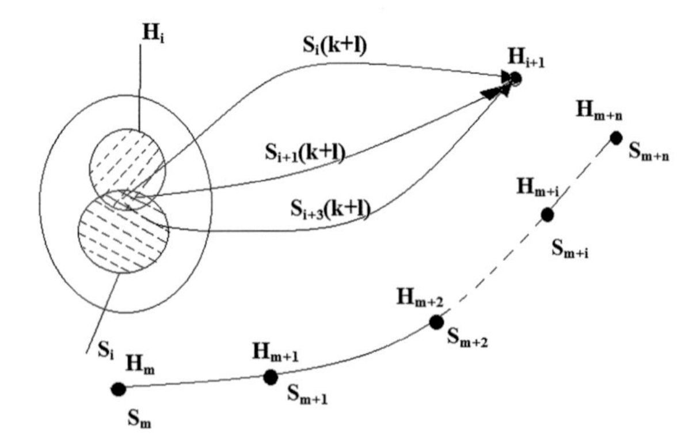


**Fig. 1** System's situation and state

The situation $H_i$ is considered as ascertainable case with six measurable attributes, which could be presented as seven-tuple:

$$H_i = \langle S, E, BC, J, B, TS, TC \rangle \tag{2}$$

where *S* is generalized system state, *E*—environment state, *BC*—business conditions, *J*—current objective function, *B*—current bound limitations, *TS*—technological settings, *TC*—technological conditions of the equipment and software.

The generalized state of the system S is considered as a dynamic case described by five-tuple:

$$S = \langle S_d, S_s, HS, PR, \Omega \rangle \tag{3}$$

where $S_d$ is the dynamic state, $S_s$—static state, *HS*—current hybrid control structure, *PR*—current tuning parameters of Building Blocks and $\Omega$—current operational area.

Dynamic state is presented in state–space from:

$$\dot{x} = f(x, u, v) \tag{4}$$

$$y = g(x, u, v) \tag{5}$$

where $x \in R^m$ is a state space variable, $u \in R^m$ is a control variable, $y \in R^m$ is a observable output, $v \in R^m$ is a disturbance, f(.), g(.) are non linear functions. $S_s$ defines borders of quasi-linearity, *HS*—describes current building blocks configuration and links among them and could be changeable during the trajectory. Disturbances *v* are considered as components of environmental state *E*.

The plant state S could belong to one of the tree operational areas, as it shown in Fig. 2.

- $\Omega_1$—area of normal functioning according prescribed constraints.
- $\Omega_2$—area of acceptable (but only for special situation and requirements) functioning, where stopping is expedient, despite of operator errors or/and some admissible technical troubles.
- $\Omega_3$—area of dangerous situations, which must be leaved as soon as possible.

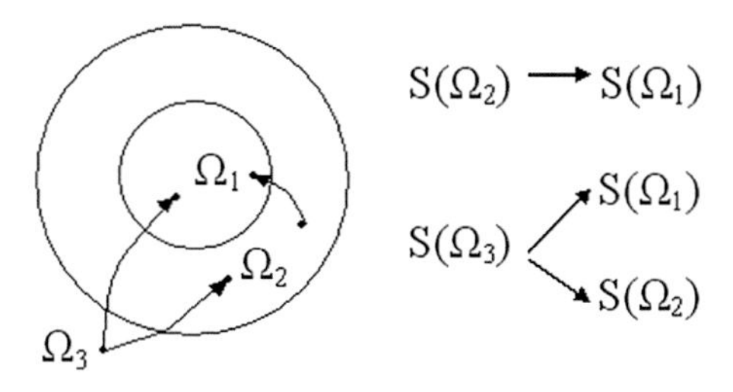


**Fig. 2** Operational area of HICS

The main goals in this investigation are:

1. To the study the integration problems in Hybrid Intelligent Control Systems (HICS) with a loose integration in the next directions: (i) Consideration the HICS with more than two intelligent building blocks, with emphases on knowledge oriented elements—Ontology (O), Case-Based Reasoning (CBR), Rule-Based Reasoning (RBR); (ii) Structural problems of HICS in dependence on control and behavioral goals, information and available knowledge; (iii) Investigation of the building blocks blending methods in order to overcome their individual limitations and disadvantages.
2. To find suitable way to combine two strategies: "Top-down" (current information and available knowledge driven) and "Bottom-up" (data-driven, acquired by learning).
3. To examine some problems of Hybrid Control Systems design in the following aspects: (i) Integration of intelligent based (FL, GAs) and knowledge based (O, CBR, RBR) building blocks with legacy part of the control system; (ii) Clarifying for the higher software degree re-using of basic hybrid structures; (iii) To construct structures easily fastable from the user.
4. To consider some programmatic tasks related to Hybrid Intelligent Control System realization, namely: (i) Using JAVA as a basic platform; (ii) Implementation of autonomous Agents (A) and Multi-Agent Systems (MAS) as structural elements/fragments; (iii) Synchronization and coordination in HICS.

# 4 Structure of Hybrid Intelligent Control Systems

## *4.1 Generalized Hybrid Structure*

The generalized structure of HICS accepted in this work is represented in Fig. 3. It contains the next main properties:

- The architecture contains functional elements, building blocks and their interconnections. The functional elements accomplish different functions and structural fragments formation, situation determination, control and operational action execution. The Building Blocks are—internally independent elements which fulfill control, intelligent local and high level data and information processing. The interconnections are intended for logical relations realization, data and information exchange and coordination.
- The accepted structure could realize loose integration of functional elements and building blocks by specification, allocation, optimizing, blending and adapting the various components saving in the same time their relative functional independence. The generalized structure used for HICS allow to fulfill three types of control behavior (Fig. 3): (i) Reactive on the basis of instrumental or human

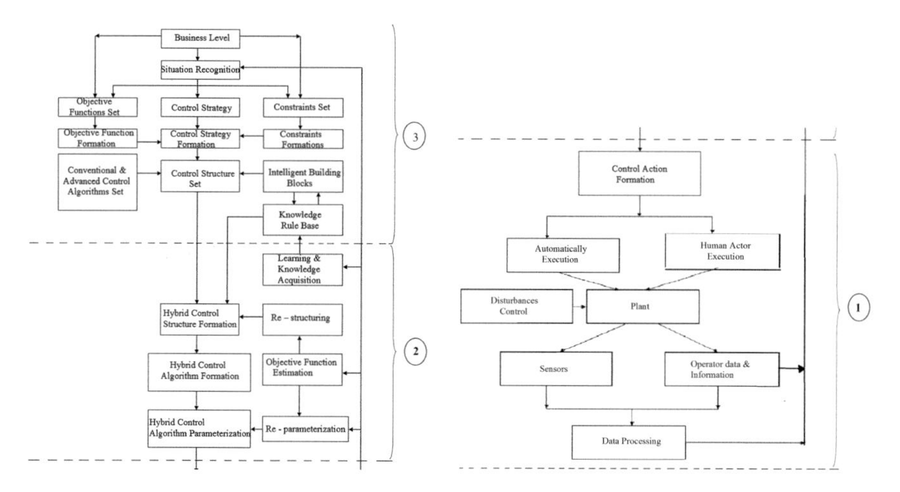


**Fig. 3** Generalized HICS structure

feedback (zone 1); (ii) Adaptive via restructuring and re-parameterization (zone 2); (iii) Proactive by posing new targets, objective functions, requirements (zone 3).

- The HICS is oriented basically for automatic operation, but it allows a human participation as well (information for monitoring and assessments generation, rules formation, control actions execution in specific situations).

The proposed generalized structure enable the user to tailored specific structures closely related to the particular control problems that he is trying to solve.

## 4.2 *Problem of Integration*

The common base for intelligent system synthesis is given in [24] on the base of Generalized Nets (GN). In the present contribution some practical aspects of the postulated theoretical auditions are considered toward of control systems design. The Building Blocks (BBs) are seen as key structural elements and without some of them the intelligent control system would suffer from structural problems. The scheme of possible integration of intelligent BBs is shown at Fig. 4.

We consider here three types of building blocks (BBs):

1. BBs for control (set C). Here are included: (i) conventional control algorithms (CC)—e.g. PID controllers; (ii) improved conventional controllers (IC)—e.g. two degree of freedom PID, feed forward control, cascade control; (iii) advanced control (AC)—e.g. multivariable, model predictive, fuzzy and neural, adaptive, state space controllers and etc. [3, 4]. As it is well known [28], the application of this BBs in industrial process control systems are in relation 100:10:1. The

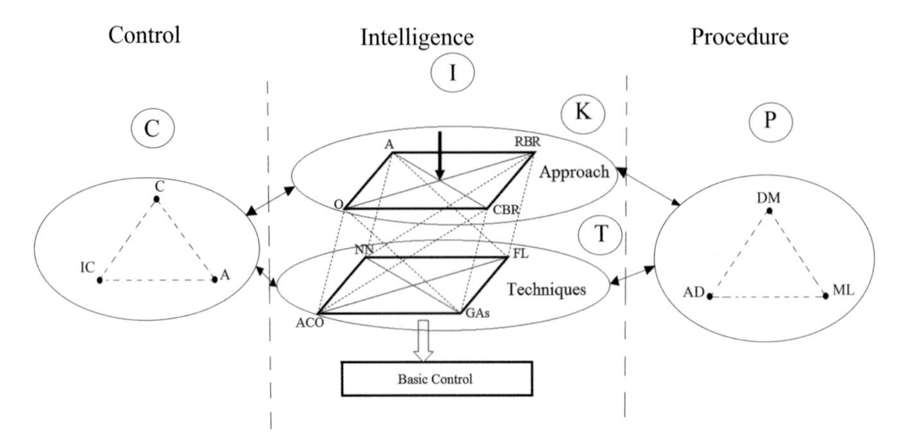


**Fig. 4** The structure of BBs integration

controllers could be new components into the structure, but quite often they represent a legacy part, which must be taken into account in order to be re-ingenerated in an optimal way.

2. Intelligent BBs (set I, (i) subset K with knowledge based elements). They are divided in two groups—with Ontologies (O), Case-Based Reasoning (CBR), Rule-Based Reasoning (RBR) and autonomous Agents (A) and Multi-Agent Systems (MAS); (ii) subset T from the computational Intelligence techniques—Neural Networks (NN), Fuzzy Logic (FL), Genetic Algorithms (GAs) and bio-inspired methods (in our case Ant Colony Optimization (ACO)).
3. BBs for a high level integration processing (set P)—Adaptation (AD), Machine Learning (ML) and Data Mining (DM).

The emphases of the present contribution are toward problems of integration among Intelligent building blocks (K and T subsets). The integration could be:

- Horizontal, when:

$$k_{ij} = k_i \cup k_j \text{ or } T_{ij} = T_i \cup T_j \quad i \neq j \quad (i, j = 1, 4) \tag{6}$$

- Vertical, when:

$$k_i \cup T_j \tag{7}$$

- Common horizontal–vertical, when:

$$k_{ij} \cup T_{ij} \tag{8}$$

**Table 1** Qualitative information about possible BBs integration

| | Knowledge processing approach | | | | Computational intelligence techniques | | | |
|---|---|---|---|---|---|---|---|---|
| | A | O | CBR | RBR | NN | FS | GAs | ACO |
| A | | *A* | *F* | *A* | *A* | *A* | *F* | *A* |
| O | *A* | | *F* | *F* | *L* | *A* | *L* | *F* |
| GBR | *F* | *F* | | *A* | *A* | *M* | *L* | *F* |
| RBR | *A* | *F* | *A* | | *A* | *M* | *F* | *A* |
| XV | *A* | *L* | *A* | *A* | | *M* | *M* | *F* |
| FS | *A* | *A* | *M* | *M* | *M* | | *M* | *A* |
| GAs | *F* | *L* | *L* | *F* | *M* | *M* | | *F* |
| ACO | *M* | *F* | *F* | *A* | *F* | *A* | *F* | |

The results published of binary integration between intelligent building blocks are extremely non-uniformed. Asserting our best knowledge the qualitative estimation of the corresponding binary combination available results are given in Table 1, where the lack (L), few (F), average (A) and many (M) notions for estimates are used. It can be seen that the integration of pairs with knowledge based Building Blocks (A, O, CBR, RBR) are still studied purely. Investigations with triple integration (e.g. $k_i \cup k_j \cup k_l$) are examined very scare. In more general approach we construct a Hybrid Intelligent Control Systems (HICS) structure using building blocks (BBs) from all sets—K, T, C, P:

$$ST = \langle k \cup T \cup C \cup P \rangle \tag{9}$$

The investigation of this type of aggregation, are extremely rare. Thus our contribution is oriented to study double, triple and more sophisticated structures of integration.

## 5 HICS Building Blocks

The emphases in this contribution is directed to intelligent building blocks, which are knowledge oriented (subset K in Fig. 4)—O, CBR, A, RBR. The BBs from the set C (control algorithms) will be not considered bellow in details because they are accepted as “Legacy Part”, both in a case of a new design or re-engineering of control systems as well. The BBs from the set P (Procedures) strongly depend on specific conditions of the system—the control tasks, level of the plant properties and behavior understanding, the uncertainness.

### 5.1 Fuzzy Systems (FS), Neural Networks (NN) and Genetic Algorithms (GAs)

The FS, NN and GAs are very well studied in all their aspects [3–5, 27]. In this work they will be considered only as parts of the hybrid systems. Thus here of a dominant interest are their type ($\phi$), structural ($\psi$) and parametrical ($\lambda$) properties. Each of the computational BBs could be presented in the form

$$T_i = T_i(\phi_i, \psi_i, \lambda_i) \tag{10}$$

where $T_i$ are elements of the sub set T of Computational Intelligence (CI) techniques [NN (i = 1), FS (i = 2), GAs (i = 3)] (Fig. 4). In Table 2 the components of $\phi_i$, $\psi_i$ and $\lambda_i$ are represented for the most often used intelligent building blocks in Hybrid Control Systems.

### 5.2 Ontology

The Ontology (O) is considered as one of the most powerful tool in the knowledge intensive systems [5, 18]. Quite often the O is integrated in the HICS in order to link the domain knowledge with conventional or advanced control BBs in situations

**Table 2** Properties of CI BBs

| Building block | Structural properties | | |
|---|---|---|---|
| | Type ($\phi$) | Structural parameters ($\Psi$) | Learning (tuning) parameter ($\lambda$) |
| Neural networks (NN) | Multi layer perceptron (MLP) | • Number of hidden layers (HL)<br>• Number of RBF in HLs | Weight metrics, bias |
| | Redial basic function (RBF) | Number of nodes in HLs | • Centers of RBF ($C_i$)<br>• Size ($\sigma_i$)<br>• Output weights |
| Fuzzy systems (FS) | Fuzzy system Type 1 T1–FS | • Dimension<br>• Type of member of shape functions (MF)<br>• Number of single tones in MF | Parameters of MF |
| | Interval Type 2 Fuzzy logic system T2–FS | • The same as in T1–FS<br>• Model of linguistic uncertainty representation | • The same as in T1–FS<br>• The size of uncertainty |
| Genetic algorithms (GAs) | Genetic algorithms (GAs) | • Chromosome structure<br>• Selection method | • Population size<br>• Mutation rate<br>• Crossover rate |

$H_i$, where some logical decisions are needed for relevant reaction. There exist a lot of ontology formal models. One simple, but representative enough model for static ontologies (SO) has the form:

$$O = \langle X, R, \Sigma \rangle \tag{11}$$

where $X$ is set of concepts, $R$ describes the relationships between the concepts, $\Sigma$ represents the interpretations, e.g. in the form of axioms. In many cases the triple $\langle X, R, \Sigma \rangle$ is adequate to represent static knowledge. But in some cases to the triple must (Eq. 11) be added temporal attributes $A(k)$ in order to take into account concepts changes or system evolving. The model (Eq. 11) should be extended and dynamic ontology (DO) could be presented as follows:

$$DO = \langle X, R, \Sigma, A(k) \rangle \tag{12}$$

The main structural characteristics of the ontology as a building block in HICS are classified in Table 3. In the HICS could be used specific individual Ontologies for knowledge sharing among the subsystems. The advantages are simplicity and flexibility, combining Ontologies with common structure, language and knowledge patterns. This gives a lot of advantages and thus is preferable [12].

The composite as well as individual but interconnected Ontologies could be converted in different forms for use in multiple target control systems. By the "bottom-up" procedures (Fig. 3) new concepts could be built, composed from those already existing and checked for consistency, allocating it at a correct position in the ontology structure. A variety of Ontology transformations are available in order to improve the knowledge representation using feedback information.

The main advantages of the Ontologies integration in HICS are: (i) A better semantic description of the local (domain) and global (upper) knowledge; (ii) The

**Table 3** Properties of Ontology based BBs

| | Type ($\phi$) | Structural parameters ($\Psi$) | Learning tuning parameters ($\lambda$) | Operation on ontologies (–) |
|---|---|---|---|---|
| Static ontology O | Upper ontology<br>Domain ontology | • Semantic structure<br>• Knowledge patterns<br>• Parameters of hierarchical structure<br>• Strategy of extension | • Level of confidence<br>• Threshold | • Extension<br>• Happing<br>• Merging<br>• Mapping<br>• Matching<br>• Adaptation<br>• Fuzzy fication<br>• Reconciliation<br>• Rectification<br>• Refining<br>• Evaluation |
| Dynamic ontology O | Domain ontology | • The same as O<br>• Structures of dynamic patterns | • The same as O<br>• Temporal horizon for adaptation<br>• Dynamic constraints in models and/or time series | |

Ontology concepts, attributes and relationships are consensual among corresponding experts; (iii) The Ontology strongly support the automatic classification based on search, querying or browsing; (iv) A scalability of Ontology is possible; (v) A number of standards, tools and languages supporting the ontology engineering have been developed. The researches in Dynamic Ontologies are promising way to incorporate the temporal attributes in the knowledge presentation.

Some disadvantages of the Ontologies implementation as building blocks are: (i) An Ontology creation takes a lot of time and money; (ii) The maintenance of the Ontology is not easy; (iii) the disambiguation and cleaning knowledge from diverse sources is still a serious problem in order to reach cross-level agreement.

## 5.3 Case-Based Reasoning (CBR)

The more sophisticated control tasks accepted in this contribution (including dynamic performance, condition based maintenance, imposing different economical constraints and goals) make reasonable integration of the HICS building blocks, which contain existing retrospective operational experience. It turn out that the Case-Based Reasoning (CBR) could be useful in many complex control systems, because: (i) For the most of the new occurred situations $H_i$, the existing "problem-solving" situations stored in the Case Base could given appropriate decision based on the principal of "similar-situation/similar decision"; (ii) The contemporary Decentralized Control Systems (DCS) contain a huge volume of the similar reiterated information from which the current situations could be recognized and represented with all necessary details in accepted form; (iii) The Human-operator experience is embedded into the solution part of the cases; (iv) The cases could be evaluated according to all accepted objective criteria; (v) The CBR theory is very well developed [5, 13].

In the CBR approach we can use examples of past "problem-solutions" to solve arising problem directly in contrast of the Rule-Based systems, where we must firstly convert this examples into general rules.

The cases are usually presented in the established "problem-solution" form:

$$C = C(P, S) \tag{13}$$

where $C$ is a case, $P$—problem, $S$—solution.

The problem $P_i$ is structure represented via attributes $a_i$ and corresponding values $v_{ij}$ for each of them:

$$P_i = P_i(a, v) \tag{14}$$

$$a = (a_1, a_2, \ldots, a_r) \tag{15}$$

$$v = \begin{vmatrix} v_{11} & v_{12} & \cdots & v_{1q_1} \\ v_{12} & v_{22} & \cdots & v_{2q_2} \\ \cdots & \cdots & \cdots & \cdots \\ v_{r1} & v_{r2} & \cdots & v_{rq_r} \end{vmatrix} \quad (16)$$

The solution $S$ in this contribution is accepted to consider two main vector-components:

$$S = S(G, M(L)) \quad (17)$$

where $G$ is the type of candidate-control algorithm

$$G = (g_1, g_2, \ldots, g_l), \quad (18)$$

$M$—maintenance action depending on diagnosis state ($L$) of the control system in corresponding operational area $\Omega_i$ (Fig. 2).

$$M = (m_1, m_2, \ldots, m_d) \quad (19)$$

$$L = (l_1, l_2, \ldots, l_v) \quad (20)$$

The current operational area $\Omega_i$ is defined using a threshold classifier

$$\text{IF } l_1 < l_1^t \text{ and } l_2 < l_2^t \text{ THEN } \Omega \in \Omega_i \quad (21)$$

where $l_i^t$ are accepted threshold values of the respective component $l_i$ of the diagnostic state $L$. The possible maintenance actions $m_j$ could be "immediate inspection", "tuning", "small current repair", "repair", "replacing", "stop".

The well established $R_e^4$ CBR cycle [5, 13] is accepted in this work using the principle of local-global closeness [13] by a applying similarity measure in the form:

$$sim(P_n, P) = \sum_{i=1}^{r} \omega_i sim_i(a_n, a_j) \quad (22)$$

where $P_n$, $a_n$ are a new problem and its attributes. The part $sim_i(a_n, a_i)$ characterizes the local system similarity between attributes $a_n$ and $a_i$ and contains mainly specific knowledge of a given local area. The weights $\omega_i$ represent the relative importance of this attributes on the common closeness between $P_n$ and $P$.

The principal structural characteristics of CBR building blocks as a component of HICS are presented in Table 4.

Using CBR as a knowledge oriented building blocks gives the next advantages: (i) Application of a direct experience, which is implicitly presented in the solution part of the model; The lack of a part of the all data in attributes is not critical; (ii) If the density of the cases around the new arising problem is acceptable it is unlikely

**Table 4** Properties of CBR based BBs

| | Type ($\phi$) | Structural parameters ($\Psi$) | Learning tuning parameters ($\lambda$) | Operation on CBR (–) |
|---|---|---|---|---|
| CBR | Static case based reasoning | • Structure of the problem<br>• Structure of the solution<br>• Similarity measure<br>• Strategy for operation | • Parameters of the problem's attributes<br>• Parameters of the solution's attributes | • CBR fitting up with cases<br>• Cases replacement<br>• CB clustering<br>• Meta case formation<br>• CB adaptation |

to allow a big mistake in the solution; (iii) The quality of solution according the adopted criteria could be improved by replacing the old cases with new ones; (iv) Dynamization of CBR (Dynamic CBR) with introduction of temporal attributes and/or presence of episodes in a significant way could improve CBR representative abilities in cases with fast situations $H_i$ changing; (v) There exists a well accepted free software for CBR (e.g. jColibry [29]).

The disadvantages of the CBR building blocks are: (i) Quite often the whole operating area is not covered by cases with enough space density; (ii) The solution part of the case gives satisfactory but not optimal solution; (iii) The query of applicable cases in a large Case-Base (CB) may be complicated and time consuming.

## 5.4 Rule-Based Reasoning

A Rule-based systems are a kind of knowledge based systems, where the knowledge is represented in the form of a set of explicit rules. The rules can be used without specifying how and when to apply them. It is possible to add new rules, to modify the existing and to create new combination of rules. The rules are usually represented in IF…THEN form:

$$R : IF\ x_1(a_1)\ and\ x_2(a_2)\ and\ldots.\ THEN\ y_1(b_1)\ and\ y_2(b_2)\ \ldots. \tag{23}$$

where $x_i$ are conditions, $y_i$—conclusions or actions, $a_i$, $b_j$ are tunable values (constants or variables).

The rules could be deep-based on theoretical knowledge, or shallow-based on a knowledge derived from human experts.

A typical Rule-Based systems (RBS) contains three main structure elements—knowledge base (KB), working memory (WM) and inference machine (IM). The rule based systems use only generalized rules, which are organized in KB. All specific information (constant values $a_i$, $b_j$) is situated into WM. The Inference

**Table 5** Properties of RBR based BBs

| | Type ($\phi$) | Structural parameters ($\Psi$) | Learning/tuning parameters ($\lambda$) | Operation on rule-based systems (–) |
|---|---|---|---|---|
| Rule based systems (RBS) | Production systems | • Search strategy<br>–Data driven or goal driven<br>–Dept or breadth<br>• Strategy for RBS extension<br>–Size<br>–Scope<br>• Strategy for conflict resolving | • Rule based refinement by rule<br>–Adding<br>–Modifying<br>–Refactoring<br>• Values into the rules adaptation<br>–Threshold<br>–Constants | • Matching<br>• Updating<br>• Extending<br>• Evaluating<br>• Reset |

Machine (IM) realizes "recognize-act" cycle using a sequential procedure for conflicts resolution.

The Rule-Based Systems (RBS) are domain specific and knowledge intensive problem solvers. The rules make possible easy to express suitable pieces of knowledge. These allow the domain experts to criticize and improve the rule base. The major structural properties of the rule-based systems are presented in Table 5.

The choice of the search strategy in the Rule Base strongly depends on possible conclusion/action number. In case of many control alternatives e.g. variety of situations, trajectory planning, the direct rule-based control, the data driven (forward) reasoning is preferable. When the number of outputs of inference chain is limited e.g. in switching control, diagnosis, equipment maintenance, the goal driven (backward) reasoning is more suitable. Extending the size and scope of RBS by only adding more rules could be unsuccessful, because of possible rule interactions arising. This makes large rule-bases inefficient and hard to update. The RBS extension must be incremental one. Thus the sequential steps of size and scope exposition should be solved as optimization problem of the evolving control system.

The matching of the condition parts of the IF-THEN rules in the rule base produce a subset of applicable rules which could be from conflicted sub set. To select on adequate rule, which will be fired, some of the conflicted resolutions have to be fulfilled. The proper choice of the corresponding strategy is critical for the reasoning based processes.

The tuning parameters could be changed during the time of system prototype testing and in the time of a real operation using a feedback information (Fig. 3). In the RBS this includes adding, modifying and refinement the rules from the rule base and re-parameterization of thresholds and other tuning constants into the working memory (WM).

The rule based Building Blocks (BBs) have a number of advantages : (i) Easy to formulate and directly to use experimental knowledge; (ii) The design process is iterative; (iii) The corresponding software (XCON, OPSS, KEE, CLIPS, DROOLS) could be incorporated into the HICS.

The disadvantages are: (i) The RBS are very hard task depending from the system conditions; (ii) They have acceptable performance in a limited specific domain; (iii) There are difficulties to handle the missing information; (iv) Near to the boundaries of domain knowledge the fired rules lose the consistency; (v) The RBS has a restricted power of the knowledge representation in cases when the IF-THEN patterns are not adequate enough.

## 5.5 *Software Agent and Multi Agent Systems*

The software Agents (A) are suitable building blocks in a HICS design due to their broad functionality [10, 11, 15]. They could fulfill a number of functions, not allowable for conventional computer programs—to be autonomous, persistent, collaborative, reactive and proactive. The HICS could be designed or not into agent-based framework, but MAS implementation gives a number of advantages. The agents in the HICS can perform technological tasks (communication, interface, synchronization) or to be problem oriented as well. In this role they are able to emulate the computational intelligence based BB like NN, FL, GAs as well as to carry out their own intelligent functions—decision making, perception, prioritization, negation. This flexibility is impossible for the other intelligent techniques, especially in an autonomous regime.

The agents classification in this work is presented in Table 6. It corresponds to their role as Building Blocks (BBs) in a HICS design and author's experience in using more than 50 types of agents in complex control systems [10–12]. As a main design characteristics are accepted their type $\phi$, structural parameters $\psi$ and tuning parameters $\lambda$.

An individual agent (A) could be described as eight-tuple:

$$A = \langle S_a, \phi, \psi, \lambda, K_a, K_o, T_s, F_a \rangle \tag{24}$$

where $S_a$ is Agent internal state, $K_a$—own knowledge about system behavior, $K_o$—knowledge about the state of the other agents, $T_s$—the task set, $F_a$—the description of the functions thought the agent system, when the agents can influence to the behavior to the other agents presented into the system.

Multi Agent System (MAS) can be presented as:

$$MAS = \langle S_{MAS}, E_{MAS}, A_{MAS}, \Lambda, K_s, TS_{MAS}, C_{MAS}, B_{MAS} \rangle \tag{25}$$

where $S_{MAS}$ is MAS internal state, $E_{MAS}$—external environment description (e.g. situation $H_i$), $A_{MAS}$—the set of Agents belonging to MAS, $\Lambda$—structure of MAS, $K_s$—internal system knowledge (Ontologies (O), Knowledge Based (KB), Rule Based (RB), Case Based (CB)), $TS_{MAS}$—the set of tasks that MAS is faced with,

**Table 6** Properties of agents BBs

| | Type (ϕ) | Structural parameters (Ψ) | Learning/tuning parameters (PR) (λ) |
|---|---|---|---|
| Agents | *Technological agents* | • Modes of action<br>–Reactive<br>–Deliberative<br>• Choice of algorithms in problem oriented agents<br>• Learning<br>–Supervised<br>–Unsupervised<br>• Functionality<br>–Roles<br>–Goals<br>• Knowledge processing<br>–Representation<br>–Manipulation<br>–Type of used logic<br>–Knowledge allocation | • Thresholds<br>–Procedural<br>–Performance<br>–Diagnosis<br>• Weights<br>–CBR metrics<br>–Data fusion<br>–Reactivity/deliberatively balancing<br>–Decisions blending<br>–Models parameters<br>• Control algorithms tuning parameters |
| | • Interface<br>• Communication<br>• Information (“Yellow pages”)<br>• Synchronization<br>• Management (MA) | | |
| | *Problem-oriented agents* | | |
| | • Data processing<br>–Preprocessing<br>–Data manning<br>–Softsensing<br>–Case retrival<br>• Monitoring<br>–Situation recognition<br>–Performance assesment<br>–Fault status<br>–Alerts/Alarms<br>• Control<br>–Reconfiguration<br>–Tasks allocation<br>–Constraints satisfaction<br>• Performance improvement<br>–Learning<br>–Optimization<br>–Simulation<br>–Re-parameterization | | |
| | *Primitive agents* | | |

$C_{MAS}$—coordination mechanisms among the different Agents in MAS, $B_{MAS}$—description of MAS behavior.

If the HICS is designed completely in a MAS framework it could be presented as shown in Fig. 5 via aggregated agents: $A_1 - A_5$.

The advantages of MAS implementation in a HICS are: (i) The possibility of autonomous relevant behavior in a case of unanticipated situations; (ii) Easy realization of interconnections in a HICS—communication information exchange, coordination; (iii) Shearing common system knowledge containing KB, RB, CB.;

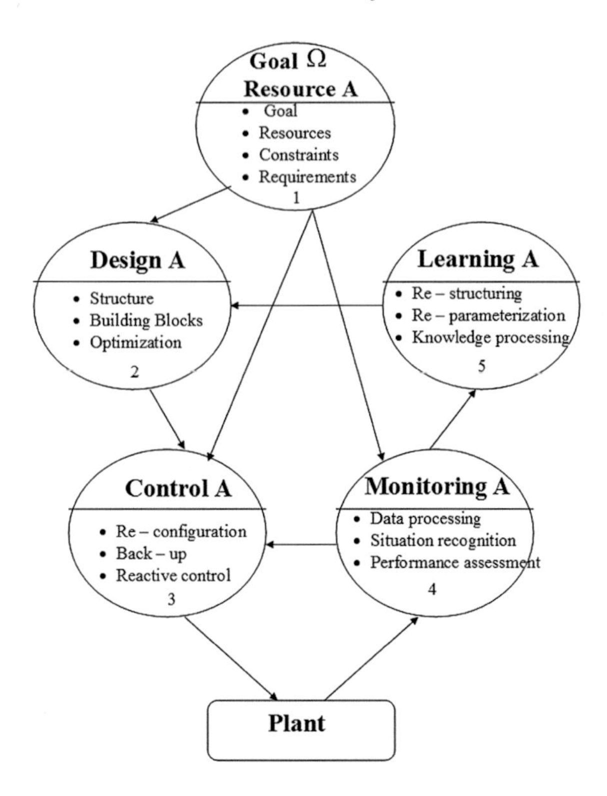


**Fig. 5** HICS design as MAS

(iv) Re-usability of individual BBs and full architecture as well; (v) Easy testing and verification.

The disadvantages are: (i) The creation of full MAS-based HICS is very hard, time consuming and costly task; (ii) The peculiarities of the MAS in complex HICS are still very poorly investigated; (iii) Variety of problems arises with the integration of the existing legacy part in MAS-based HICS.

## 5.6 Ant Colony Optimization (ACO)-Based Approach

The convincing results received in our previous works [10, 11] have stimulated further implementation of ACO in given contribution.

According the Fig. 4 the Agents (A) and Ontologies (O) interact each other in order to retrieve the knowledge, to update the pheromones in MAS and to adapt the system environment [14]. The ants communicate each other through the system environment represented by ontology. The control action is chosen according the probability the control to be "bad" or "good" [14]:

$$P_u(m) = \frac{(u_m + g)^b}{(u_m + g)^b + (l_m + g)^b} \tag{26}$$

$$P_l(m) = 1 - P_u(m) \tag{27}$$

where: $u_m$ is the number of ants accepted decision for "good"; $l_m$—number of ants accepted decision for "bad"; $P_u(m)$—the probability of decision to be "good"; $P_l(m)$—the probability of decision to be "bad"; g and b-parameters; m-number of all ants in the colony.

When the knowledge of the systems is updated, the ants spread the information about their believes for a "good" or "bad" decision and the quality of approved decisions in the current time k can be describe by the relation:

$$Q(k) = \frac{TP(k)}{TP(k) + FN(k)} \frac{TN(k)}{FP(k) + TN(k)} \tag{28}$$

$$\tau_{ij}(k+1) = \tau_{ij}(k) + \tau_{ij}(k)Q(k) \tag{29}$$

where: TP(k) is the number of the ants accepted $P_u(m)$ for "good" decision; TN(k)—number of the ants accepted $P_u(m)$ for "bad" decision; FP(k)—number of ants accepted $P_l(m)$ for "good" decision; TN(k)—number of the ants accepted $P_l(m)$ for "bad" decision; $\tau_{ij}(k+1)$ is the quantity of the pheromones; $Q(k)$—the approved decision into system.

The Eq. (28) gives the possibilities to define the quantity of pheromones which are necessary to update the information in the system $\tau_{ij}(k+1)$. That could be represented in the form of Eq. (29), where $\tau_{ij} = \sum_{i=1}^{a} b_i^{-1}$ is the quantity of dispersed pheromones; a—number of attributes included in decision making; b—the possible attribute values.

From Eqs. (28) and (29), the change of knowledge can be given as follows:

$$SK(k+1) = SK(k) + SK(k)Q(k) \tag{30}$$

where: SK(k), SK(k + 1)—are current and foregoing system knowledge.

Advantages: (i) Flexible; (ii) Possible to self-decision making; (iii) Working into co-operative way (conflict resolving); (iv) Adaptive to the current system state; (v) Spread the knowledge through the Ant Colony Optimization; (vi) Decision co-ordination into colony; (vii) Asynchronous way of work; (vix) Functionality re-placement.

Disadvantages: (i) A large rank of communication; (ii) A lot of rank of ants interaction; (iii) Necessity of time of knowledge receiving; (iv) Necessity of a management level; (v) Slow adaptation of the colony; (vi) Mish-match of knowledge about the current system state (Table 7).

**Table 7** Properties of ACO

| | Type ($\phi$) | Structural parameters ($\Psi$) | Learning/tuning parameters ($\lambda$) | Operation on rule-based systems (–) |
|---|---|---|---|---|
| Ant colony optimization (ACO) | • Active ants (decision making ants)<br>• Re-active ants (working computational) ants)<br>• Follower ant colony decision ants<br>• Management ants (queen Ants)<br>• Knowledge ants (spreading the knowledge into colony)<br>• Conflict resolution ants (warriors ants)<br>• Passive ants (spreading system knowledge ants) | • Autonomous acting ants<br>• Co-operation between ants—acting together to achieve the common goals<br>• Associations of ants—completing each-other functionality in a way to accomplish the common system targets<br>• Manager ants—manage the process of knowledge sharing, conflict resolution, behavior coordination and decision making | • Parameters of the searching the shortest way algorithms<br>• Numbers of the ants composing an association<br>• Number of the functions disposed between the ants<br>• Rank of the communications (relationships between ants)<br>• Rank of the interactions between ants | • Extension<br>• Merging<br>• Mapping of functionality<br>• Controlling the rank of communication<br>• Controlling rank of knowledge sharing<br>• Forming the Ants association based of the ants interests<br>• Evaluation of ACO behavior<br>• Fusion of ACO common work |

# 6 Reconfigurable Hybrid Intelligent Control System (HICS)

## 6.1 Design Principles

The main goal is to design a supervisor control system, which could face all possible operational situations in an effective way. Tree main design principles are accepted: Situation based; Hybridization via integration of variety of Building Blocks; Intelligent usage of knowledge, reasoning and learning.

As it was shown above the notion "situation" according to Eq. (2) includes except usually used internal state S, as well as main real business and managerial circumstances: the external state E (disturbances, environmental changes), the business conditions (productivity, dead lines, quality etc.), the set of actual criteria J (profit, accuracy, priority, experience, maintenance etc.), the variety of limitations (B) (time limits, resources, investments, ecological requirements etc.), the

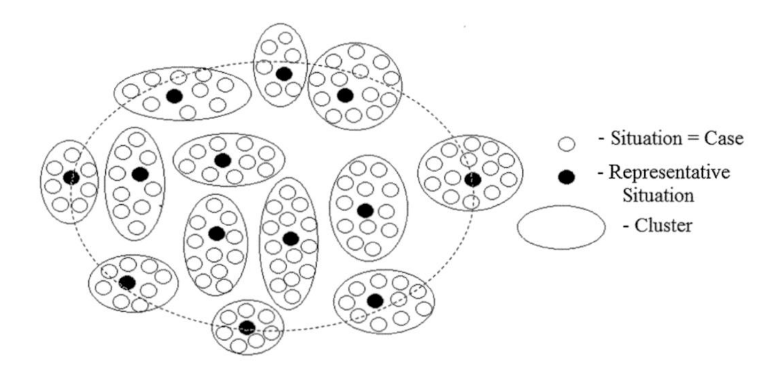


**Fig. 6** Set of simulations, corresponding clusters and representative situations

technological settings (TS) of different units (optimal temperature, pressure, recycle, degree, composition)., the technical condition of equipments (operational area Ω), Remain Useful Life (RUL) etc.). The situation is considered as a problem part P of a case in the CBR approach (Eq. 13):

$$H \equiv P = \langle S, E, BC, J, B, TS, TC \rangle \tag{31}$$

All attributes in Eq. (31) are presented in the form of Eq. (14) and it is assumed that they are measurable. The situations H are acquired from the archive trends, expert information, simulation and literature. They could be presented as a set of situations $H_i \in H$, as shown in Fig. 6. Using some cluster technology a number of representative situations $\tilde{H}_i$ could be received.

The solution parts of the case according to Eq. (13) is accepted to be design of the Hybrid Intelligent Control System (part D) presented as four-tuple:

$$S_l \equiv D = \langle ST, HS, AL, PR \rangle \tag{32}$$

where ST is a strategy, HS—Hybrid Structure, AL—Algorithms of Building Blocks (BBs), which forms HS, PR—parameters of BBs. The structure of the Hybrid Intelligent Control System (HICS) is considered as a number of BBs and interconnections (IC) between them:

$$HS_i = \langle BB_i, IC_i \rangle \tag{33}$$

Each Building Block is described using algorithms (AL) and corresponding parameters (PR):

$$BB_i = \langle AL_{ij}, PR_{ij} \rangle \tag{34}$$

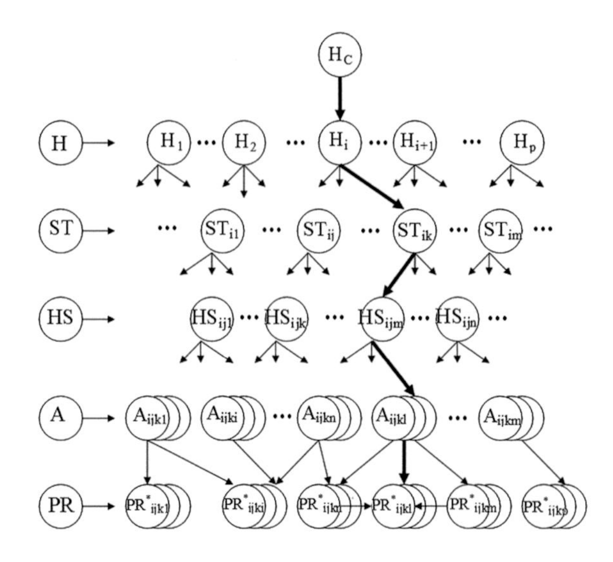


**Fig. 7** The scheme of situation oriented design

The objective function J in Eq. (31) is multi component:

$$J = \langle J_1, J_2, J_3 \rangle \tag{35}$$

The partial objective functions are as follow: $J_1$ is a estimation of the control system accuracy, $J_2$ is assessment of the unit economical efficiency, $J_3$ is evaluation of availability and safety of the unit.

The goal of the design stage is to derive for the each representative situation $\bar{H}_i$ an optimal control system $\bar{D}_i^*$:

$$\bar{D}_i^* = \bar{D}_i^*(\bar{H}_i) = \arg\min_{\bar{H}_j} J(D) \tag{36}$$

Each optimal design $\bar{D}_i^*(\bar{H}_i)$ contains optimal strategy $ST_i^*$, structure $HS_i^*$, algorithms $AL^*$ and respective parameters $PR^*$ according Eqs. (32)–(34). This is shown in Figs. 7 and 8.

The main goal is to use available information of the peculiarities of the plant, environment, management and in off-line regimes to derive a set of optimal designs $\bar{D}_i^*$, corresponding to representative situations $\bar{H}_i$, individual $BB_{ij}$, their optimal algorithms $AL_{ijk}$ and respective parameters $PR_{ijk}$ are reused in different situations as it is shown in Fig. 7.

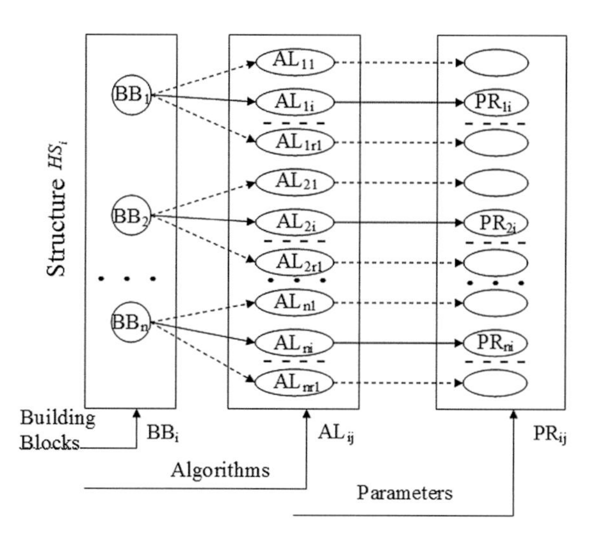


**Fig. 8** Structure $HS_i = \langle BB_i, AL_{ij}, PR_{ij} \rangle$

## 6.2 *Control Principals*

In the paper a situation—based control algorithm is proposed, containing the next steps (Fig. 9):

1. Using input (reference and plant data) to recognize the current situation $H_C$.
2. By retrieval in representative Situation Design Case Base (DCB) find the closest situation $\bar{H}_{NN}$ and respective design $\bar{D}_{NN}$.
3. After the analysis of the situation $H_C$ to make (if necessary) reconfiguration of the HICS using all information for the design $\bar{D}_{NN}$.

$$\bar{D}^*_{NN} = \langle ST^*_{NN}, HS^*(BB), AL^*(BB), PR^*(BB) \rangle \tag{37}$$

This is a "fast" reactive mode control, which provide supervisory formed references for the basic control level $A^C_C$ and maintenance $A^m_C$. The performance of reconfigured control system ($J^C$) will be not optimal because: (i) the current situation $H_C$ is not identical with the representative one $\bar{H}_{NN}$, (ii) the unknown characteristic of the plant and environment (e.g. models) are probably different from this used in the design procedure to receive $\bar{D}^*_{NN}$, (iii) $\bar{H}_{NN}$ and $\bar{D}_{NN}$ could be a result of human operation action, which are heuristic. Thus a new step is needed.

Improving the HICS behavior using data driven procedures. After the performance estimation towards the current objective function $J_C$ Eq. (35), a re-parameterization and (if necessary) a re-structuring should be fulfilled. This will be "slow" deliberative operation (Fig. 10).

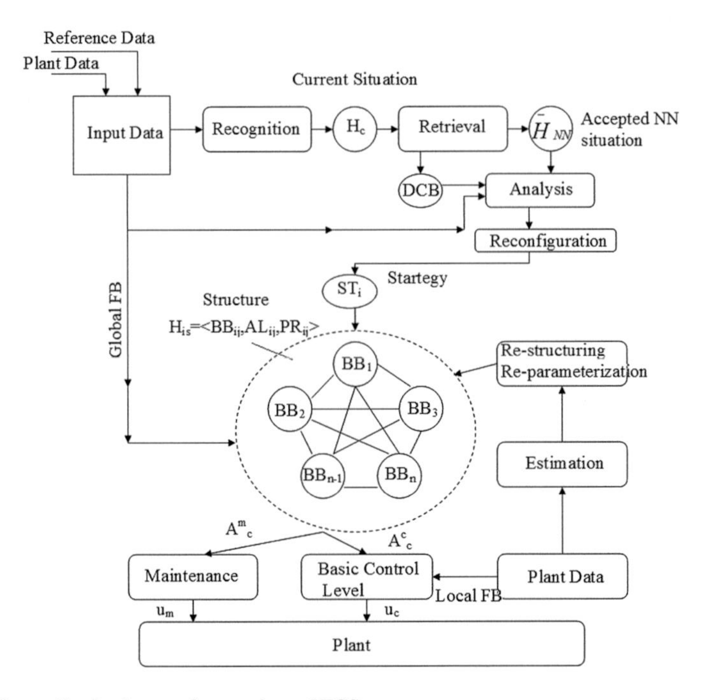


**Fig. 9** Generalized scheme of supervisory HICS

**Fig. 10** Nearest neighbor finding $H_C \rightarrow H_{NN}$

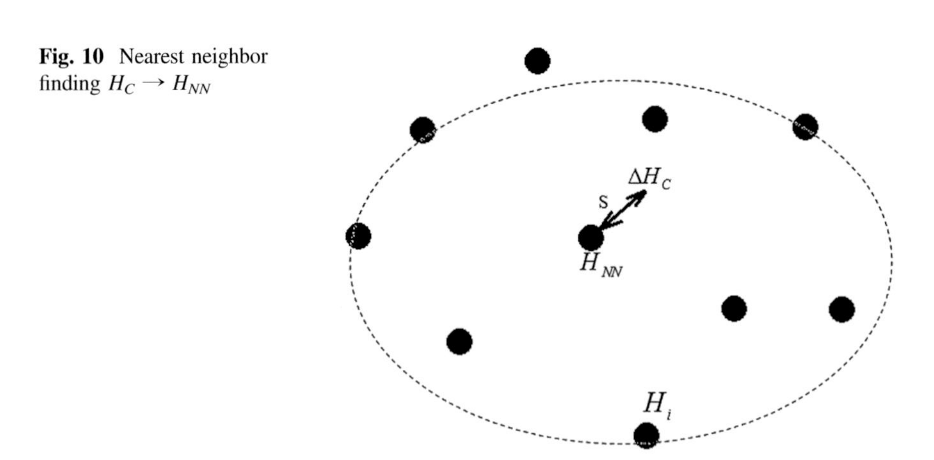


## 6.3 *Situation Based Design*

In the Fig. 11 is presented the accepted generalized situation based approach for Hybrid Intelligent Control System design. The goal is to create a correct and complete enough Design Case Base (DCB) corresponding to the set of representative situations $\bar{H}$ as a subset of feasible situation set $H_i$:

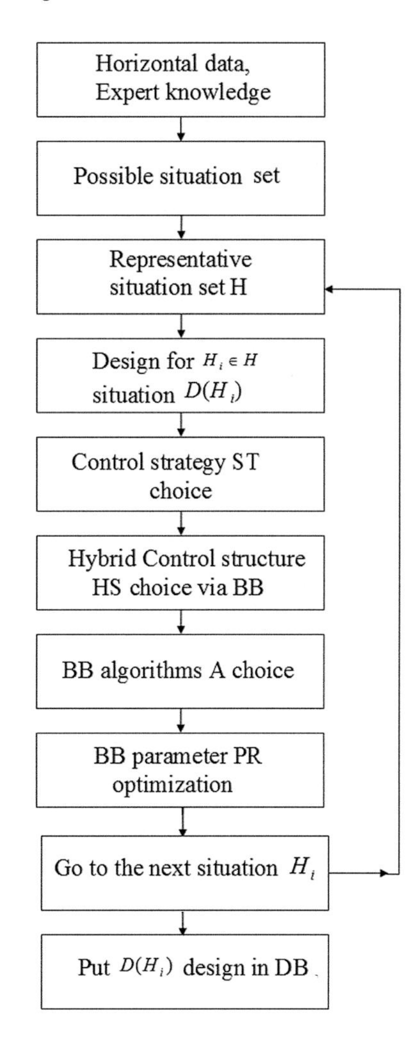


**Fig. 11** Situation based design scheme

$$C_{DCB} = \langle \bar{H}, \bar{D} \rangle \tag{38}$$

This goal is decomposed in four interrelated tasks for off-line solving the optimization problem Eq. (36). The used procedure is "top-down": Strategy (ST) → Hybrid Control Structure (HS) via Building Blocks (BBs) choice → $BB_i$ algorithms choice (AL) → Algorithms parameterization (PR). This is shown with more details in Fig. 8.

## 6.4 Reconfiguration Based Control of HICS

The scheme presented in Fig. 12 gives additional details of the reconfiguration control extending Figs. 8 and 9. After the current situation $H_C$ recognition a necessity of a reconfiguration is estimated. If the distance between new $H_C(k)$ and previous $H_C(k-1)$ situation exceeds given set of thresholds a reconfiguration is accomplished. If performance of the HICS using the design $\bar{D}_{NN}(\bar{H}_{NN})$ is not acceptable in accordance with the current objective function $J_C$, system stability

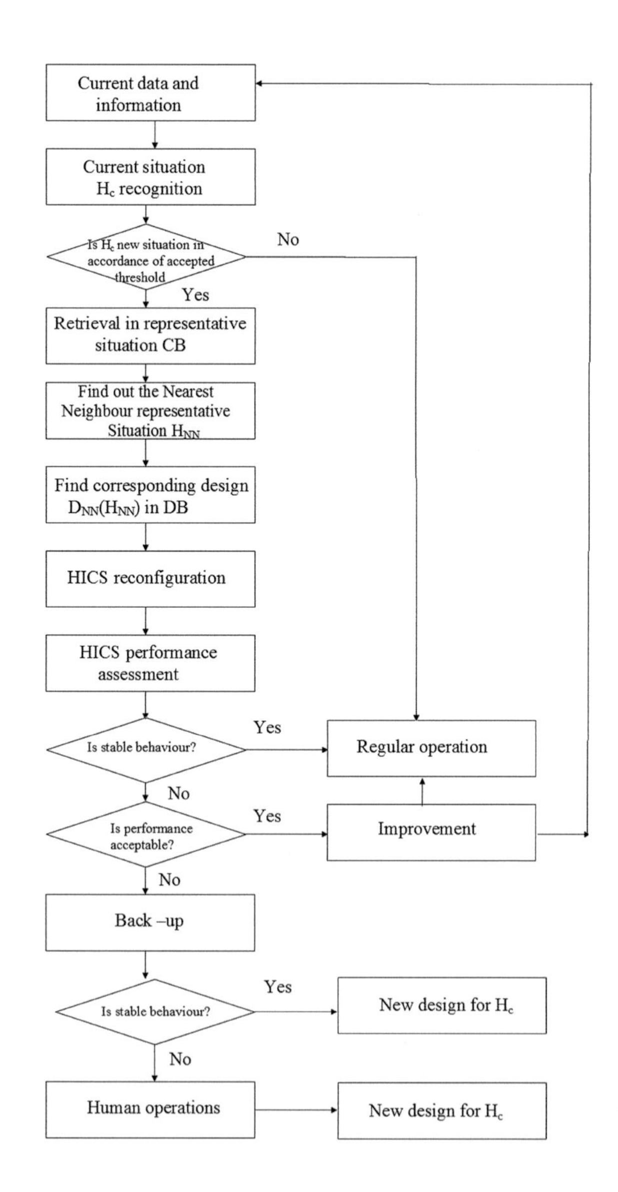


**Fig. 12** Reconfiguration based control of HICS

must be checked. In a case when the behavior of the HICS go to be unstable a "back-up" system or at the end a human operator must be switched on. In this case a re-design is necessary.

## *6.5 Improvement of HICS*

The improvement should be undertaken when the behavior of the HICS is not acceptable, but system is still stable in the new situation. Four sequential stages of improvement are foreseen: (i) Re-parameterization (PR changes in BB's algorithms); (ii) BB's algorithms alteration; (iii) Re-structuring (BB changes); (iv) Strategy alteration. This is shown in Fig. 13.

A transition forward the next stage is only in a case when the possibilities of the lower stage are exhausted.

The main problem of the HICS composition is to integrate different system functional components using basic Building Blocks. The problems are provoked from the specific system behavior into the current system state. Some of the functions of the HICS can be spread among different BBs. This is a complex task, which try to optimize the system work according to the system objective functions.

## *6.6 Joint Operation Problems*

The joint operations in HICS aim to achieve improvement of the system behavior in each possible situation $H_c$ satisfying a numerous established criteria—dynamic and static accuracy and robustness $J_1$, economical efficiently $J_2$, availability and safety $J_3$ (Eqs. (35) and (36)). Bellow will be considered the next joint operation problems: the system functions allocation; communication; synchronization.

### 6.6.1 System Functions Allocation

The functions allocation in hybrid control systems is not trivial but is still a poorly studied problem. In this paper we will consider for instance the integration of Legacy System (LS) with Agents (A), and Ontology (O) building blocks.

Instead the global criteria $J_1$, $J_2$, $J_3$ some local criteria estimating the control system performance in three aspects have been examined: $Q_1$—Dynamic accuracy of the control system; $Q_2$—Volume of the inter task communication; $Q_3$—Degree of the knowledge utilization.

To fulfill above requirements three sets of functions connected with LS, A, and O building blocks $F_{LS}$, $F_A$ and $F_O$ are formed. The degree of partial participation of Legacy System (LS), Agents (A) and Ontology (O) are defined by the relations:

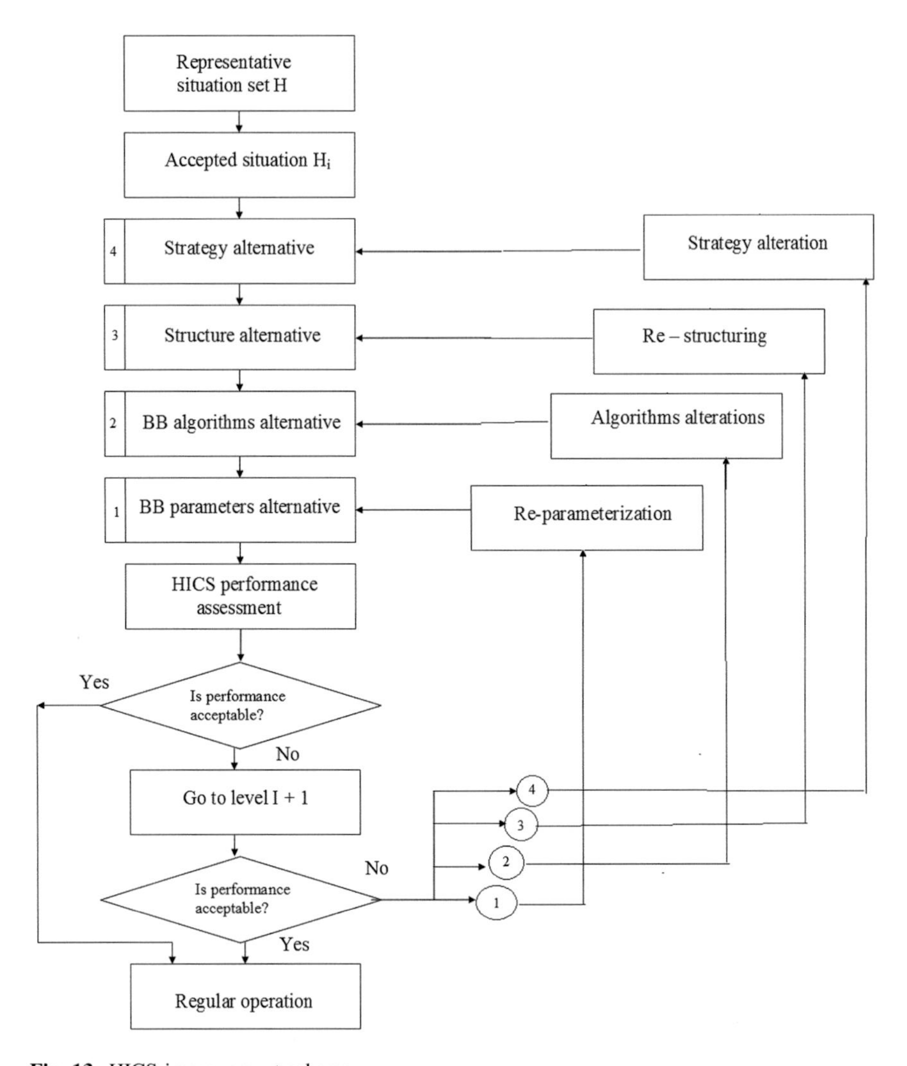


**Fig. 13** HICS improvement scheme

$$\alpha_{LS} = \frac{n_{LS}}{n}, \quad \alpha_A = \frac{n_A}{n}, \quad \alpha_O = \frac{n_O}{n} \tag{39}$$

where $n$ is the number of all functions in integrated LS-A-O system, $n_{LS}$, $n_A$, $n_O$ are the corresponding numbers of functions in LS, A and O parts.

The considered three dimensional cases could be presented graphically as a simplex (Fig. 14). If some of the individual functions are obligatory, a sub simplex MPN is formed. Since often the cross set $F_A \cap F_O$ is significantly lager than $F_A \cap F_{LS}$ and $F_O \cap F_{LS}$ the hybrid system with fixed LS functions could be presented as it is

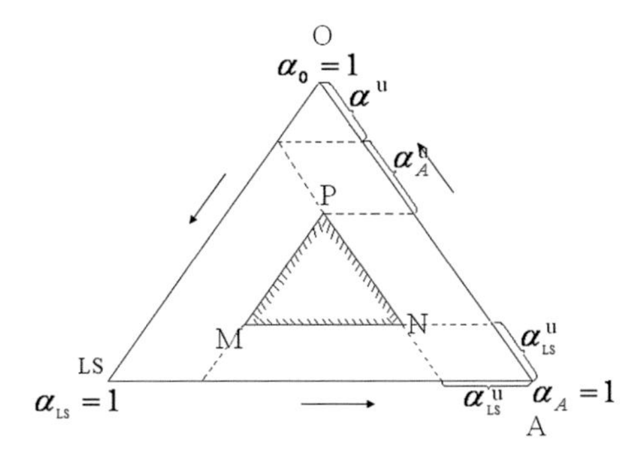


**Fig. 14** Tree component simplex diagram of partial function distribution

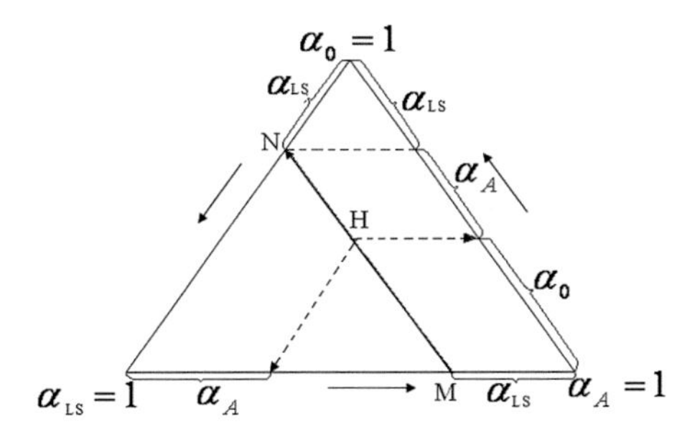


**Fig. 15** Restricted distribution among part A and part O (segment MN)

shown at Fig. 15 by segment MN. In this particular case of integration a scalar index could be accepted for relative distribution of functions $F_A \cap F_O$ among BBs A and O:

$$\beta = \frac{\tilde{n}'_O}{\tilde{n}_A + \tilde{n}_O} \tag{40}$$

where $\tilde{n}_A, \tilde{n}_O$, are the number of common functions of the subset $F_A \cap F_O$, belonging respectively to BBs A and O.

The common objective function, which combines a number of control indexes such as accuracy, time for communication between the agents and different control blocks and knowledge elements is accepted in sacralized form using weights $\alpha_i$:

$$Q = \sum \alpha_i Q_i \to \max \tag{41}$$

The system function distribution rate into the HICS can be defined as a rating of the common system functionality distributed among the different BBs. The common formation of the problem could be presented as:

$$DR = \frac{HICS_f}{A_f + O_f + CBR_f + LS_{fs}} \tag{42}$$

where: $HICS_f$—all functions of HICS; $A_f$—functions which are implemented from agent BBs; $O_f$—functions which are implemented from ontological BBs; $CBR_f$—

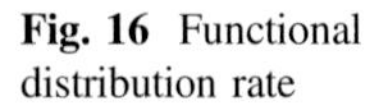
Fig. 16 Functional distribution rate

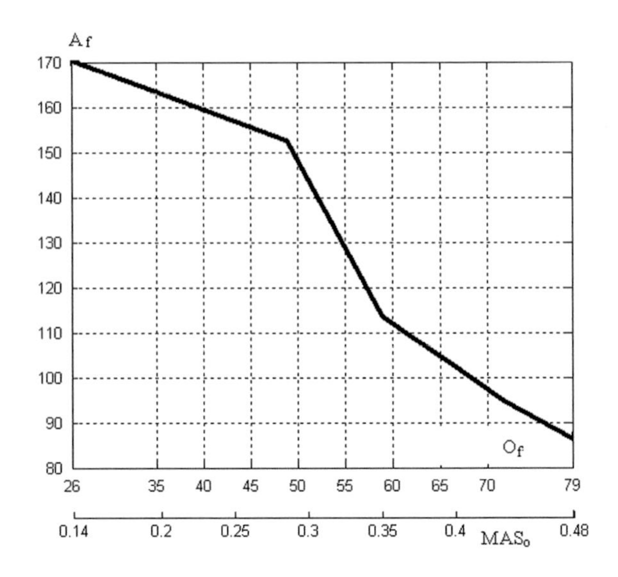


functions which are distributed to the BBs, containing information about the system behavior and stored as previous real(or simulated) cases; $LS_{fs}$—legacy system functions.

In a new design or re-engineered control systems the functions distribution among the two of the basic HICS components—Agents and Ontology has a dominant importance. The function distribution rate of specific HICS realization as Multi Agent System following Eq. (42) can be present in the form:

$$MAS_o = \frac{O_f}{A_f + O_f} \tag{43}$$

where: $MAS_O$—distribution of functions in the MAS; $O_f$—functions taken from the ontological part; $A_f$—agent functions.

When the distribution rate $MAS_O \rightarrow 0$—this means that the agents takes the whole system functionality. If $MAS_O \rightarrow 1$—the ontological part fulfill all of the system's functions. Figure 16 shows that the distribution functions rate $MAS_O$ between 0.14 and 0.48 guaranties system stability in a particular studied case of HICS (Fig. 16).

It was found out [10, 11] that when the index $\alpha = 0$ the optimal distribution rate of the system must be not less than 0.4 in order to assure the planed system functionality. When the index is $\alpha = 1$, the distribution of the functions among the other intelligent knowledge based blocks must be less than 0.5 (taking into account the possibility of the agents to work with a local knowledge). When the constraints become "soft" the distribution rate level should be between 0.35 and 0.5.

### 6.6.2 Communication in HICS

The communication rate into the HICS can be representing as:

$$R_c = \frac{n_{BB}}{M_S + M_R}/n \qquad (44)$$

where: $R_C$—the rate of all communications into the system; $n_{BB}$—number of the BBs actions performed into the HICS; $M_S$—number of the sent messages between the BBs; $M_R$—number of the received messages from the different BBs; $n$—number of the iterations between the BBs.

Figure 17 shows that the system communications rate strongly depends on the HICS BBs. The communication distribution between different BBs into the HICS is based on the communication model [30]:

$$R_c = \{(n, M_S) \in n_{BB}^2 | d(n, M_S) \leq R_c^{\max}\} \qquad (45)$$

where: $R_c$—level of the communications rate between BB; $V$—number of BBs composing the HICS; $R_c \in n_{BB}^2$—the area of a possible communication which can be presented into HICS; $n$—the number of system node represented as a part of BBs, which can sent information to other system elements $M_S$; $(n, M_S)$—part of $R_c$ represented the number of nodes (n) and number of the sent messages (Ms); $R_c^{\max}$ is a maximum communication capacity which theoretical is possible to all BBs composing HICS—$d(n, M_S)$.

The HICS simulation time $HICS_S$ depends on the number of the BBs and their functionalities, and could be present as:

$$HICS_S = (\frac{n_{BB}}{T_S})/n \qquad (46)$$

where: $T_S$—is the computation time of the simulation; $n_{BB}$—number of the BBs implemented in HICS; $n$—number of the iterations between BBs. The HICS

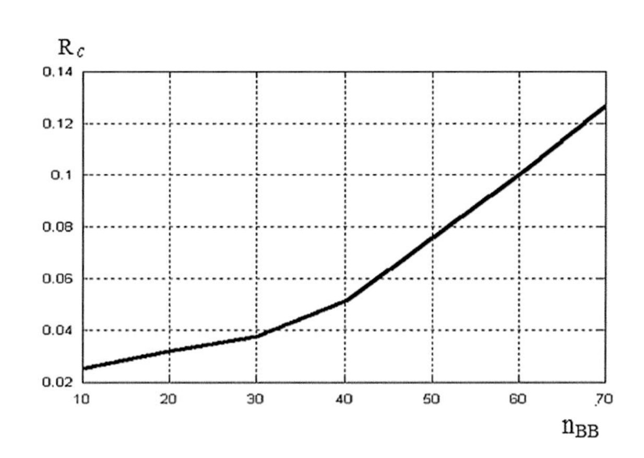


**Fig. 17** The level of communications rate in dependence on BBs number

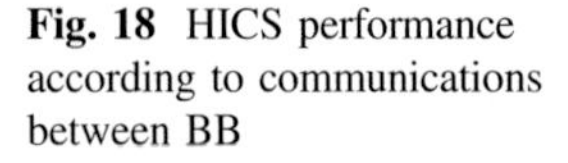
**Fig. 18** HICS performance according to communications between BB

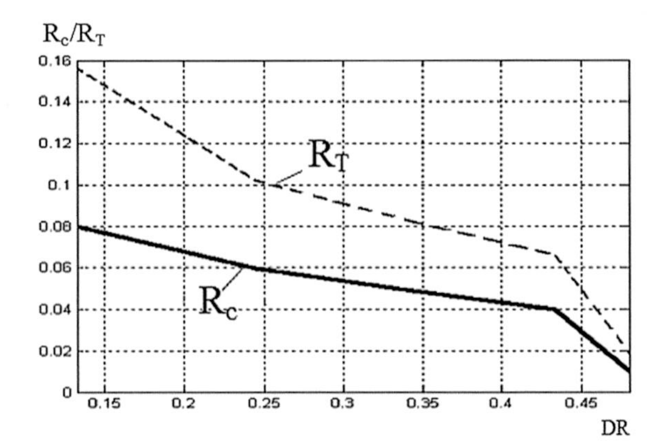


performance taking into account the communication rate $R_c$ and the necessary time to reach the optimal behavior $R_T$ is presented in Fig. 18 according the relation:

$$J_P = 1 - (1 - \beta)(R_c + R_T) \tag{47}$$

where $J_P$ is the objective function depending on the current level of the control system performance and the necessary time for execution the accepted decision from Agent knowledge; $\beta$—weighting coefficient; $R_T$—necessary time of HICS to achieve the optimal control.

### 6.6.3 HICS Synchronization

Into the HICS there is large flow of information, which posses the problem of synchronization of the communication between the BBs. The HICS can works into two modes: synchronous or asynchronous one. The synchronization between different integrated BBs, leads to a big internal rank of communications. That could provokes a bad performance lying out of the necessity to wait each building block to accomplish its function. Thus it couldn't react on time in fast and proper way in cases of large system disturbances, alarming situation and etc. This is shown in Fig. 19.

In order to compare MAS with different mode of communication the average rank of received messages, sent messages and performed by the agent actions are represented in Fig. 20. The MAS-S have a big communication rank, caused from the received and transmitted between agents useless information while awaiting the other agents to reach their targets.

The communications into the Multi-Agent Systems (MAS) depends on the agent's number and their relationships. When the agent's number increases the communication rank increases sharply ("curse of dimensions"). To reduce the communication rank the additional elements should be integrated in a way to perform the simple tasks which don't need a synchronization.

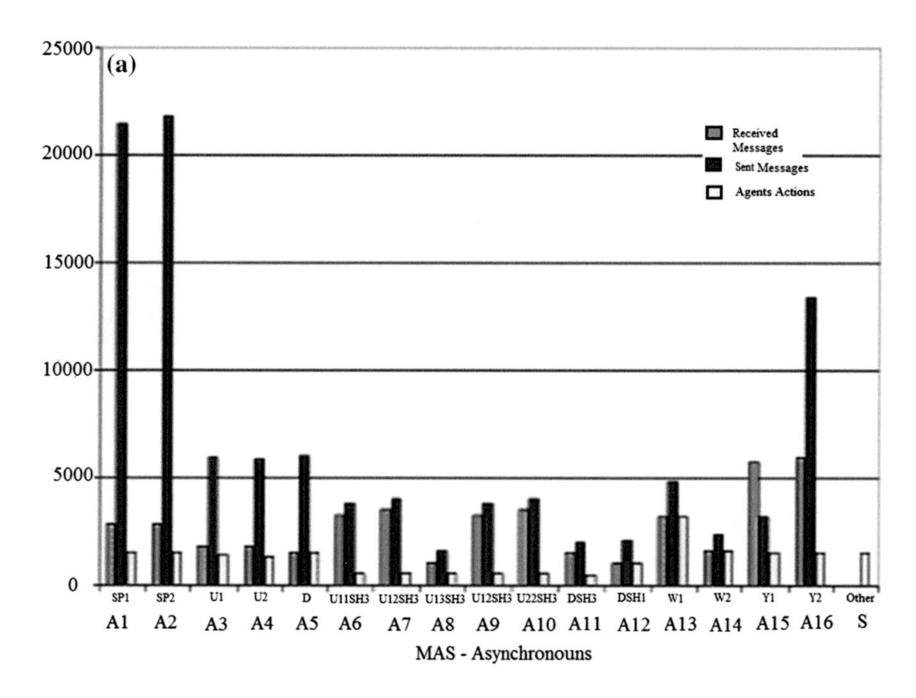


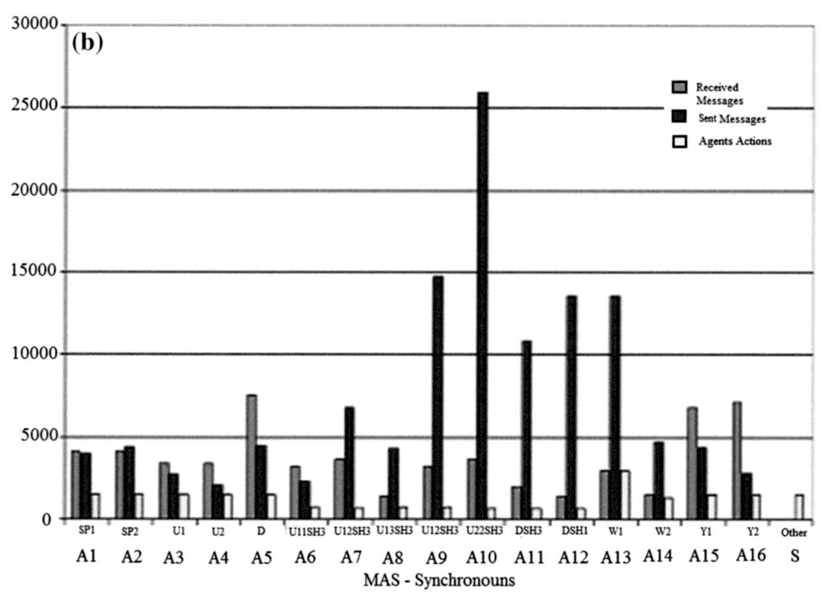


**Fig. 19** Communication rate MSR for asynchronous (**a**) and synchronous (**b**)

Fig. 20 Synchronization of BBs into HICS

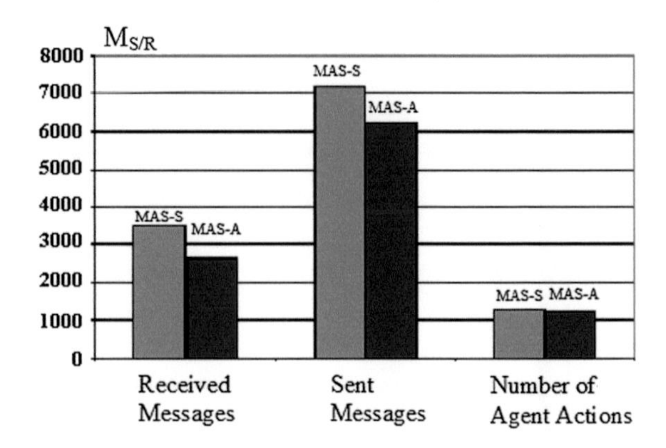


## 6.7 Software Realization

In the previous works of the authors developing industrial Hybrid Intelligent Control Systems (Power Plants [8, 26]), HVAC Systems [10, 11] have been combined the advantages of the own developed software code and its integration with the free open source code distributed as JAVA third part libraries. The main reason to accept JAVA as a base programming language is the possibility to use open intelligent techniques and algorithms (Fig. 21). All knowledge oriented building blocks (presented as set K in Fig. 4) are realized in JAVA (Fig. 22). Taking into account that there exists [28, 30, 33, 34] a variety of software realizations of computational intelligence building blocks (set T), a combination of different software languages and interfaces has been used in order to integrate variety of program techniques as well (Fig. 21).

The basic models represented as knowledge oriented building blocks (O, CBR, RBR, A, ACO) can be easy integrated with the core JAVA technologies. They are based on the implementation of a program language (www.oracle.com [34]) which provides different developing environments running in separated JAVA Virtual Machine—JVM. The JVM contains the basic JAVA libraries which interpret in runtime the program logic. JAVA could use Operational Systems (OS) such Windows, Linux, Unbutton, Mac and etc., acting as a cross-platform interpretation

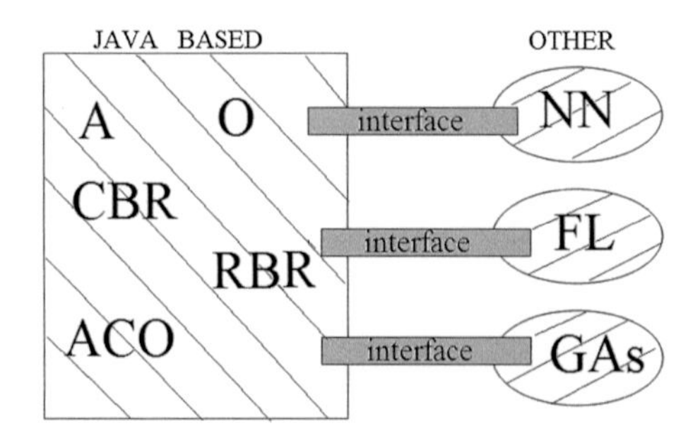


Fig. 21 Integration of BBs by using JAVA

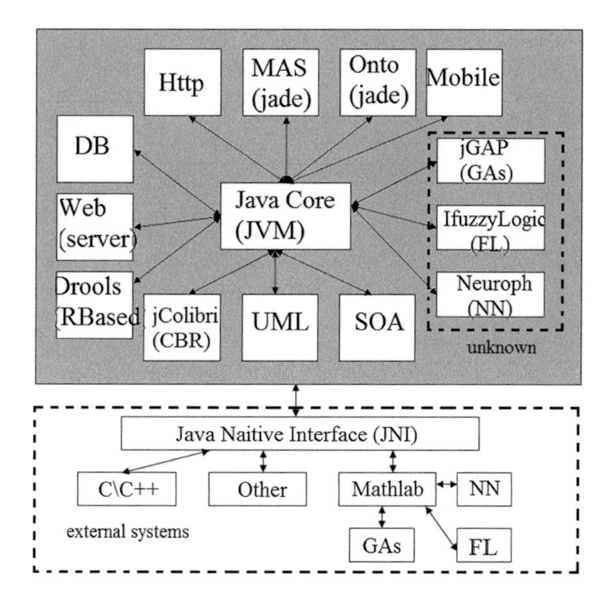


**Fig. 22** Integration of different BBs JAVA interfaces

provided as JAVA Development Kit—JDK. This environment is based on the program languages independence. It incorporates JAVA based components and various interfaces of complex HICS. According our experience using the JAVA programming and the free JAVA based frameworks—JADE, jColibri, UML, Onto about 60 % of the integration in the HICS could be covered.

The JAVA based specific open source software provides easy way to integrate basic different technologies implemented into building blocks. Some of the open source frameworks developed on JAVA are: Fuzzy Systems (FS) by using of IFuzzyLogic; Neural Networks (NN) with Neuroph framework and jGAP for implementation of Genetic Algorithms (GAs). In this way the JAVA language can cover more than 90 % of the software for the HICS realization. An alternative is to use the "external systems", which can be integrated into JAVA based software platform by using the provided mechanisms of core JAVA program language and an implementation of different adapters between the software components (Fig. 22).

# 7 Structure of HICS

## *7.1 Legacy System–Agents–Ontology (LS, A, O)*

As illustration in Fig. 23 is presented a conventional Legacy System (a) and hybrid (LS, A, O) realization (b) in a case of SITO plant where a control of a plant with non square matrices is a goal. A reconfigurable control algorithm is accepted in both designs. The main advantage of intelligent agents as BBs implementation is

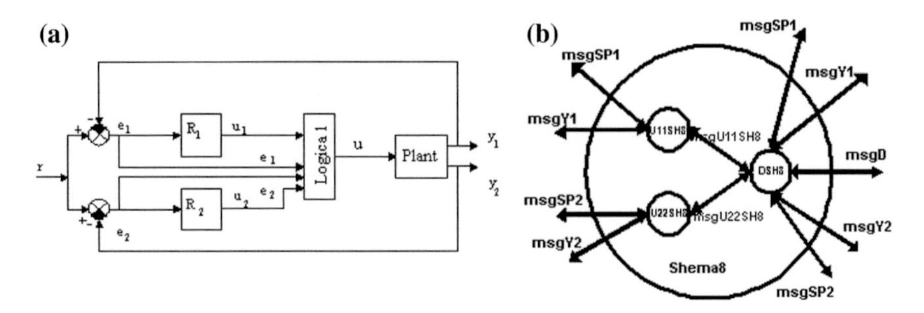


**Fig. 23** Non-square plant control using legacy system (**a**) and hybrid realization (**b**)

**(a)**

```
if (msg != null){
    restart();
    msg.getByteSequenceContent();
       msg.getLanguage();
       msg.getOntology();
    msg.getSender();
       try{
           ContentElement ce = manager.extractContent(msg);
           if (ce instanceof abssclassC1){
              abssclassC1 abssd = (abssclassC1) ce;

              p = abssd.getRabotenC1();
              u = abssd.getUC1();
             // //System.out.println(" W1 ==== " + u);
             iteu++;
          T = 20;
          arru[iteu] = u;
          arrw1[iteu]=w1;
          w1 = (0.9*arru[iteu-2] + T*arrw1[iteu - 1])/(T + 1);



            }
```

**(b)**

```
if (msg != null){
    restart();
    msg.getSender();
           try{

            ContentElement ce = manager.extractContent(msg);
            if (ce instanceof abssclassW1M){
               abssclassW1M absswl = (abssclassW1M) ce;

               p = absswl.getSoupW1M();
               w11 = absswl.getW1M();
            // //System.out.println("Y1 Y1 Y1 W11 received +++++
               w1 =  w11 + w12 + w1n;
               ite++;
             arrw1[ite] = w1 ;

             w1=arrw1[ite];

             e1=sp1-w1;
              myAgent.addBehaviour(new ConstraintsY1M(myAgent,w1

            if(ite%5 == 0){

               i1++;
```

**Fig. 24** Software fragments realization of an Agent-based (**a**) and Hybrid (LS–A–O) control (**b**)

the easy sharing the expert knowledge for each current operational situation $H_i$ (Fig. 7) and forming a relevant trade-off control action $u$.

Some of the control system algorithms [such as realization of the system with or without adaptation of control loops, which are a part of the whole structure (Table 6)] are implemented as pure JAVA-based agent functional components (Fig. 24a). A software fragment of a hybrid LS–A–O control algorithm is presented at Fig. 24b.

## 7.2 *Fuzzy Logic–Ontology Integration (FL–O)*

The function distribution among the BBs gives the possibility to separate the global system knowledge and the knowledge presented into some of the BBs as a local one consolidating all knowledge for the current system situation ($H_C^i$). The software realizations of the knowledge representations as a part of BBs the HICS based composing blocks could be carried out via separate software components, which allow removal of a agent's part knowledge. These BBs can be applied as a part of the Ontology knowledge structure. According to the possible functions distribution presented in Sect. 6.6.1 above, some of the HICS functionalities can be implemented as a part of the Ontology block realization and the rest part need to be represented by the other BBs. In order to prevent possible conflicts between different knowledge segments presented into Ontology based BBs it was accepted that, the available knowledge about the control system can be defined by fuzzification of the common HICS knowledge. The knowledge referred to the BBs can be re-presented as a similarity between the knowledge fragments, rank of fuzzification between terms describing relations and interpretations defined into the BBs. The fuzzyfication of the system knowledge is mainly based on merging the human experience via the combination of ACO–Ontology Knowledge (Fig. 25).

In Figs. 26 and 27 are given parts of (FL–O) software realization. They illustrate the flexibility of control structure creation according to the $HS_i$ choice independent on the current situation $H_i$ and the possible consequential re-parameterization (PR) (Fig. 7).

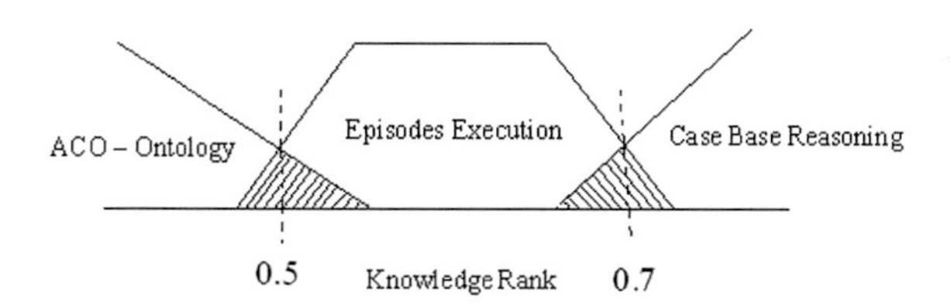


**Fig. 25** System structure $HS_i$ fuzzy fication

```
if (ce instanceof absclassSP1) {
    absclassSP1 absspl = (absclassSP1) ce;
    sp1 = absspl.getSP1();
    askForCBR();
    if (cbrUsing.equals("false")) {
        SwitchingControl1();
        myAgent.addBehaviour(new SendtoUC1(myAgent, u1));
        myAgent.addBehaviour(new SendtoUC2(myAgent, u2));
    } else {
        SwitchingControl1();
        if (countSim > 2) {
            if (countSim > 30) {
                u1 = su1 / 1000;
                u2 = su2 / 1000;
            }
        }
        myAgent.addBehaviour(new SendtoUC1(myAgent, u1));
        myAgent.addBehaviour(new SendtoUC2(myAgent, u2));
    }
}
if (ce instanceof absclassY1) {
    absclassY1 abssy1 = (absclassY1) ce;
    countSim++;
    y1 = abssy1.getY1();
    if (countSim < 100) {
        y1 = 0;
```

**Fig. 26** The fragment of source code implemented stigmergy A

```
public ConstraintsY2(Agent myAgent, double e2,double w2, double sp2, dou
    super(myAgent);
    this.w2=w2;
    this.sp2=sp2;
    this.sp2min=sp2min;
    this.sp2max=sp2max;
    this.e2=e2;
}

public void action() {
    try{
        restart();
            System.runFinalization();

    if ((w2 > (sp2max))|(w2 < (sp2min))){

        y2p="Y2-BAD!";
    }
    else {
        y2p="Y2-OK!";
    }
    myAgent.addBehaviour(new SendY2(myAgent,y2p,w2,e2,sp2));

    }catch(Exception e){
        e.printStackTrace();
    }
}
```

**Fig. 27** Part of source code describing decision making

The system merge knowledge ($SK_{12}$) retrieved from the first and second MAS in case when for the fuzzy ontology can be formed from the first one ($SK_1$) has a priority. The merge part ($SK_{21}$) shows that the knowledge retrieved from the second system ($SK_2$) is with higher priority. The destination among the present knowledge in two MAS and the common knowledge could be estimated as follow:

$$D(k+1) = \frac{(SK_{12}(k) - SK_1(k)) + (SK_{21}(k) - SK_2(k))}{SK_1(k) + SK_2(k)} \tag{48}$$

In case when the couple $\{SK_{12}, SK_{21}\} > \{SK_1, SK_2\}$ the D(k) is always positive. That assures the knowledge covering and stable new merge ontology

The rank of similarity of the knowledge D(i) has been defined on the basis of the new present knowledge versus old one into the HICS according the next relationship:

$$K_{fuzzy}(i) = \mu_{SK_{12}}(i)D(i) \tag{49}$$

where: $K_{fuzzy}$—rank of the fuzzyfication; i—number of Ontology building block; $\mu_{SK_{12}}(i) = \frac{S_{12}(i)}{S_{total}}$ —rank of the relation between the knowledge allocated into the Ontologies; $S_{12}$ is the number of common related terms and concepts presented into Ontologies; $S_{total}$—the total known relations between the BBs into the HICS.

The proposed algorithm for fuzzification of the present knowledge into the HICS according to different system situations $\bar{H}_C$ and possible changes into the control structures $HS_i$ is illustrated in Fig. 28. The part of the software realization of the Fuzzy Ontology is presented as a source code realization for specific HICS implementation in the developed HVAC Control System [10] at Fig. 29.

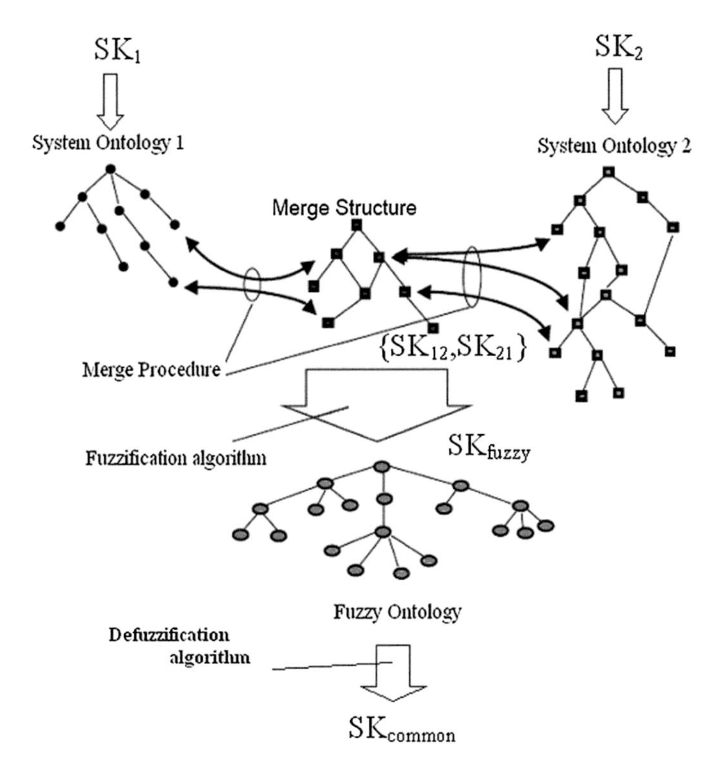


**Fig. 28** Ontology merge and fuzzyfication

## *7.3 CBR–Ontology Integration (CBR–O)*

The hybridization of the CBR and Ontology BBs is based on the relevant combinations between some parts of Ontologies and Case Base representation techniques. The Ontology blocks can be separated into different layers according to the specific needs of the system: Global Knowledge—describes the common knowledge shared into different systems for the total plant control; Domain Knowledge—presents a specific knowledge for the unit's control; Process Knowledge—contains a specific knowledge for the technological processes in the plant units (flow rate and ratio, temperature, pressure, degree of recirculation and etc.) (Fig. 30).

The layered knowledge could be represented as a component in various BBs: The Agent based representation is preferable for keeping the temporal knowledge about the current system situation. The Ontology representation is mainly used for keeping the global knowledge about the HICS optimal design ($\tilde{D}_i^*(\tilde{H}_i)$), including strategies ($ST_i^*$) for the structure ($HS_i^*$), control algorithms ($AL^*$) and respective parameters ($PR^*$) (Figs. 7, 11 and 12); The Case-based knowledge representation is focused on the retrospective information about the historical system states, situations expert solutions and "simulation-solution" results. We use as a hybridization

**Fig. 29** Merge Fuzzy Ontology representation procedure

```
<owl:Class rdf:ID="HVACControl">
    <rdfs:subClassOf>
        <owl:Restriction>
            <owl:onProperty rdf:resource="#hasAirControl"/>
            <owl:cardinality rdf:datatype="&xsd;nonNegativeIn
 </owl:cardinality>
        </owl:Restriction>
        <owl:ObjectProperty rdf:ID="forAirControl">
            <rdfs:domain rdf:resource="#Control"/>
            <rdfs:range rdf:resource="#AirConotrol"/>
            <rdfs:parameter rdf:resource="#fuzzyRank"/>
        </owl:ObjectProperty>
    </rdfs:subClassOf>
    <rdf:HVACControl rdf:ID="computeVentilationControl">
        <owl:differentFrom rdf:resource="#control1"/>
        <owl:differentFrom rdf:resource="#control2"/>
        <rdfs:parameter rdf:resource="#fuzzyRank"/>
    </rdf:HVACControl>
    <rdf:AirControl rdf:ID="setpoint1">
        <owl:sameAs rdf:resource="#control1"/>
        <owl:diferentFrom rdf:resource="#control1second"/>
        <rdfs:parameter rdf:resource="#fuzzyRank"/>
    </rdf:AirControl>
    <rdf:AirControl rdf:ID="setpoint2">
        <owl:sameAs rdf:resource="#output1"/>
        <owl:diferentFrom rdf:resource="#output1second"/>
        <rdfs:parameter rdf:resource="#fuzzyRank"/>
    </rdf:AirControl>
    <owl:intersectionOf rdf:parseType="Control">
        <owl:Class rdf:about="#HVACControl"/>
        <owl:Restriction>
            <owl:onProperty rdf:resource="#hasAirControl"/>
            <owl:hasValue rdf:resource="#control1"/>
        </owl:Restriction>
    </owl:intersectionOf>
</owl:Class>
```

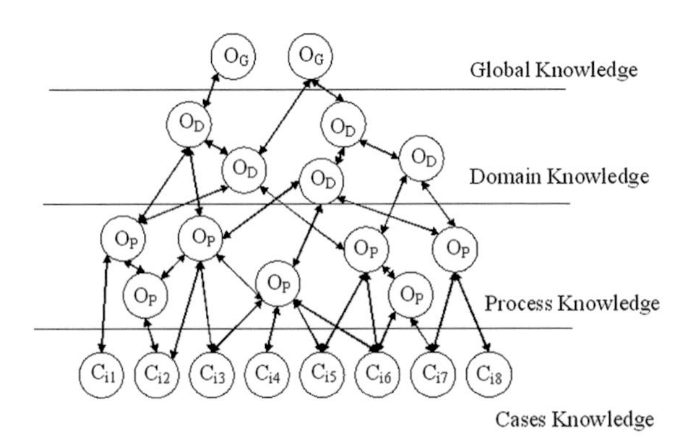


**Fig. 30** The Ontology–CBR integration composition

technology a configuration files, which maps the common system properties and relations between them. Some of the system properties as attributes and solutions can be represented via the Ontology part and the Case Base one. It is preferable to put the global system properties into the Ontology part and map them with the specific ones presented as part of the Case Based description. Such kind of implementation of CBR–O hybridization makes the HICS flexible against

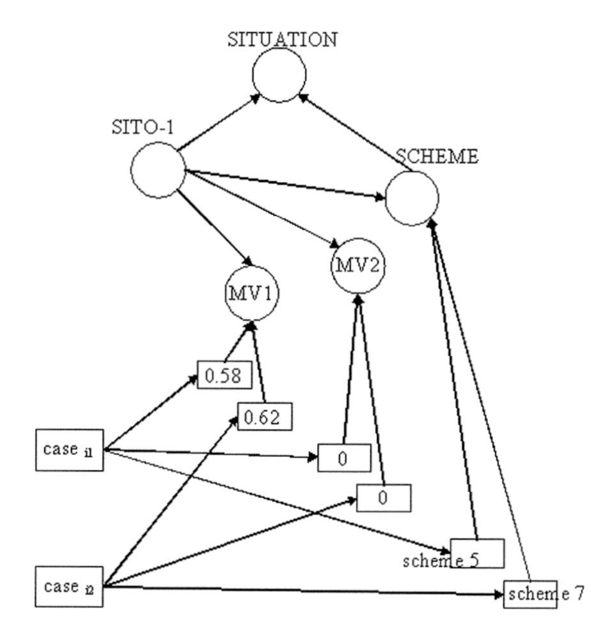


**Fig. 31** Structure case and ontology mapping

unpredictable fast operational changes. The scheme of Case-Base Reasoning and Ontology mapping is presented in Fig. 31 for a particular TITO non-linear control system [10].

A part of CBR–O configuration fail, which combines the ontology knowledge sharing into system mapped into the present system with Case-Base reasoning (CBR) approach is shown in Fig. 32. The CBR realization of BBs as a part of the HICS uses the jColibri free software code [29]. The small fragment of the software realization of the integration between CBR–O is presented into Fig. 33.

## 7.4 CBR–Agent–Ontology Integration (CBR–A–O)

In Fig. 34 is shown a common structure of Multi Agent Systems (MAS) integrated with a CBR system using a specific Similarity Agent estimated the similarity of the retrieved case to the recognized current system situation ($H_i^c$). According to the present system knowledge the HICS can adapt the case or to create a new one (Fig. 35).

Some parts of the code realizing the (CBR–A–O) integration are presented in Figs. 36 and 37. These examples show the Data Base integration as a present part of HICS representation (Fig. 36) and its absorption from the Ontology structure as a part of (CBR–A–O) implementation (Fig. 37). The present two examples are focused on loose coupling into CBR–A–O hybridization approach.

**Fig. 32** A fragment of a CBR–O integration file software structure

# 8 Simulation Results

## *8.1 Function Distribution into the HICS*

The simulation results confirm the importance of function distributions between different BBs (Ontology and Agents). When the rank of the function distribution implemented from Ontology part exceeds 50 % of the HICS functions the system can tend to instability (Fig. 38). When more functions are implemented in Agent BBs, the communication rate increases rapidly (Fig. 39) which brings the HICS to bad performance as it was mention of above. The simulations results show that the optimal rank of the function distribution among Agents and Ontologies is between 14 and 48 % (Fig. 40). The Legacy Part of the software components can cover some of the system functionality, but the reactivity of the system in whole operational space will be loosed. As illustration of a SITO control system performance is presented in Fig. 41.

```xml
Java EE - INTELLIGENT_SYSTEMS_2012_CBR_version2/src/jcolibri/cbrsystem/CBRDescription.hbm.xml - Eclipse
File Edit Source Navigate Search Project Run Window Help
CBRSolution.hbm.xml | CBRDescription.hbm.x | CBRSolution.java | *hibernate.cfg.xml | Adaptation.java | Test.java | int

<?xml version="1.0"?>
<!DOCTYPE hibernate-mapping PUBLIC "-//Hibernate/Hibernate Mapping DTD//EN" "http://hibernate.sc
<hibernate-mapping default-lazy="false">
    <class name="jcolibri.cbrsystem.CBRDescription" table="INTELLIGENTSYSTEMS2012_CBR">

        <id name="caseId" column="caseId">
            <generator class="native" />
        </id>
        <property name="sp1" column="sp1" />

        <property name="mv1" column="mv1" />
        <property name="mv1_1" column="mv1_1" />
        <property name="mv1_2" column="mv1_2" />
        <property name="mv1_3" column="mv1_3" />

        <property name="cv1" column="cv1" />
        <property name="cv1_1" column="cv1_1" />
        <property name="cv1_2" column="cv1_2" />
        <property name="cv1_3" column="cv1_3" />

        <property name="noise1" column="noise1" />
        <property name="noise1_1" column="noise1_1" />
        <property name="noise1_2" column="noise1_2" />
        <property name="noise1_3" column="noise1_3" />

        <property name="rank" column="rank" />
    </class>
</hibernate-mapping>
```

**Fig. 33** A software structure describing a CBR

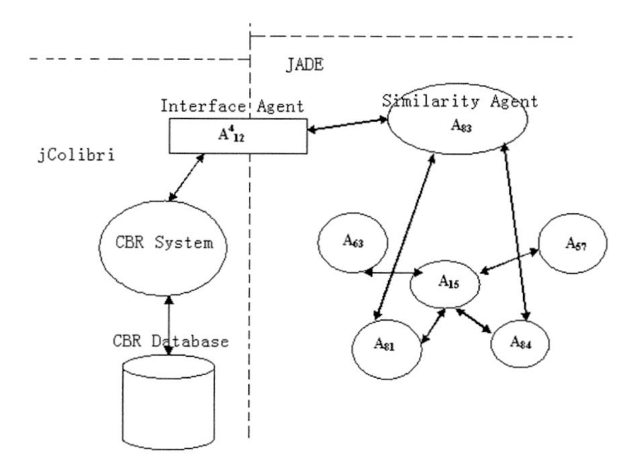


**Fig. 34** The common scheme of interaction between different BB and implemented integration interfaces

## 8.2 *Legacy Systems–Multi-agent Systems (LS–MAS)*

Some of the simulation results are based on investigation of a reconfigurable control system of technological plant with transfer functions presented in Fig. 42. There exists seven possible control structures provoked from different system's situations (Table 8). In Figs. 43 and 44 the performance of SITO1 and SITO2 control

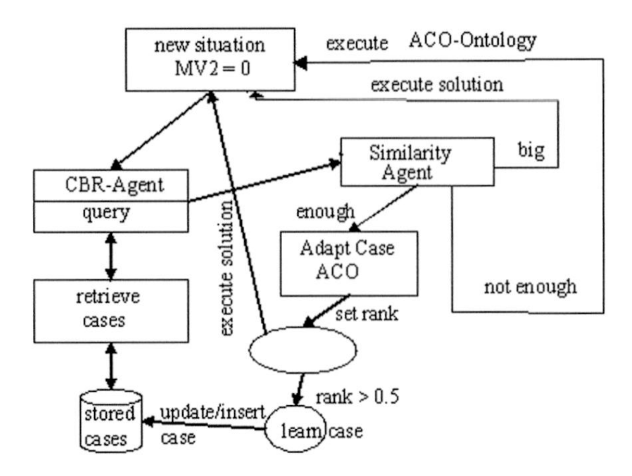


**Fig. 35** Retrieving and adaptation of cases

```
<?xml version="1.0"?>
<!DOCTYPE hibernate-mapping PUBLIC "-//Hibernate/Hibernate Mapping
DTD//EN" "http://hibernate.sourceforge.net/hibernate-mapping-3.0.dtd">
<hibernate-mapping default-lazy="false">
      <class name="jcolibri.cbrsystem.CBRDescription"
table="CBRSYSTEM">

                <id name="caseId" column="caseId">
                      <generator class="native" />
                </id>
                <property name="sp1" column="SP1" />
                <property name="sp2" column="SP2" />
                <property name="mv1" column="MV1" />
                <property name="mv2" column="MV2" />
                <property name="cv1" column="CV1" />
                <property name="cv2" column="CV2" />
                - - - - - - - - - - - - - - - - - -
                <property name="error1" column="ERROR1" />
                <property name="error2" column="ERROR2" />

      </class>
</hibernate-mapping>
```

**Fig. 36** Case description mapping in jColibri

**Fig. 37** Case description object

```
import jcolibri.cbrcore.Attribute;
import jcolibri.datatypes.Instance;

public class CBRDescription implements jcolibri.cbrcore.CaseComponent{

        int caseId;
        Integer sp1, sp2, mv1, mv2, cv1, cv2, error1, error2;
        Instance DomainOntology;

        public Instance getDomainOntology() {
                return DomainOntology;

        public int getCaseId() {return caseId;}

        public void setCaseId(int caseId) {this.caseId = caseId;}

        public Integer getSp1() {return sp1;}

        public void setSp1(Integer sp1) {this.sp1 = sp1;}

    - - - - - - - - - - - - - - - - - -
```

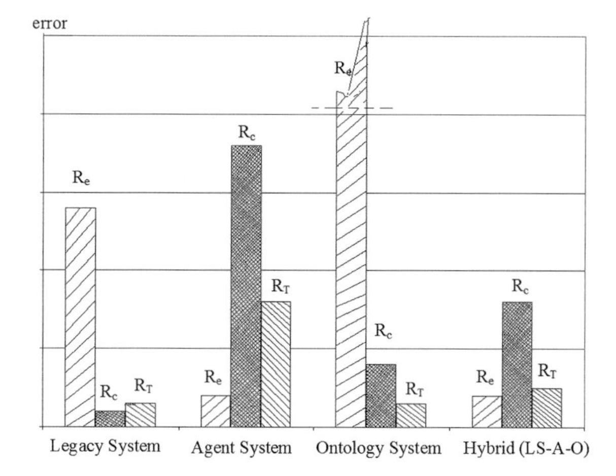


**Fig. 38** Comparison of different ways of hybrid control based on functional distribution between the knowledge components

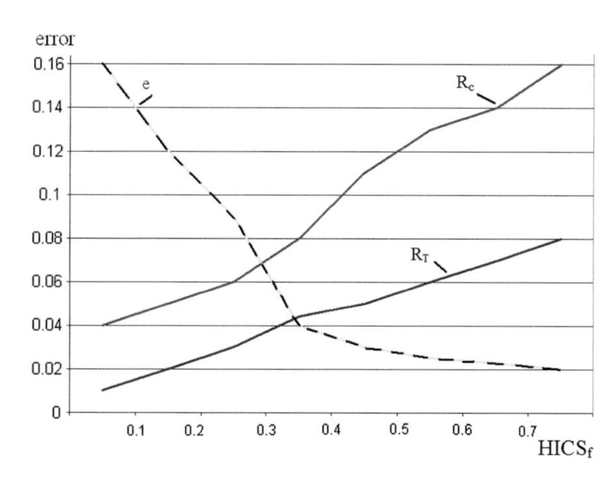


**Fig. 39** Indexing for system accuracy with MAS

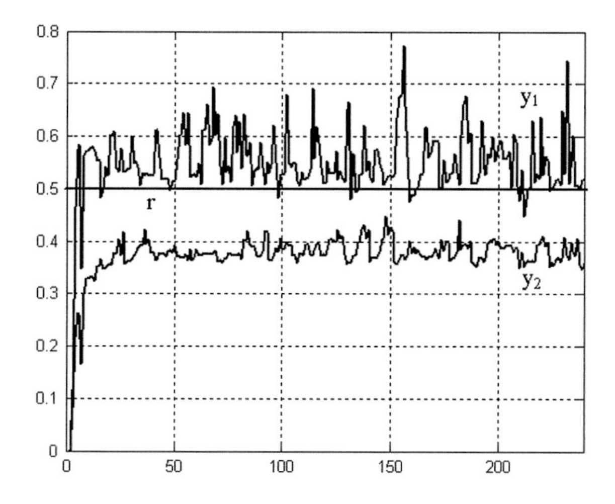


**Fig. 40** SITO system behavior when $MAS_{onto} = 0.48$

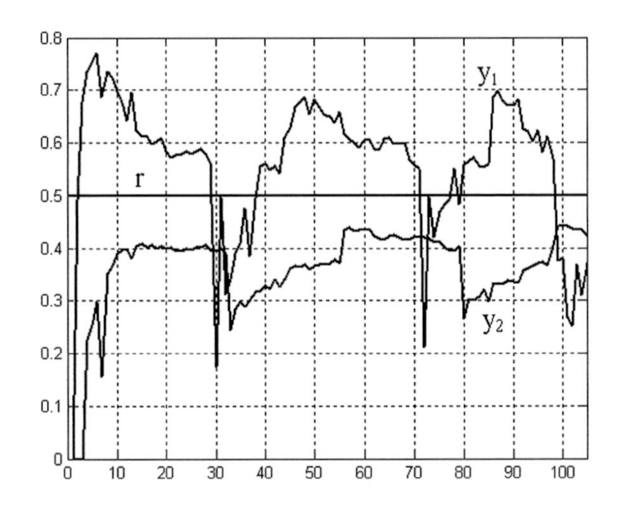


**Fig. 41** SITO system behavior when $MAS_{onto} = 0.57$

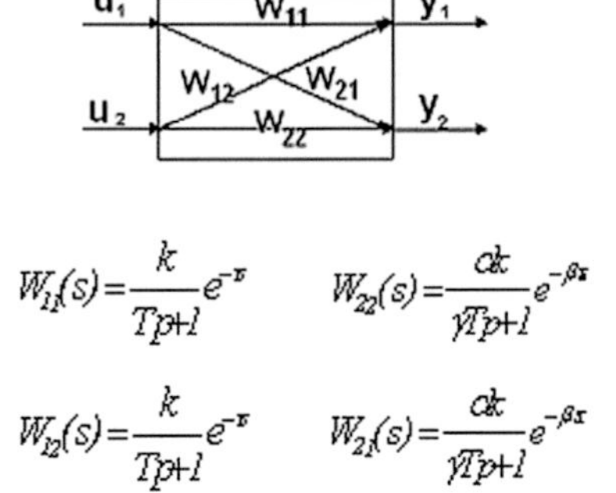


**Fig. 42** Plant control structure

**Table 8** The possible HICS structures—HSi

| № | Configuration | MVs | CVs | Figure | Transfer functions | Controler(s) |
|---|---|---|---|---|---|---|
| 1 | TITO | $u_1$, $u_2$ | $y_1$, $y_2$ | 7a | $W_{11}$, $W_{12}$, $W_{21}$, $W_{22}$ | $R_1$, $R_2$ |
| 2 | TISO-1 | $u_1$, $u_2$ | $y_1$ | 7b | $W_{11}$, $W_{12}$ | $R_1$, $R_2$ |
| 3 | TISO-2 | $u_1$, $u_2$ | $y_2$ | 7c | $W_{21}$, $W_{22}$ | $R_1$, $R_2$ |
| 4 | SITO-1 | $u_1$ | $y_1$, $y_2$ | 7d | $W_{11}$, $W_{12}$ | $R_1$, $R_2$ |
| 5 | SITO-2 | $u_2$ | $y_1$, $y_2$ | 7e | $W_{21}$, $W_{22}$ | $R_1$, $R_2$ |
| 6 | SISO-1 | $u_1$ | $y_1$ | 7f | $W_{11}$ | $R_1$ |
| 7 | SISO-2 | $u_2$ | $y_2$ | 7g | $W_{22}$ | $R_2$ |

structures are shown of investigation of the system is made with respect to synchronous and asynchronous control system mode. Some of the results are presented in Figs. 45 and 46.

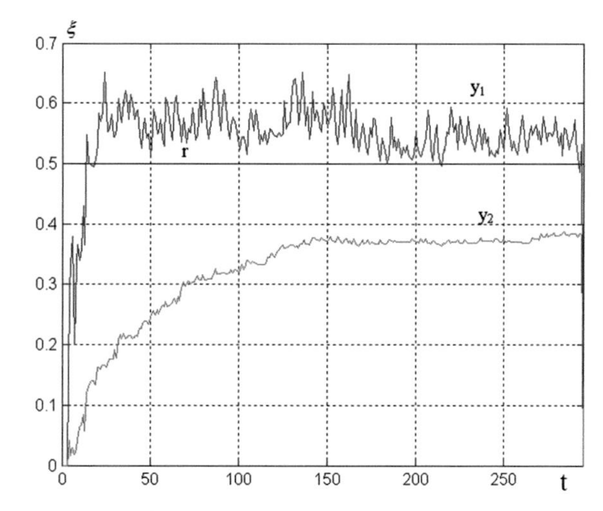


**Fig. 43** SITO1 configuration performance

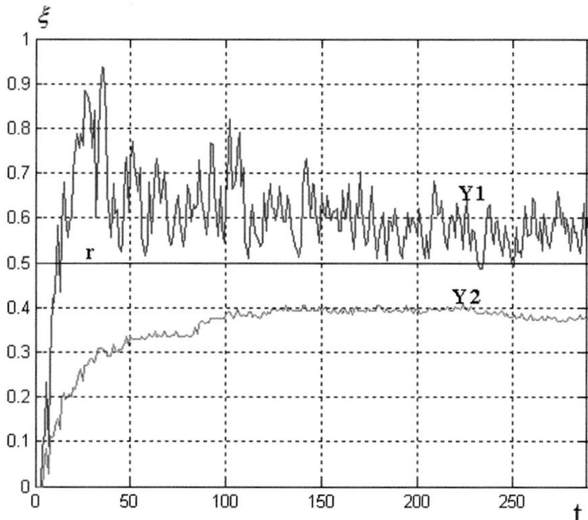


**Fig. 44** SITO2 configuration performance

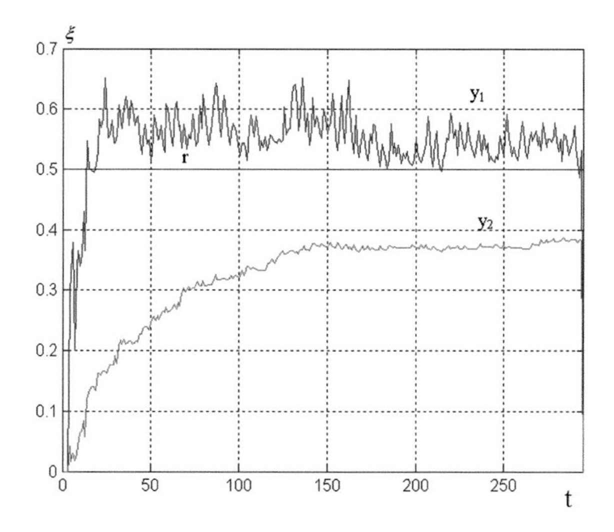


**Fig. 45** Synchronous LS–MAS mode

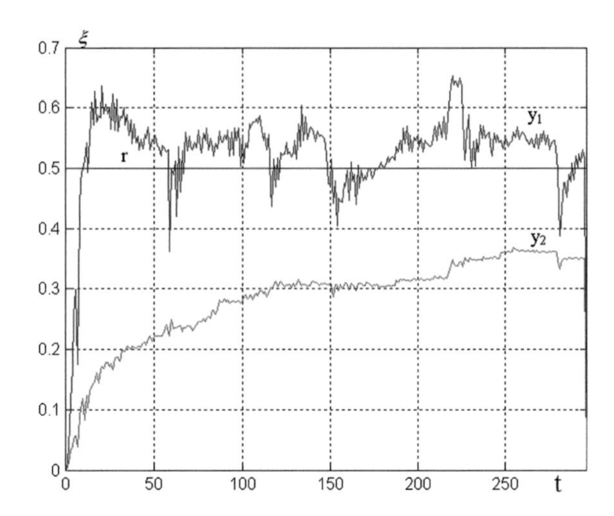


**Fig. 46** Asynchronous LS–MAS mode

## 8.3 Multi-agent Systems–Dynamic Ontology (MAS–DO)

In the Hybrid Intelligent Control System the concurrent work of Ant Colony Optimization (ACO) and environment ensures a stable behavior when the function distribution rank is $MAS_{onto} \geq 0.35$ (Eq. 43), the resulting settling time is large, because after the ants find out the optimal control actions, they must change the state by updating the environment. Thus each separate agent waits to receive messages from the other agents and then takes their own decisions [8] as it is shown in Fig. 47. In order to separate the ants and the environment the latter could be represented by dynamic ontology (DO). Using dynamic ontology directly for the

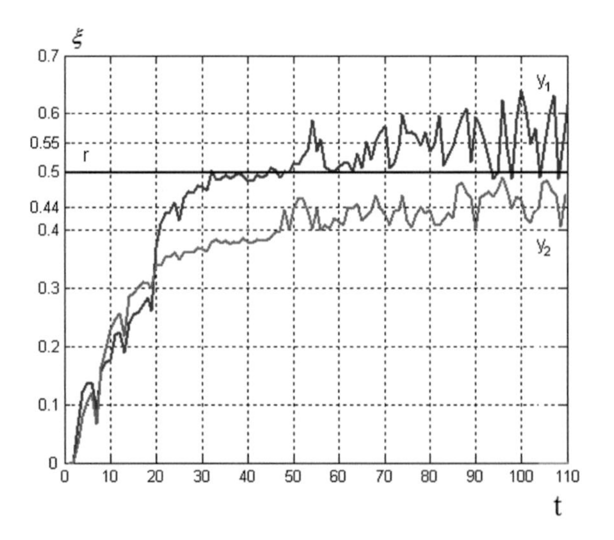


**Fig. 47** HICS control via MAS

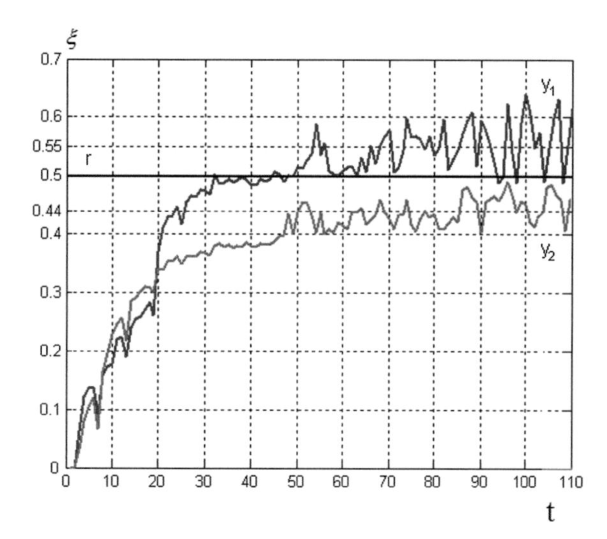


**Fig. 48** HICS control via DO

control action formation the system becomes unstable (Fig. 48). This is caused by the lack of knowledge refreshing and no effective information feedback in stigmergy [14].

## 8.4 Legacy–Fuzzy–Ontology–Neural Network–Multi-agent System (LS–FS–O–NN–MAS)

Using the plant model from the Fig. 42 and the operational situations listed in Table 8, a variety of control structures (HS) with different BBs and algorithms (AL) was examined according the "top-down" and "bottom-up" procedures (Fig. 12). Some of the simulation results including Neural Networks, Fuzzy Logic, MAS, Ontology and some part of Legacy System are presented at Figs. 49 and 50. The main idea of such kind of techniques incorporation is to estimate the possibility to achieve a higher level of the HICS performance (Eq. (36)). The simulation results show that the inserting of NN into the FS–O–A structure can additionally improve the control system behaviour (Fig. 50).

## 8.5 Multi-agent–Ontology–Case–Base–Reasoning Systems (MAS–O–CBR)

Initially a MAS–O system has been studied with Ant Colony Optimization (ACO) BBs incorporating in order to find out relevant control of the system shown in Fig. 23 under the possible situations listed in Table 8. The system responses,

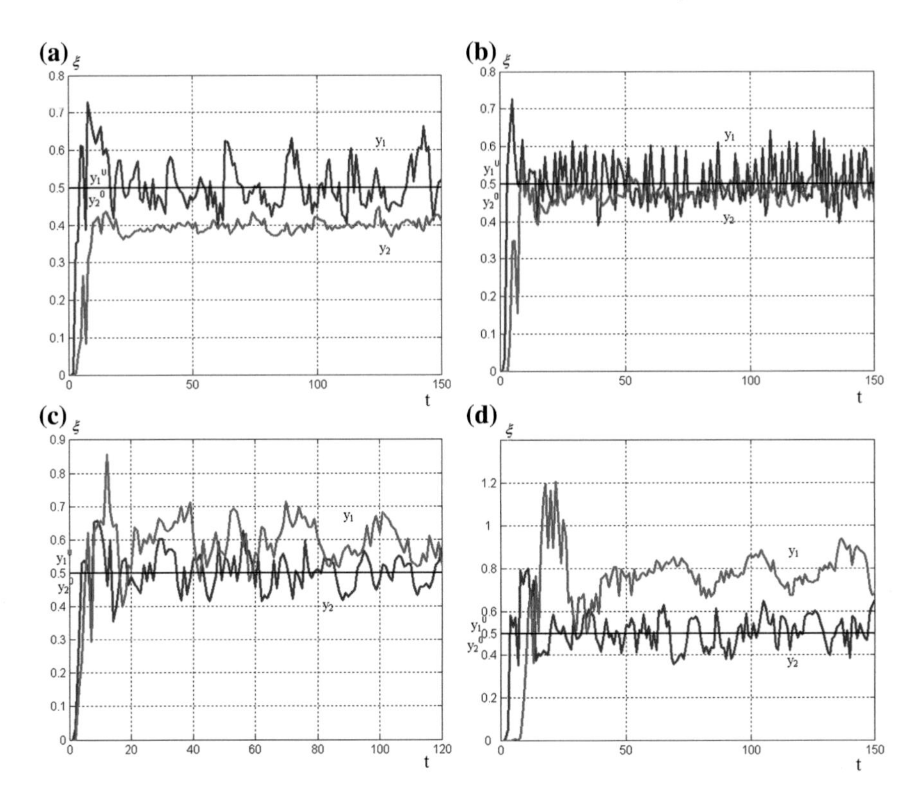

Fig. 49 LS–FS–O–MAS control. **a** SISO, **b** SITO 1, **c** SITO 2, **d** TITO

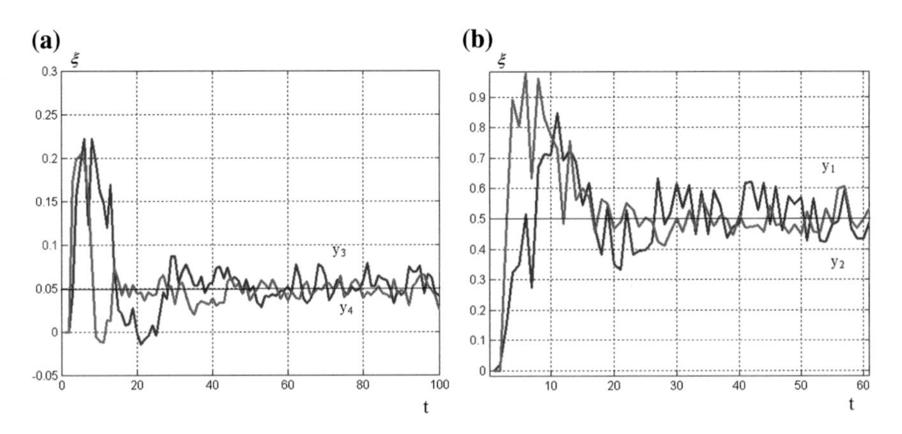

Fig. 50 LS–FS–NN–O–MAS control. **a** SITO, **b** TITO

presented in Fig. 51 show that the bounds of dangerous/alert control (Fig. 2) are reached and quasi–normal regime runs with large deviations. In Fig. 52 the similar situations have been simulated but after adding CBR-based BBs into the previous

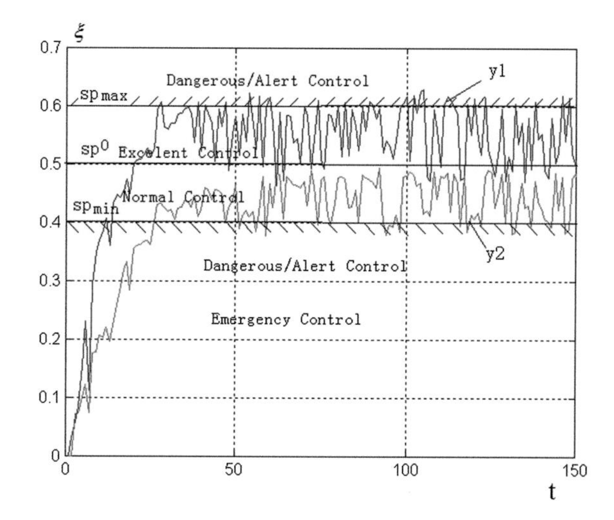


**Fig. 51** MAS–O simulation result

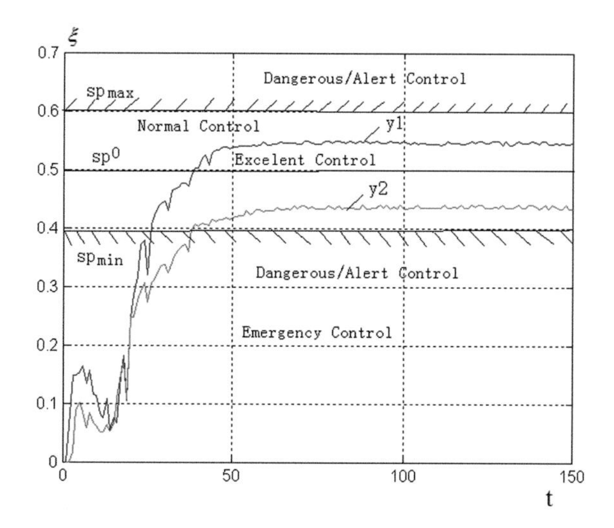


**Fig. 52** CBR–A–O simulation results with execution

structure and it becomes MAS–O–CBR-structure using again the ACO as an optimizer during the retrieving the nearest available cases from Case Base (CB) is found. If the reached degree of similarity is big enough the system behaviour becomes quite satisfactory. When the autonomous agents detect trend-off in some tracing points the control action is selected among the ACO proposals if there isn't adequate cases available.

## 9 Conclusions

The design of successful hybrid intelligent control systems by integration variety of Building Blocks (BBs) achieve an efficient solution of complex real-world problems is still an open field of research. In this contribution we basically consider structural and interconnection problems in the framework of loose integration approach. The choice of elements, interactions and algorithms strongly depends on the domain specificity and the control tasks. Instead to demonstrate as a successful application of a few ad-hoc designed systems it is more constructive to consider some unified results of investigation addressed to a set of representative cases. The reported results in this work show that a wide number of designed strategies could be relevant to make up the control system in order to be knowledge-based, agent-based, case-based or rule-based. At the same time in engineering design, not only functionality, but a lot of complementary aspects should be considered in dept also—for instance building blocks software realization, integration of intelligent part with legacy existing conventional part, maintenance, Human Machine Interface (HMI).

To take advantages of loose coupling integration of different knowledge components and to facilitate the design of Hybrid Intelligent Control System (HICS), we have made an attempt to generalized some of the results received for a variety of the studied control structures addressed to sequential design steps—intelligent techniques choice, task and knowledge allocation, blending of recommended actions learning and adaptation.

The results achieved using the accepted holistic intelligent techniques and knowledge representation can be considered as promoting way for creation of large number of new hybrid intelligent control systems with increasing problem-solving capabilities, improved performance and robustness.

**Acknowledgment** This work was supported in part from the National Science Fund of Republic of Bulgaria, Project No. DVU-10-0267/2010.

## References

1. Bemporad, A., Morari, M.: Control of systems, integrating logic, dynamics and constraints. Automatica **35**(3) (1999)
2. HYCON: Second Review Meeting, Brussels (2005)
3. Castiliio, O., Melinand, P., Kasprzyk, J. (eds.): Recent Advantages on Hybrid Intelligent Systems. Springer (2013)
4. Ramay, M., Corchando, G.E., Garcia-Sebastian, M.T. (eds.): Hybrid artificial intelligence systems. In: 5-th International Conference on HIAS, 2010, San Sebastian, Spain, Proceedings, Springer (2010)
5. Negnevitski, M.: A Guide to Intelligent Systems. Addison Wesley (2005)
6. Ruan, Da (ed.): Intelligent Hybrid Systems: Fuzzy Logic, Neural Networks and Genetic Algorithms. Springer, London (2012)
7. Hadjiski, M., Doukovska, L.: CBR approach for technical diagnostic on mill fan system. Comptes rendus de l'Academie bulgare des Sciences (2012)

8. Hadjiski, M., Boishina, V.: Hybrid supervisory control of complex systems incorporating case based reasoning. In: Proceedings of the International Conference on Complex Systems, COSY 2011, Ohrid, Macedonia (2011)
9. Hadjiski, M., Sgurev, V., Boishina, V.: Intelligent control of uncertain complex systems by adaptation of fuzzy ontologies. In: Intelligent Systems: From Theory to Practice, SCI 299, Springer (2009)
10. Hadjiski M., Sgurev, V., Boishina, V.: Multi-agent intelligent control of centralized HVAC systems. In: Proceedings of the IFAC workshop on energy saving control in plants and buildings, Bansko, Bulgaria, 3–5 Oct 2006
11. Hadjiski M., Sgurev, V., Boishina, V.: Adaptation of fuzzy ontology for cascade multi-agent system. In: 4th International IEEE Conference on "Intelligent Systems", Varna (2008)
12. Hadjiski, M., Boishina, V.: Functions selection and distribution in hybrid computational-agent-ontology control system. In: Hadjiski, M., Petrov, V. (eds.) Ontologies-Philosophical and Technological Problems, Marin Drinov, Sofia (2008)
13. Pal, S., Shin, S.: Foundation of Soft Case-Based Reasoning. Wiley, New York (2004)
14. Dorigo, M., Birattari, M., Stützle, T.: Ant colony optimization. In: IEEE Comput. Intell. Mag. **1**(4) (2006)
15. Wooldridge, M.: An Introduction to Multi–Agent Systems. Wiley, New York (2002)
16. Abraham, A.: Hybrid intelligent systems: evolving intelligence in hierarchical layers. In: Gabris, B., Leiviska, K., Strackeljan, J. (eds.) Structures of Fuzzyness and Soft Computing, vol. 173. Springer (2004)
17. Jain, L.C.: Fusion of neural networks, fuzzy sets and generic algorithms in industrial automations. IEEE Trans. Ind. Electron. **46**(6) (1999)
18. Gao, J., Deng, G.: Semi-automatic construction of ontology-based CBR systems for knowledge integration. Int. J. Electr. Electron. Eng. **4**(4): 2010
19. Berka, P.: NEST: a compositional approach of rule-based and case-based reasoning. Adv. Artif. Intell. **1** (2011)
20. Kowalski, M., Zelewski, S., Bergenrodt, D.: Applying of an ontology—driven case-based reasoning systems in logistics. Int. J. Comput. Technol. **3**(2) (2012)
21. Bragaglia, S., Chesani, F., Ciampolini, A., Mello, P., Montontani, M., Sottara, D.: Intelligent Architecture Integrating to with Fuzzy Ontological Reasoning. Springer, Berlin (2010)
22. Man, X., Haigan, Y., Jiang, S.: New algorithm for CBR–RBR fusion with robust threshold. Chin. J. Mech. Eng. **25**: 2012
23. Prentzas, J., Hatzilygeroudis, I.: Categorizing approaches combining rule-based and case-based reasoning. Expert Syst. **24**(2): 2007
24. Atanassov, K., Hadjiski, M.: Generalized nets and intelligent systems. Int. J. Gen. Syst. **39**(5) (2010)
25. Hadjiski, M., Doukovska, L., Koprinkova-Hristova, P.: Intelligent diagnostics of mill fan system. In: Proceedings of the 6-th IEEE International Conference on Intelligent Systems IS'12, Sofia, Bulgaria (2012)
26. Vachtsevanos, G., Lewis, F., Roemer, M., Hess, A., Wu, B.: Intelligent Fault Diagnosis and Prognosis for Engineering Systems. Wiley, New York (2006)
27. Kano, M., Ogava, M.: The State of the Art in Advanced Chemical Process Control in Japan. IFAC ADCHEM, CD-ROM, Istanbul (2009)
28. jColibri: http://gaia.fdi.ucm.es/research/colibri/jcolibri (2014)
29. Praccreditation: http://www.praccreditation.org/secure/documents/APRSG_Comm_Models.pdf (2014)
30. JADE: http://jade.tilab.com (2014)
31. Oracle: http://oracle.com (2014)
32. Reyes, N., Barczak, A., Susnjak, T.: A reconfigurable hybrid intelligent system for robot navigation. Res. Lett. Int. Math. Sci. **25** (2012)
33. FIPA Specification: http://www.fipa.org (2006)
34. W3C: http://www.w3.org

# Learning Intelligent Controls in High Speed Networks: Synergies of Computational Intelligence with Control and Q-Learning Theories

**Georgi M. Dimirovski**

**Abstract** The phenomena of congestion and packet-drops in high-speed communications and computer networks do affect the quality-of-service and overall performance more than ever existing time-delays with uncertain variations. Their control and, possibly, prevention are subject to extensive research ever since the internet is available. Because of the uncertainties and time-varying phenomena, obtaining the accurate and complete information on the network traffic patterns, especially for the multi-bottleneck case, is rather difficult hence learning intelligent controls are needed. One such control is a multi-agent flow controller (MAFC) based on Q-learning algorithm in conjunction with the theory of Nash equilibrium of opponents' strategies. The other is a model-independent Q-learning control (MIQL) scheme having focus on the flow with higher priority, which also does not need prior-knowledge on communication traffic and congestion. The competition of communication flows with different priorities is considered as a two-player non-cooperative game. The Nash Q-learning algorithm control obtains the Nash Q-values through trial-and-error and interaction with the network environment so as to improve its behaviour policy. The MAFC can learn to take the best actions in order to regulate source flows that guarantee high throughput and low packet-loss ratio. The MIQL control, through a specific learning processing, does achieve the optimum sending rate for the sources with lower priority while observing the sources with higher priority. Designed intelligent controls achieve superior performances in controlling the flows in high-speed networks in comparison to the standard ones and avoid communications congestion.

G.M. Dimirovski (✉)
School of Electrical Engineering & Information Technologies, St Cyril and St Methodius University, MK-1000 Skopje Karpos 2 BB, Republic of Macedonia
e-mail: dimir@etf.ukim.edu.mk; gdimirovski@dogus.edu.tr; dimir@feit.ukim.edu.mk

G.M. Dimirovski
School of Engineering, Acibadem, Dogus University of Istanbul, Zeamet Sk 21, TR-34722 Istanbul, Republic of Turkey



V. Sgurev et al. (eds.), *Innovative Issues in Intelligent Systems*,
Studies in Computational Intelligence 623, DOI 10.1007/978-3-319-27267-2_4

**Keywords** Communication networks · Computer networks · Control theory · Fuzzy-neural systems · Lyapunov stability · Learning systems · Nonlinear systems · Q-learning theory · Strategy synergies

# 1 Introduction

The interests and research activities oriented towards understanding and building large network-interconnected, complex systems for various applications ranging from electrical power to telecommunications and to transportation [1, 2] have increased rapidly during the last decade of the previous and the first one of this century. In fact, this trend has emanated with the seminal monographs on Cybernetics [3, 4] but it begun to expand rapidly ever since it was understood that Earth's real-world space represents a dynamical interplay below the light-speed of the fundamental natural quantities and that information is the third fundamental quantity next to energy and matter [5–8]. For, it was understood the information differs from energy and matter only in that it requires human mind to sense and transcend it, unlike the energy and matter which are sensed and transcended by biological human sensors.

On the other hand, it was also well understood that solely information has the potential to re-shape both the energy and the material processes via man-made decision and control systems [1, 5, 7], which are integrating the information signal processing, decision making, and supervision coordination accompanied by various executive control actions. In turn, indeed rather complex networked systems, largely viewed as cyber-physical systems, have emerged in which information processing propagates in cycles from decision computation through communication transmission to executive controlling actions re-shaping the useful dynamical processes towards the human posed goals [2, 9]. And yet, this propagation can be crucially deteriorated or even disrupted within the telecommunications network integrating the entire complex systems because of the potentially present, rather serious, phenomenon of communications congestion [10, 11]. This kind of congestion can cause network-interconnected complex systems to loose their essential features that make up the very reasons for which they have been built. Thus, indeed various techniques ranging from analytical control and game theories, Nash non-cooperative games in particular, to machine learning theory have been applied in order to prevent or cope with the network congestion problem (e.g., see [10, 12–24]).

The flow control in high-speed networks is rather difficult owing to the uncertainties and different traffic patterns that are highly time-varying. The flow control mainly checks the availability of bandwidth and buffer space necessary to guarantee the requested quality of service (QoS). A major problem here is the lack of information related to the characteristics of the source flows indicating patters of communication traffic [1, 19] that must be rediscovered and learned by some pattern recognition methodology (e.g. see [25, 26]). Naturally, deriving a mathematical model for the source flow is the fundamental issue, of course. But it has been found to be a rather difficult task, especially for the case of broadband sources, and often impossible in the real-time of communications such as in the TCP/ATM networks. In order to overcome the above-mentioned difficulties, the control schemes with learning capability have been introduced into applications to flow controls [18, 20, 27, 28]. The basic advantage gained in this way is the ability to learn the source traffic characteristics from sufficiently big and representative data samples. Yet, it is obvious that the accurate data, that are needed to train the parameters, are hard to obtain because of disturbances and errors in measuring instruments [18].

It should be noted, the crucial feature of network-interconnected complex systems is that essentially they are high-speed networks [29–31]. For, it may readily be argued that their functioning (and malfunctioning too) propagates with half light-speed at least as so happens the information processing at large. Thus, the growing interest on congestion problems in high-speed networks has arisen from the need to control the sending rates of traffic sources because of the ever existing potential for congestion induced collapse in communications hence functional disintegration of the complex system [14]. Congestion problems result from a mismatch between the offered load and the available link bandwidth between network nodes. Such problems can cause high packet loss ratio (PLR) and long time delays (LTD), and even can break down the entire system. Therefore high-speed networks must have an appropriate flow control scheme not only to guarantee the quality of service (QoS) for the existing links but also to achieve high system utilization in terms of maximum throughput and allowable time delays while preventing congestion blocking [10, 12, 16].

Recently, certain breakthrough progress has been to cope with the congestion in high-speed networks by employing synergies of computational intelligence control and learning theories. Certain contributions to such advanced intelligent controls, verified by extensive simulations (among other standard package for telecommunications, also MathWorks Matlab [32]; Simulink [33] were used), are presented in this chapter. Clearly, regardless the computational intelligence (either fuzzy-system or neural-net based) the stability property at the network's operating equilibrium has to be investigated in the sense of the recently established stability theories for hybrid and switched nonlinear dynamic systems [34–40] to the latter category also the switched fuzzy systems belong to [41]. Without dwelling into details of this stability topic it should be noted that this is a requirement sine-qua-non hence must be guaranteed for the network's system dynamics within the proximity of it operating equilibrium [14, 42]. Nonetheless, it also should be noted that this stability topic appears beyond the main goal and purpose of the present contribution study hence it shall not be address in more detail.

In their recent work Hwang et al. [31], unlike in works by Chatovich et al. [27] and by Hsiao et al. [30], they have extended the operating environment to multi-bottleneck network and then proposed a cooperative multi-agent congestion controller. Yet, they missed to prove or demonstrate the convergence of the scheme under their controller. On the other hand, Li et al. [20] successfully combined the Q-learning and simulated annealing to solve the problems of ABR flow control in

ATM networks. It is on these grounds that Yuan Wei Jing [43] and co-authors as well as Xin Li and their co-authors [19–21] managed to contribute further advances into learning intelligent flow control for high-speed networks within the setting of Nash non-cooperative games enhanced by using Q-learning algorithm [44].

In this contribution, the author has revisited the papers he co-authored with Jing and Li (2007, 2008, 2009) as well as his own monographs chapter [14] thereby expanding and finalizing those studies in a unified contribution to this monograph. The focus is placed on the two main case studies of network with multiple and with single bottleneck nodes. The intelligent control for networks with multiple bottleneck nodes was derived as a multi-agent flow controller (MAFC) based on Q-learning algorithm in conjunction with the theory of Nash equilibrium of opponents' strategies. The second one, networks with single bottleneck nodes, was derived as a model-independent Q-learning control (MIQL) scheme not needing a prior-knowledge on communication traffic and congestion but rather extracting it online following the flow with higher priority. The underlying idea in both designs is to regard the competition of communication flows with different priorities as a non-cooperative game involving the opponent players.

## 2 Case-Study on Nash Q-Learning Multi-agent Flow Control for High-Speed Networks

Especially in the case studies considered in here, the employed pattern recognition methodology is the reinforcement learning [44–46]. For, it has proven its operating superiority since it just needs a very simple, yet critical information, such as "right" or "wrong" [47, 48]. For, it is independent of mathematic model and priori-knowledge of system dynamics. It does obtain the needed knowledge through trial-and-error and interaction with environment to improve its own behavioural policy. Thus it possesses the ability of self-learning. Because of these advantages, it has been played a very important role in the flow control in high-speed networks (e.g., see [15, 27, 30, 31]). Furthermore, the Q-learning technique of the reinforcement learning class is fairly easy to apply and has a firm foundation in the theory [47, 49]. It is therefore that in here the creation of the synergy between computational intelligence and systems control theory evaolves around the idea of employing the Q-learning methodology [44].

In here, on the grounds of the Q-learning algorithm, a multi-agent intelligent flow controller (intelligent MAFC; denoted as MFC in figures for simplicity) for high-speed networks is proposed. The fuzzy inference system is adopted to generate the reward signal in Q-learning scheme. In this proposed controller design, each learning agent in a bottleneck node has a separate memory structure in order to explicitly implement its own objectives towards achieving Nash equilibrium Q-values. It is the Nash equilibrium solution that serves as the optimum sending rates of traffic flow sources. By means of learning procedures, the proposed controller adjusts the source sending rate to the optimal value so as to reduce the average length of queue in the buffer. The convergence of the proposed learning algorithm is demonstrated in due course of the analysis. A set of simulation results shows that the proposed controlling strategy can effectively avoid the occurrence of congestion with the features of high throughput, low PLR, low end-to-end delay and high utilization.

## 2.1 *Controller Strategy and Learning Intelligent Control Design*

Nowadays, there is widely employed into high-speed networks the well-known Adaptive-Increase-Multiplicative-Decrease (AIMD) flow controlling strategy (e.g., see [50]). In the well-known AIMD case, the agent senses the network system's states and makes a decision based on a rate control scheme to avoid packet losses and increase the utilization of multiplexer's output bandwidth [15].

However, it is hard to achieve high system performance by reactive AIMD scheme because of the propagation delay and the very dynamic nature of high-speed networks. Thus, it may well be inferred that not only an innovated pattern recognition learning [25, 26] is needed but also novel system architectures of controllers in communication networks become prerequisite too [19, 20]. Therefore in here the Multi-Agent-Flow-Controller (MAFC) initially developed by Li et al. [19, 20] is revisited and further improved within the fuzzy-neural setting of Dimirovski [51].

### 2.1.1 The Architecture of the Intelligent MAFC

The here presented intelligent MAFC can operate in an optimum mode solely by relying on the interaction with unknown environment and provide the best action for a given state. The architecture of MAFC in the $n$ node case is shown in Fig. 1.

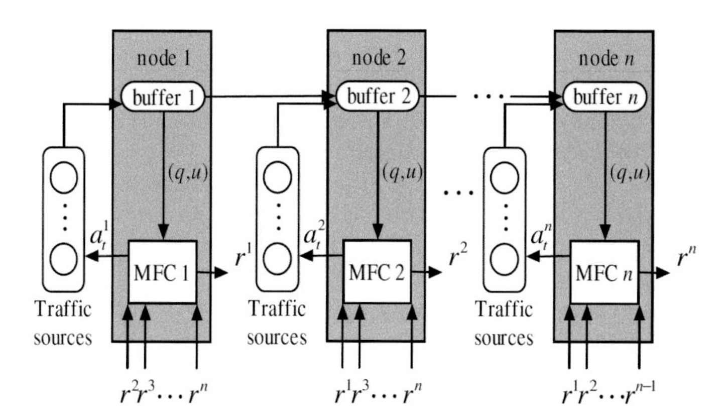


**Fig. 1** Architecture of the intelligent MAFC in the $n$ node case

The high-speed network is assumed to have $n$ bottleneck nodes that possess controllable sources. Each MAFC has its own state variables (S), which is composed of the current queue length $q$, the current rate of queue length change $\dot{q}$, and the current rate of source sending rate change $u$ irrespective of the operation of other individual MAFCs within the network. The output of each MAFC is the feedback signal $a$ to the traffic sources, which is the ratio of the sending rate. It determines the sending rate $u$ of traffic sources. However, all agents have incomplete but perfect information, meaning agents do not know other agents' reward functions and state transition probabilities, but they can observe other agents' immediate rewards and actions taken previously. The sample for each MAFC is same. By means of multi-agent strategy-search learning, the proposed MAFC is endowed [17, 20] with convergence property to a Nash equilibrium balance [52]. In contrast to single-agent flow controller (SAFC), MAFC uses joint-action learning algorithms to learn values for joint actions to avoid the bias in individual action decision.

The sending rate is controlled by the feedback control signal *at* periodically activated actions. The controlled sending rate is defined by the equation

$$u_t = a_t FL \tag{2.1}$$

where: $a_t \in [0.2, 1.0]$ is the feedback signal by the flow controller; $F$ is a relative value in the ratio of source offered load to the available bit rate (ABR) output; and $L$ denotes the outgoing rate of link. Notice that in here $u_t \in [0.2FL, FL]$ is the controlled sending rate at sample time $t$ [47].

### 2.1.2 The Reward Signal

Q-learning is to learn what to do and how to map situations to actions, so as to maximize the reward signal $r$. The reward $r$ is the only information for the controller to judge whether the sending rate taken is good or bad, so it is vital to choose an appropriate $r$. $r$ is in the range [0, 1], the larger $r$ is, the better control affects.

In high-speed networks, the fuzzy reward evaluator (FRE) (see [47]) evaluates the reward for environmental states. FRE relies on three parameters $(q, \dot{q}, \dot{u})$ to generate a reward or a punishment for the action (the ratio of the source sending rate) in a state. If the state of the network is enhanced toward positive evolution, the action will be rewarded; otherwise punished.

The term set should be determined at an approximate level of granularity to describe the values of linguistic variables. For queue length, the term set is defined as $f(q)$ = {Low (L), Medium (M), High (H)}, which is used to describe the degree of queue lengths. The term set for the rate of queue length change is defined as $f(\dot{q})$ = {Decrease (D), Increase (I)}, which describes the rate of queue length change as "Decrease" or "Increase". The term set for the rate of source transmission rate change is defined as $f(\dot{u})$ = {Negative (N), Positive (P)}, which describes the rate of source transmission rate change as "Negative" or "Positive". In order to provide a precise graded reward in various states, the term set for reward is defined as

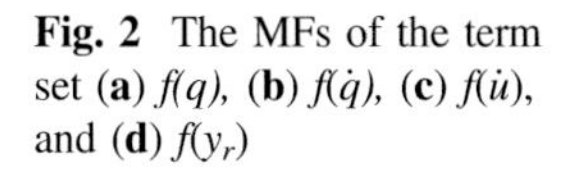

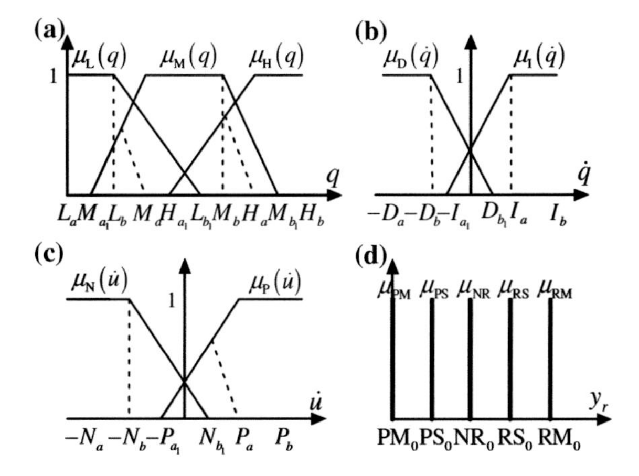


**Fig. 2** The MFs of the term set (**a**) $f(q)$, (**b**) $f(\dot{q})$, (**c**) $f(\dot{u})$, and (**d**) $f(y_r)$

**Table 1** Rule table of fuzzy reward evaluator

| Rule | $q$ | $\dot{q}$ | $\dot{u}$ | $y_r$ | Rule | $q$ | $\dot{q}$ | $\dot{u}$ | $y_r$ |
|---|---|---|---|---|---|---|---|---|---|
| 1 | L | D | N | PM | 7 | M | I | N | RS |
| 2 | L | D | P | PS | 8 | M | I | P | RM |
| 3 | L | I | N | NR | 9 | H | D | N | PS |
| 4 | L | I | P | NR | 10 | H | D | P | PM |
| 5 | M | D | N | RS | 11 | H | I | N | PS |
| 6 | M | D | P | PM | 12 | H | I | P | PM |

$f\,(y_r)$ = {Penalty More (PM), Penalty Slightly (PS), No Reward (NR), Reward Slightly (RS), Reward More (RM)}. The membership functions (MFs) of the term set are shown in Fig. 2.

The fuzzy rule base is a reward knowledge base, characterized by a set of linguistic statements in the form of "if-then" rules that describe the fuzzy logic relationship between the input variables and the reward (penalty) $y_r$. According to fuzzy set theory, the fuzzy rule base forms a fuzzy set with dimensions $3 \times 2 \times 2 = 12$. Table 1 shows a total of twelve inference rules in the fuzzy rule base under various system states. For example, rule 1 can be linguistically started as "if the queue length is low, the queue length change rate is decreased, and the transmission change rate is negative, then give more penalty."

The proposed FRE adopts the max-min inference method for the inference engine because it is designed for real-time operation.

### 2.1.3 The Nash Q-Learning Multi-agent Flow Controller

The Q-learning is a value learning version of the model-free reinforcement learning that learns utility values (Q-values) of state and action pairs (see [45]). The objective of Q-learning is to estimate Q-values for an optimal strategy. The proposed flow controller in each bottleneck node acts as a learning agent. During the learning process, an agent uses its experience to improve its estimate by blending new information into its prior experience [44].

The difference between single-agent and multi-agent system exists in the environments. In multi-agent systems, other adapting agents make the environment no longer stationary and violate the Markov property that traditional single-agent behaviour learning relies upon. Based on the framework of stochastic games, the single-agent Q-learning is extended to multi-agent systems.

In general form, an n-agent is defined by a "(2n + 3)-tuple" of special mathematical objects <$n$, S, $A^1$,..., $A^n$, $r^1$,..., $r^n$, $p$>. In there notice the following: $n$ is the number of agents, or in another word the number of bottleneck nodes in high-speed networks; S is a set of discrete state space of high-speed networks composed of $(q, \dot{q}, \dot{u})$; $A^1$,..., $A^n$ is a collection of actions (feedback control signal to traffic sources) available to each agent ($A^i$ is the discrete action space available to agent $i$); $p$ is the reward function for agent $i$; $p$ is the transition probability map.

In Nash Q-learning flow controller, the objective of each agent is to maximize the discounted sum of rewards, with discount factor $\beta \in [0, 1)$. Let $\pi^i$ be the action strategy of agent $i$. For a given initial state $s$, agent $i$ tries to maximize

$$v^i(s, \pi^1, \ldots, \pi^n) = \sum_{t=0}^{\infty} \beta^t E(r_t^i | \pi^1, \ldots, \pi^n, s_0 = s) \tag{2.2}$$

A strategy $\pi = (\pi_0, \ldots, \pi_t, \ldots)$ is defined over the whole course of learning process. $\pi_t$ is called the decision rule at sample time $t$.

The Nash equilibrium solution is a tuple-set of $n$ strategies $(\pi_*^1, \pi_*^2, \ldots, \pi_*^n)$ such that for all $s \in S$ and $\pi^i \in \Pi^i$,

$$v^i(s, \pi_*^1, \ldots, \pi_*^n) \geq v^i(s, \pi_*^1, \ldots, \pi_*^{i-1}, \pi_*^i, \pi_*^{i+1}, \ldots, \pi_*^n) \tag{2.3}$$

where $\Pi^i$ is the set of strategies available to the agent $i$. The definition of Nash equilibrium requires that each agent's strategy is a best response to the opponent agent's strategy.

The Nash Q-function for the $i$th agent is defined over $(s, a^1, \ldots, a^n)$ as the sum of its current reward plus its future rewards when all agents follow a joint Nash equilibrium strategy. That is,

$$Q_*^i(s, a^1, \ldots, a^n) = r(s, a^1, \ldots, a^n) + \beta \sum_{s\prime \in S} p(s'|s, a^1, \ldots, a^n) v^i(s', \pi_*^1, \ldots, \pi_*^n) \tag{2.4}$$

where $(\pi_*^1, \ldots, \pi_*^n)$ is the joint Nash equilibrium strategy, $r^i$ $(s, a^1, \ldots, a^n)$ is agent $i$'s one-period reward in state $s$ under joint action $(a^1, \ldots, a^n)$, $v^i(s', \pi_*^1, \ldots, \pi_*^n)$ is agent $i$'s total discounted reward over infinite periods starting from state $s'$ given that agents follow the equilibrium strategies.

The learning agent, indexed by *i*, learns about its Q-values by forming an arbitrary guess at time 0. One simple guess would be letting $Q_0^i(s, a^1, \ldots, a^n) = 0$ for all $s \in S$, $a^1 \in A^1, \ldots, a^n \in A^n$. At each time *t*, agent *i* observes the current state, and takes its action. After that, it observes its own reward, actions taken by all other agents, others' rewards, and the new state $s'$. It then calculates a Nash equilibrium $\pi^1(s')\ldots\pi^n(s')$ for $(Q_t^1(s'), \ldots, Q_t^n(s'))$, and updates its Q-values according to

$$Q_{t+1}^i(s, a^1, \ldots, a^n) = (1 - \alpha_t)Q_t^i(s, a^1, \ldots, a^n) + \alpha_t[r_t^i + \beta\, NashQ_t^i(s')] \tag{2.5}$$

where $\beta \in [0, 1)$ is the discount factor, if $\beta$ is large, systems will easily tend to follow the current strategy so that it will not have more opportunities to find a better strategy; if $\beta$ is small, systems will not easily follow a strategy so that it will do explorations all the time. This will cause the convergence rate to be slow. On the other hand, $\alpha \in [0, 1)$ is the learning rate. The convergence rate is determined by the value of $\alpha$. If $\alpha$ is small, the convergence rate will be slow but it will easily tend to stabilize. If $\alpha$ is large, the convergence rate will be fast but it will not easily tend to stabilize.

The $NashQ_t^i(s')$ is defined as

$$NashQ_t^i(s') = \pi^1(s')\ldots\pi^n(s')Q_t^i(s'). \tag{2.6}$$

In order to calculate the Nash equilibrium $(\pi^1(s')\ldots\pi^n(s'))$, agent *i* would need to know $Q_t^1(s'), \ldots, Q_t^n(s')$. Information about other agents' Q-values is not given, so agent *i* must learn about them too. Agent *i* forms conjectures about those Q-functions at the beginning of learning, for example, $Q_0^j(s, a^1, \ldots, a^n) = 0$ for all *j* and all $s, a^1, \ldots, a^n$. As the learning process, agent *i* observes other agents' immediate rewards and previous actions. That information can then be used to update agent *i*'s conjectures on other agents' Q-functions. Agent *i* updates its beliefs about agent *j*'s Q-function according to the same updating rule [1] and it applies to its own

$$Q_{t+1}^i(s, a^1, \ldots, a^n) = (1 - \alpha_t)Q_t^j(s, a^1, \ldots, a^n) + \alpha_t[r_t^j + \beta\, NashQ_t^j(s')] \tag{2.7}$$

Notice that $\alpha_t = 0$ for $(s, a^1, \ldots, a^n) \neq (s_t, a_t^1, \ldots, a_t^n)$. Therefore [53] does not update all the entries in the Q-functions. It updates only the entry corresponding to the current state and the actions chosen by the agents.

### 2.1.4 The Convergence of the Proposed Controller Design

In this section, we would like to demonstrate the convergence of $Q_t^i$ to the Nash equilibrium $Q_*^i$ for agent *i*. The value of $Q_*^i$ is determined by the joint strategies of all agents. That means the agent has to learn Q-values of all the agents and derive strategies from them. The learning objective is $(Q_*^1, \ldots, Q_*^n)$, and we have to show the convergence of $(Q_t^1, \ldots, Q_t^n)$ to $(Q_*^1, \ldots, Q_*^n)$

The convergence proof requires a basic assumption that every state and action should be visited infinitely often [49] or, equivalently, almost so. The proof relies on the following lemma given, which is Corollary 5 in [48], that establishes the convergence of a general Q-learning process updated by a pseudo-contraction operator. Let Q be the space of all Q functions.

**Lemma** *If the mapping $P_t : Q \to Q$ satisfies*

$$\|P_t Q - P_t Q_*\| \leq \|Q - Q_*\| + \lambda_t \tag{2.8}$$

*for all $Q \in$ Q and $Q_* = E[P_t Q_*]$, where $0 < \gamma < 1$ and $\lambda_t \geq 0$ zero with the probability 1.0, then the iteration defined by*

$$Q_{t+1} = (1 - \alpha_t) Q_t + \alpha_t [P_t Q_t] \tag{2.9}$$

*converges to $Q_*$ with the probability 1.0.*

*Notice that if the computed Nash optimal Q-value $Q_*$ matches the Lemma with an existing operator $P_t$, then we can draw the conclusion that the iteration converges to Nash optimal Q-value $Q_*$ with probability 1.0. For the present Nash Q-learning algorithm, we define the operator $P_t$ as follows. Let $Q = (Q^1, \ldots, Q^n)$ where $Q^k \in$ Q$^k$, for k = 1,..., n, and Q = Q$^1 \times \cdots \times$ Q$^n$. $P_t Q = (P_t Q^1, \ldots, P_t Q^1)$, where*

$$P_t Q^k(s, a^1, \ldots, a^n) = r_t^k(s, a^1, \ldots, a^n) + \beta \pi^1(s') \ldots \pi^n(s') Q^k(s') \tag{2.10}$$

*Let suppose further that $v^k(s\prime, \pi_*^1, \ldots, \pi_*^n)$ are the agents k's Nash equilibrium reward in order for $(Q_*^1(s'), \ldots, Q_*^n(s'))$, and $(\pi_*^1(s'), \ldots, \pi_*^n(s'))$ be the Nash equilibrium solution. The analysis consideration along this line of thinking, in turn, does yield the following evaluation result*

$$\begin{aligned} & Q_*^k(s, a^1, \ldots, a^n) \\ & \quad = r^k(s, a^1, \ldots, a^n) + \beta \sum_{s' \in S} p(s'|s, a^1, \ldots, a^n) v^k(s', \pi_*^1, \ldots, \pi_*^n) \\ & \quad = r^k(s, a^1, \ldots, a^n) + \beta \sum_{s' \in S} p(s'|s, a^1, \ldots, a^n) \pi_*^1(s'), \ldots, \pi_*^n(s') Q_*^k(s') \end{aligned}$$

that in fact represents the respective mathematical expectation

$$\begin{aligned} & Q_*^k(s, a^1, \ldots, a^n) \\ & \quad = \sum_{s\prime \in S} p(s'|s, a^1, \ldots, a^n)(r^k(s, a^1, \ldots, a^n) + \beta \pi_*^1(s') + \pi_*^n(s') Q_*^k(s')) \\ & \quad = E\left[P_t^k Q_*^k(s, a^1, \ldots, a^n)\right] \end{aligned}$$

*for all* $s, a^1, \ldots, a^n$. *Thus* $Q_*^k = E\left[P_t^k Q_*^k\right]$ *since this holds for all k,* $E[P_t Q_*] = Q_*$. *During the learning process of MAFC in high-speed networks, every stage game* $(Q_t^1(s), \ldots, Q_t^n(s))$, *for all t and s, holds Nash equilibrium, and the agents' rewards in this equilibrium are used to update their Q-functions. At the Nash equilibrium, each agent effectively holds a correct, common sense and mathematical, expectation about other agents' behaviours hence it acts rationally with respect to this expectation.*

*For every agent acting rationally it does mean the agent's strategy is a best response to the opponent agent's strategies. In contrast, any deviation would make that agent worse off. This is to say that:*

$$\pi^k(s)\pi^{-k}(s)Q^j(s) \geq \hat{\pi}^k(s)\pi^{-k}(s)Q^j(s) \tag{2.11}$$

$$\pi^k(s)\pi^{-k}(s)Q^j(s) \leq \pi^k(s)\hat{\pi}^{-k}(s)Q^j(s) \tag{2.12}$$

*for all k, and all* $\hat{\pi}^k \in \pi(A^k), \hat{\pi}^{-k} \in \pi(A^{-k})$, *where*

$$\pi^{-k}(s) = \pi^1(s)\ldots\pi^{k-1}(s)\pi^{k+1}(s)\ldots\pi^n(s) \tag{2.13}$$

*Also it should be noted, for all* $Q, \hat{Q} \in Q$ *there is fulfilled:*

$$\begin{aligned} \left\|P_t Q - P_t\hat{Q}\right\| &= \max_j \left\|P_t Q^j - P_t\hat{Q}^j\right\|_{(j)} = \max_j \max_s \left\|P_t Q^j(s) - P_t\hat{Q}^j(s)\right\|_{(j,s)} \\ &= \max_j \max_s \left|\beta\pi^1(s)\ldots\pi^n(s)Q^k(s) - \beta\hat{\pi}^1(s)\ldots\hat{\pi}^n(s)\hat{Q}^k(s)\right| \end{aligned}$$

*Hence we have*

$$\left\|P_t Q - P_t\hat{Q}\right\| = \max_j \max_s \beta\left|\pi^k(s)\pi^{-k}(s)Q^k(s) - \hat{\pi}^k(s)\hat{\pi}^{-k}(s)\hat{Q}^k(s)\right| \tag{2.14}$$

*On the grounds of the known property of Nash equilibrium* [52], *if it is valid* $\pi^k(s)\pi^{-k}(s)Q^j(s) \geq \hat{\pi}^k(s)\pi^{-k}(s)Q^j(s)$, *then we have*

$$\begin{aligned} &\pi^k(s)\pi^{-k}(s)Q^k(s) - \hat{\pi}^k(s)\hat{\pi}^{-k}(s)\hat{Q}^k(s) \\ &\leq \pi^k(s)\pi^{-k}(s)Q^k(s) - \pi^k(s)\hat{\pi}^{-k}(s)\hat{Q}^k(s) \\ &\leq \pi^k(s)\hat{\pi}^{-k}(s)Q^k(s) - \pi^k(s)\hat{\pi}^{-k}(s)\hat{Q}^k(s) \leq \left\|Q^k(s) - \hat{Q}^k(s)\right\| \end{aligned}$$

*where*

$$\left\|Q^k(s) - \hat{Q}^k(s)\right\| = \max_{a^1,\ldots,a^n} \left|Q^k(s, a^1, \ldots, a^n) - \hat{Q}^k(s, a^1, \ldots, a^n)\right| \tag{2.15}$$

*If it be* $\pi^k(s)\pi^{-k}(s)Q^j(s) - \hat{\pi}^k(s)\hat{\pi}^{-k}(s)\hat{Q}^j(s)$, *then the proof is rather similar to the above presentation and thus omitted here.*

*Therefore we con conclude that*

$$\left|\pi^k(s)\pi^{-k}(s)Q^k(s) - \hat{\pi}^k(s)\hat{\pi}^{-k}(s)\hat{Q}^k(s)\right| \leq \left\|Q^k(s) - \hat{Q}^k(s)\right\| \tag{2.16}$$

*Moreover, on the grounds of this result and* (2.14) *above, it follows at once that*

$$\left\|P_tQ - P_t\hat{Q}\right\| \leq \max_j \max_s \beta\left\|Q^k(s) - \hat{Q}^k(s)\right\| = \beta\left\|Q - \hat{Q}\right\|$$

*Since it is also true that* $\hat{Q} \in \mathrm{Q}$, it *follows that the Nash optimal Q-value* $Q_*$ *does satisfy the inequality* (2.8), *which completes the proof.*

In summary of the above presented discussion, apparently, we may well infer the needed conclusion that the proposed Nash Q-learning process converges to the optimum Nash Q-values thus to the equilibrium status that keeps on the operating equilibrium state.

## 2.2 Simulation Investigations and Performance Comparisons with Other Controllers

The standard assumptions on simulation investigations have been adopted. It is assumed that all packets have a fixed length of 1000 bytes, and a finite buffer length of 20 packets in each node is available. On the other hand, the offered loading in the simulations is assumed to vary in the range between 0.6 and 1.2 as corresponding to the systems' dynamics around the equilibrium state. Therefore, the higher loading appears in conjunction with the heavier communication traffic and vice versa.

From the knowledge of the evaluating system performance, the parameters of the membership functions for input linguistic variables in the FRE are selected as given in the sequel. Namely, for $\mu_L(q)$, $\mu_M(q)$, and $\mu_H(q)$, $L_a = 0$, $L_b = 6$, $L_{b1} = 10$, $M_{a1} = 2$, $M_a = 8$, $M_b = 12$, $M_{b1} = 20$, $H_{a1} = 9$, $H_a = 14$, $H_b = 20$, and $H_{b1} = 20$; for $\mu_D(\dot{q})$ and $\mu_I(\dot{q})$, $D_a = 4$, $D_b = D_{b1} = 2$, $I_{a1} = I_a = 2$, and $I_b = 4$; for $\mu_N(\dot{u})$ and $\mu_p(\dot{u})$, $N_a = 0.8$, $N_b = 0.4$, $N_{b1} = 0.2$, $P_{a1} = 0.2$, $P_a = 0.4$, and $P_b = 0.8$. Also, the parameters of the membership functions for output reward linguistic variables are given by means of $PM_O = 0$, $PS_O = 0.25$, $NR_O = 0.5$, $RS_O = 0.75$, and $RM_O = 1$.

In the simulation experiments, three schemes of flow control agent, AIMD, MAFC with general Q-learning algorithm and the Nash Q-learning algorithm proposed here are individually implemented in high-speed networks. The first scheme AIMD increases its sending rate by a fixed increment (0.11) if the queue length is less than the predefined threshold; otherwise the sending rates are decreased by a multiple of 0.8 of the previous sending rate to avoid congestion. Finally, for the second and third schemes, the sending rates are controlled by the feedback control signal periodically. For establishing the assurances the proposed

MAFC as applied to high-speed networks to be feasible and effectively achieved, comparisons among those schemes were analyzed.

Four measures, namely, throughput, PLR, buffer utilization and packets' mean delay, are used as the indices for evaluating the achievable performances. The status of the input multiplexer's buffer in each node reflects the degree of congestion, which may result in possible packet losses. For simplicity, packets' mean delay only takes into consideration the processing time at each node plus the time needed to transmit packets. The details are delineated in the following.

The investigated configuration of high-speed networks with multi-node is shown in Fig. 3. It is composed of four switches Sw1, Sw2, Sw3, and Sw4 in cascade. Three nodes Sw1, Sw2, and Sw3 are implemented with its own control agent, respectively; hence, the sending rates of sources S1 and S2, S3 and S4, and S5 and S6 are regulated by MFC1, MFC2 and MFC3, respectively. Consequently, three agents jointly interact to reach a common goal with high system performance. For case of three schemes, each node has the same training loading pattern, which is generated by a shuffle of loading pattern (0.6, 0.7, 0.8, 0.9, 1.0, 1.1, 1.2), each of which lasts for 0.6 s; i.e., a training epoch will last for a period of 4.2 s.

We have taken the Sw2 as an example to consider the performance of the flow control schemes adopted in simulation. Figure 4 shows the throughputs of Sw2 controlled by three different kinds of control agents individually. Analogously, because of the reactive control, the throughput of nodes for the AIMD method decrease seriously at loading of about 0.9. Conversely, the MAFC methods remain a higher throughput even though the offered loading is over 1.0.

The subsequent Figs. 5, 6 and 7 show the PLR, buffer utilization and mean delay of Sw2 controlled by four different kinds of control agents individually. It is obvious that the PLR of no control is high, even though that the AIMD method were adopted. However, using the MAFC with Nash Q-learning method can decrease the PLR enormously with high throughput and low mean delay. The MAFC with Nash Q-learning algorithm has a better performance over MAFC with general Q-learning in PLR, buffer utilization and mean delay. It demonstrates once again that MAFC with Nash Q-learning possesses the ability to predict the network behaviour in advance.

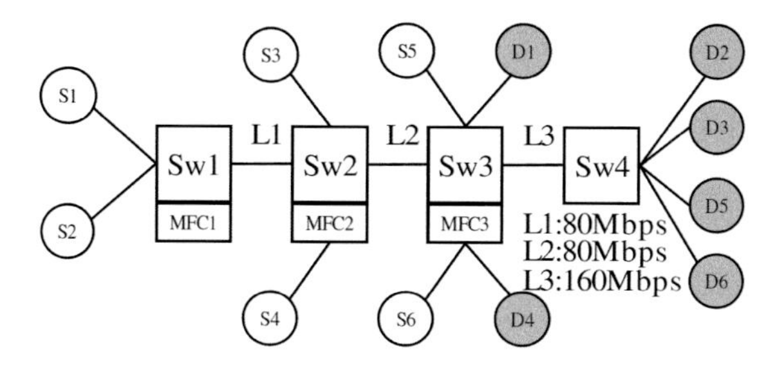


**Fig. 3** The configuration of high-speed networks with four switches

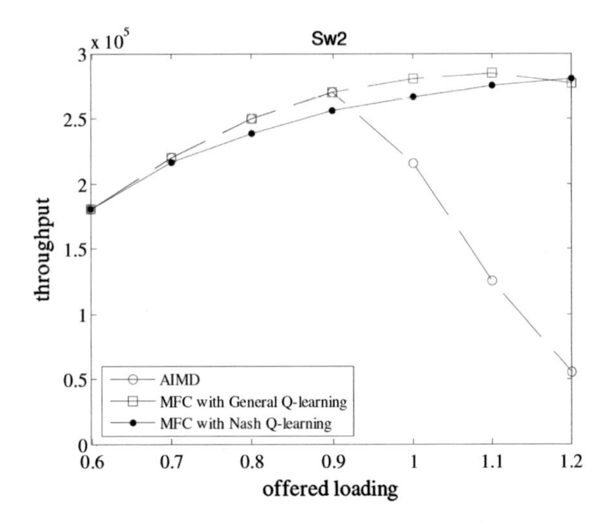


**Fig. 4** Throughput versus various offered loading at Sw2

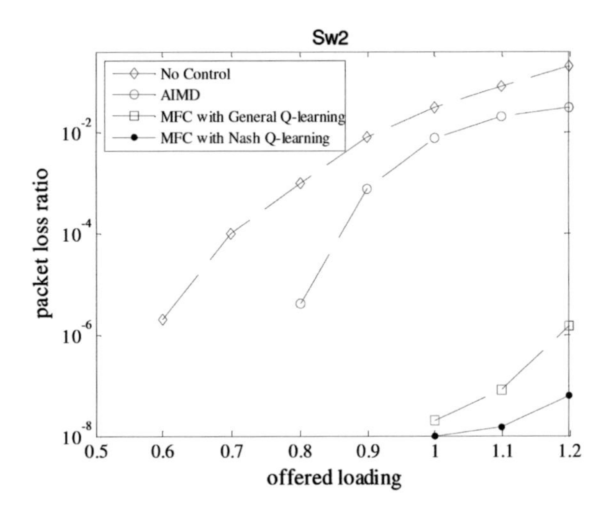


**Fig. 5** PLR versus various offered loading at Sw2

## 2.3 *Concluding Remarks for the Intelligent MAFC Design*

It is well known that most packet losses in high-speed networks result from the dropping of packets owing to the congested nodes in the network. The standard reactive scheme AIMD [22, 54] therefore cannot accurately respond to a time-varying environment because of the lack of some predicting capability. In contrast, all learning intelligent controls can cope considerably better than the AIMD precisely because being endowed with some predicting capability. In this regard, the Nash Q-learning algorithm of the reinforcement learning category due to its prediction capacity does perform even better than the one endowed with general Q-learning capacity.

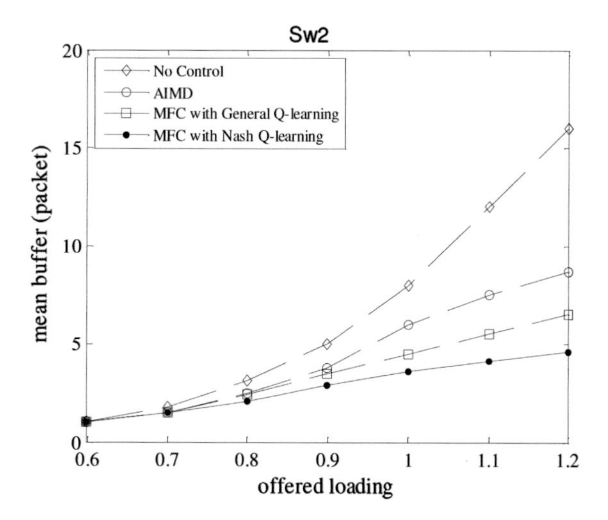


**Fig. 6** Mean buffer versus various offered loading at Sw2

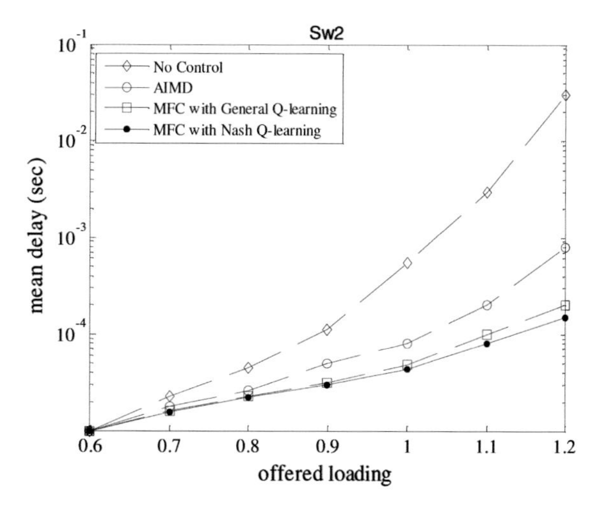


**Fig. 7** Mean delay versus various offered loading at Sw2

The proposed design of an intelligent MAFC, which was applied to a communication network having multiple bottlenecks, can respond to the networks' traffic dynamics as needed. Through a proper learning-training process, MAFC can learn empirically without prior information on the operating environmental dynamics of the network. The sending rate of traffic sources can be determined by the well-learned Nash Q-values and in the course of learning the convergence was well established and demonstrated via the simulations. Simulation results have shown that the proposed method can increase the utilization of the buffer and decrease PLR and delay simultaneously. Therefore, MAFC with Nash Q-learning algorithm not only guarantees low PLR for the existing links, but also achieves high system utilization [17].

# 3 Case-Study on Simple Q-Learning Model-Independent Flow Control for High-Speed Networks

The second case study on intelligent learning flow controls for communication networks in this chapter is a synthesis for a simple Q-learning model-independent flow controller design [19]. The here derived controller is shown to behave and operate in an optimum mode without utilizing an analytical mathematic model of the network environment. It is shown that it can cope with the unpredictably varying operating environment only by relying on its continuing interaction with the unknown environment, and yet provide the best action for a given state of the traffic dynamics. By means of the Q-learning procedures [53, 55, 56], the designed controller adjusts the source sending rate to the optimum value in the sense of reducing the average length of queue in the buffer. The obtained results in simulation experiments have demonstrated that this simple Q-learning controller can effectively avoid the congestion occurrence while ensuring the desired operating features of high throughput, low PLR, low end-to-end delay, and high network utilization. However, all these useful properties have been proven achievable in the case of single bottleneck high-speed network.

## *3.1 Controller Strategy and Learning Intelligent Control Design*

Although nowadays the AIMD control scheme [54] is the employed standard in the high-speed networks, such as TCP/ATM networks are, it is well known that is hard if not impossible to achieve high system performance because of the propagation delay and the very dynamic nature of high-speed networks (e.g., see [16, 18, 50]). This section gives the detailed architecture of the proposed flow controller as shown

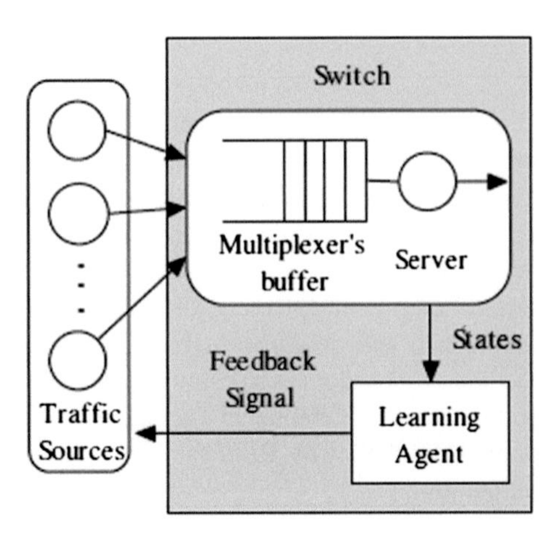


**Fig. 8** Architecture of the proposed flow controller

in Fig. 8. In high-speed networks, the proposed controller in bottleneck node acts as a flow control agent with flow control ability. The inputs of controller are state variable in high-speed networks composed of the current queue length in the buffer and the available bandwidth for the controlled traffic sources. In here the synthesis for a simple Q-learning, model-independent, flow controller design is investigated. It will be shown in due course that even such a fairly simple Q-learning [44] can cope with the unpredictably varying operating environment only through its continuing interaction with the unknown environment.

### 3.1.1 The Architecture of Flow Controller Design

This section gives the detailed architecture of the proposed flow controller as shown in Fig. 1, [19]. In high-speed networks, the proposed controller in bottleneck node acts as a flow control agent with flow control ability. The inputs of controller are state variable in high-speed networks composed of the current queue length in the buffer and the available bandwidth for the controlled traffic sources. In high-speed networks, the proposed controller in bottleneck node acts as a flow control agent with flow control ability. The inputs of controller are state variable in high-speed networks composed of the current queue length in the buffer and the available bandwidth for the controlled traffic sources. The output of controller is the feedback signal to the traffic sources we need to control, which is the determined source sending rate. The learning agent and the network environment interact continually in the learning process.

### 3.1.2 An Outline of the Theoretical Framework

For the purpose of the topic in this section, as shown in [19], we consider the following discrete-time system as an approximate model of the high-speed network operation

$$x_{k+1} = f(x_k, u_k, v_k) = Ax_k + Bu_k + Cv_k \tag{3.1}$$

It is also assumed the discrete-time feedback control is give as

$$u_k = Lx_k \tag{3.2}$$

And also that

$$v_k = Kx_k \tag{3.3}$$

In the above mathematical models the letter symbols denote: $x_k$ is the state of the high-speed networks, composed of the queue length in the buffer and the available bandwidth for the controlled traffic sources; $u_k$ and $v_k$ are the control inputs of the controller, $u_k$ means the controlled sending rate of traffic sources with lower priority; $v_k$ means the sending rate of traffic sources with higher priority. $L$ and $K$ are the control policies for $u_k$ and $v_k$ respectively. Here $A$, $B$, $L$, and $K$ are the matrices of dimensions 2 × 2, 2 × 1, 1 × 2, and 1 × 2, respectively. Matrices $L$ and $K$ are assumed to be chosen so that the matrix $A + BL + CK$ has all its eigenvalues strictly within the unit circle, and thus the local feedback dynamic stability is guaranteed.

From Eq. (3.1) it may well be inferred that, provided $A$, $B$, $C$ in network's dynamics are known, we could design the flow controllers by adopting any one of the model-dependent control methods offered by control science. But because of the difficulty to derive a reasonably good mathematic model for high-speed networks, the value of $A$, $B$, $C$ are hard to find and even harder to achieve the needed accuracy. It is therefore that we turned towards finding a model-independent control method in order to overcome the above mentioned difficulties. Again the background idea emanating from the Q-learning gives the incentive for a model-independent control method, which can achieve a good controlling performance.

Associated with the network system we define a one-step cost as follows

$$c_k = c(x_k, u_k, v_k) = x_k^T R x_k + u_k^T E u_k - v_k^T F v_k \tag{3.4}$$

where $R$ is a symmetric positive semi-definite matrix of dimensions 2 × 2. In turn, the total cost of a state $x_k$ under the control policy $(L,K)$, $V(x_k)$, is defined as the discounted sum of all costs that will be incurred by using $(L,K)$ from time $k$ onward with the following sum

$$V(x_k) = \sum_{i=0}^{\infty} \beta^i c_{k+i} \tag{3.5}$$

where $0 \leq \beta \leq 1$ represents the discount factor.

In Q-learning technique, the learning agent in the flow controller tries to optimize the function $V(x_k)$ when the sources with lower priority taking the sending rate $u_k$ while the sources with higher priority taking the sending rate $v_k$ at state $x_k$. The definition of $V(x_k)$ implies the recurrence relation

$$V(x_k) = c(x_k, Lx_k, Kx_k) + \beta V(x_{k+1}) \tag{3.6}$$

On the other hand, $V(x_k)$ is a quadratic form function in the state and therefore can be expressed as

$$V(x_k) = x_k^T P x_k \tag{3.7}$$

where $P$ is the 2 × 2 cost matrix for policy $(L,K)$. $(L^*, K^*)$ denotes the policy which is optimal in the sense that the total discounted cost of every state is minimized. $P^*$ represents the cost matrix associated with $(L^*, K^*)$.

It is a simple matter to derive $(L^*, K^*)$ if accurate models of the network dynamic behaviour and cost function are available. The problem we address, however, is how to derive $(L^*, K^*)$ without having access to such models. Thus, we follow Watkins [44] who defined the Q-function for a stable control policy $(L, K)$ as

$$Q(x,u,v) = c(x,u,v) + \beta V(f(x,u,v)) \tag{3.8}$$

The actual value of $Q(x, u, v)$ is the sum of the one-step cost incurred by taking action $(u, v)$ from state $x$, plus the total cost that would accrue if the fixed policy represented by $(L, K)$ were followed from the state $f(x, u, v)$ and all subsequent states (also see [9, 46, 56]). The function $Q$ can also be defined recursively as below

$$Q(x_k,u_k,v_k) = c(x_k,u_k,v_k) + \beta Q(x_{k+1}, Lx_{k+1}, Kx_{k+1}) \tag{3.9}$$

along with noting that

$$Q(x,Lx,Kx) = c(x,Lx,kX) + \beta V(f(x,Lx,Kx)) = V(x) \tag{3.10}$$

As it is shown further below, the Q-function can be computed explicitly. Namely, note first the following evaluation:

$$\begin{aligned}
&Q(x,u,v)\\
&\quad = c(x,u,v) + \beta V(f(x,u,v))\\
&\quad = x^T Rx + u^T Eu - v^T Fv + \beta(Ax+Bu+Cv)^T P(Ax+Bu+Cv)\\
&\quad = x^T\left(\beta A^T PA + R\right)x + u^T\left(\beta B^T PB + E\right)u + v^T\left(\beta C^T PC - F\right)v\\
&\qquad + \beta x^T A^T PBu + \beta x^T A^T PCv + \beta u^T B^T PAx + \beta u^T B^T PCv\\
&\qquad + \beta v^T C^T PAx + \beta v^T C^T PBu.
\end{aligned}$$

It is therefore that, upon a close examination of the last obtained expression, it may readily be found the next summarized formula:

$$\begin{aligned}
&Q(x,u,v)\\
&\quad = \begin{bmatrix} x\\ u\\ v\end{bmatrix}^T \begin{bmatrix} \beta A^T PA + R & \beta A^T PB & \beta A^T PC\\ \beta B^T PA & \beta B^T PB + E & \beta B^T PC\\ \beta C^T PA & \beta C^T PB & \beta C^T PC - F\end{bmatrix} \begin{bmatrix} x\\ u\\ v\end{bmatrix}\\
&\quad = \begin{bmatrix} x\\ u\\ v\end{bmatrix}^T \begin{bmatrix} H_{(11)} & H_{(12)} & H_{(13)}\\ H_{(21)} & H_{(22)} & H_{(23)}\\ H_{(31)} & H_{(32)} & H_{(33)}\end{bmatrix} \begin{bmatrix} x\\ u\\ v\end{bmatrix}\\
&\quad = \begin{bmatrix} x\\ u\\ v\end{bmatrix}^T H \begin{bmatrix} x\\ u\\ v\end{bmatrix} = y^T Hy.
\end{aligned} \tag{3.11}$$

where $y = [x^T, u^T, v^T]^T$ is the column vector concatenation of *x, u,* and *v*. *H* is a symmetric positive definite matrix of dimensions (2 + 1 + 1) × (2 + 1 + 1). Given the policy $(L_j, K_j)$ and the value function *V*, then we can find an improved policy that does not depend on *x*, say policy $(L_{j+1}, K_{j+1})$, by minimized *Q function*. That is, we can find the minimizing *u* and *v* by taking the partial derivative of *Q*(*x*, *u*, *v*) with respect to *u* and *v* respectively.

By taking the derivatives we can get

$$\begin{cases} \frac{\partial Q(x,u,v)}{\partial u} = 2(\beta B^T PB + E)u + 2\beta B^T PCv + 2\beta B^T PAx \\ \frac{\partial Q(x,u,v)}{\partial v} = 2\beta C^T PBu + 2(\beta C^T PC - F)v + 2\beta C^T PAx \end{cases} \tag{3.12}$$

Setting these derivatives to zero and solving for *u* and *v* does yield solutions

$$\begin{aligned} u = & \left(\beta B^T P_j B + E - \beta^2 B^T P_j C\left(\beta C^T P_j C - F\right)^{-1} C^T P_j B\right)^{-1} \\ & \times \left(\beta^2 B^T P_j C\left(\beta C^T P_j C - F\right)^{-1} C^T P_j A - \beta B^T P_j A\right)x = L_{j+1}x \end{aligned} \tag{3.13}$$

and

$$\begin{aligned} v = & \left(\beta C^T P_j C - F - \beta^2 C^T P_j B\left(\beta B^T P_j B + E\right)^{-1} B^T P_j C\right)^{-1} \\ & \times \left(\beta^2 C^T P_j B\left(\beta B^T P_j B + E\right)^{-1} B^T P_j A - \beta C^T P_j A\right)x = K_{j+1}x \end{aligned} \tag{3.14}$$

Since the new policy $(L_{j+1}, K_{j+1})$ does not depend on *x*, in fact, it is the minimizing policy for all states *x*. By making use of (3.11) then policy $(L_{j+1}, K_{j+1})$ can be written in the terms of *H* as follows:

$$\begin{cases} L_{j+1} = \left(H_{j(22)} - H_{j(23)}H_{j(33)}^{-1}H_{j(32)}\right)^{-1} \times \left(H_{j(23)}H_{j(33)}^{-1}H_{j(31)} - H_{j(21)}\right), \\ K_{j+1} = \left(H_{j(33)} - H_{j(32)}H_{j(22)}^{-1}H_{j(23)}\right)^{-1} \times \left(H_{j(32)}H_{j(22)}^{-1}H_{j(21)} - H_{j(31)}\right). \end{cases} \tag{3.15}$$

Notice that formulae (3.15) depend only on the *H* matrix, and they are needed to find the controller gains. However, it was shown in Bradke et al. [55] that if *H* is a known matrix, then the network model is not needed for the purpose of compute the controller gains because it can learn them via interaction with the network's flows.

### 3.1.3 The Controller's Learning Capacity

In this section, we show how the function *Q* can be directly estimated using recursive least squares (RLS). It is not necessary to identify either the network model or the one-step cost function separately.

First, we define a specific, say "over-bar", function for vectors so that $\bar{y}$ is the vector whose elements are all of the quadratic basis functions over the elements of $y$

$$\bar{y} = \left(y_1^2, \ldots, y_1 y_n, y_2^2, y_2 y_3, \ldots, y_{n-1}^2, y_{n-1} y_n, y_n^2\right) \tag{3.16}$$

where $n$ is the sum of dimensions of $x$, $u$, and $v$.

In the next step, we define the vector function $\Theta$ for square matrices. In turn, $\Theta(H)$ is the vector whose elements are the $n$ diagonal entries of $H$ and the $[n(n+1)/2 - n]$ distinct sums $(H_{ij} + H_{ji})$. The elements of $\bar{y}$ and $\Theta(H)$ are ordered so that it may well be written

$$Q(x, u, v) = Q(y) = y^T H y = \bar{y}\Theta(H) \tag{3.17}$$

Finally, we rearrange Eq. (3.9) so as to yield

$$\begin{aligned}
& c(x_k, u_k, v_k) \\
& = Q(x_k, u_k, v_k) - \beta Q(x_{k+1}, Lx_{k+1}, Kx_{k+1}) \\
& = \left[x_k^T, u_k^T, v_k^T\right] H \left[x_k^T, u_k^T, v_k^T\right] \\
& \quad - \beta \left[x_{k+1}^T, Lx_{k+1}^T, Kx_{k+1}^T\right]^T H \left[x_{k+1}^T, Lx_{k+1}^T, Kx_{k+1}^T\right] \\
& = \left[\overline{x_k^T, u_k^T, v_k^T}\right]^T \Theta(H) - \beta \left[\overline{x_{k+1}^T, Lx_{k+1}^T, Kx_{k+1}^T}\right]^T \Theta(H) \\
& = \phi_k^T \theta
\end{aligned}$$

where

$$\varphi_k = \left[\overline{x_k^T, u_k^T, v_k^T}\right] - \beta \left[\overline{x_{k+1}^T, Lx_{k+1}^T, Kx_{k+1}^T}\right] \tag{3.18}$$

and

$$\theta = \Theta(H) \tag{3.19}$$

The well known recursive Least Squares (RLS) method [57] can now be used in order to compute the estimate of $\theta$. The recurrence relations for the RLS method are given by means of the following formulae:

$$\hat{\theta}_j(i) = \hat{\theta}_j(i-1) + \frac{U_j(i-1)\phi_k\left(c_k - \phi_k^T \hat{\theta}_j(i-1)\right)}{1 + \phi_k^T U_j(i-1)\phi_k} \tag{3.20}$$

$$U_j(i) = U_j(i-1) - \frac{U_j(i-1)\phi_k \phi_k^T U_j(i-1)}{1 + \phi_k^T U_j(i-1)\phi_k} \tag{3.21}$$

$$U_j(0) = U_0 \tag{3.22}$$

where $U_0 = \alpha I$ for some large positive constant $\alpha$ to be chosen in the course of applications. Quantity $\theta_j = \Theta(H_j)$ is the true parameter vector for the function $Q_j$, while quantity $\hat{\theta}_j(i)$ is the $i$th estimate of $\theta_j$. The subscript $k$ and the index $i$ are both incremented at each time step in the algorithm's execution.

It has been shown that this algorithm converges to the true parameters if $\theta_j$ is fixed and $\phi_k$ satisfies the condition of the persistent excitation [57] as follows:

$$\varepsilon_0 I \leq \frac{1}{N}\sum_{i=1}^{N} \phi_{k-i}\phi_{k-i}^T \leq \bar{\varepsilon}_0 I \quad \text{for all } k \geq N_0 \text{ and } N \geq N_0 \tag{3.23}$$

where $\varepsilon_0 \leq \bar{\varepsilon}_0$, and $N_0$ is also a positive number.

The policy improvement starts with the network system at some initial state $x_0$ and with some stabilizing controller $(L_0, K_0)$. Then the $j$ keeps track of the number of policy iteration steps. The $k$ keeps track of the total number of time steps. The index $i$ counts the number of time steps since the last change of policy, and when it happens $i = N$, one policy improvement step is executed.

Each policy iteration step consists of two phases: estimation of the Q-function for the current controller, and policy improvement based on that estimate. Consider the jth policy iteration step. Then policy pair $(L_j, K_j)$ represents the currently acting controller. The quantity $\hat{\theta}_j = \hat{\theta}_j(N)$ is the best estimate of $\theta_j$ at the end of the parameter estimation interval. Each estimation interval is $N$ time-steps long. The algorithm is initialized at the start of the jth estimation interval by setting $U_j(0) = U_0$ and initializing the parameter estimates for the jth estimation interval to the final parameter estimates from the just previous interval, i.e., $\hat{\theta}_j = \hat{\theta}_{j-1}(N)$. The index $i$ used in Eqs. (3.20) and (3.21) to count the number of time steps since the beginning of the estimation interval.

After identifying the parameters $\Theta(H_j)$ for $N$ time-steps, one policy improvement step is taken on the grounds of the computed estimate $\hat{\theta}_j$. This produces the new controller $(L_{j+1}, K_{j+1})$, and a new policy iteration step is about to begun. Thus the Q-learning enables to evaluate and re-evaluate the acting controllers as learning the improved policies $(L_{j+1}, K_{j+1})$ at all time of the network operation.

## 3.2 *Simulation Investigations and Performance Comparisons with Other Controllers*

The simulation model of high-speed networks is shown in Fig. 9. It comprises two switches in cascade, namely, the Sw1 with a control agent and the Sw2 with no control agent. The constant output link L is 80 Mbps. The sending rates of the sources are regulated by the flow controllers individually.

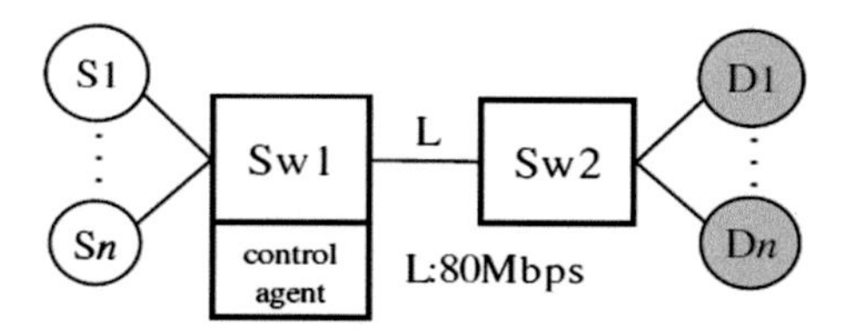


**Fig. 9** The simulation model of single bottleneck network with two switches

In the simulation experiments, we assumed that all packets are with a fixed length of 1000 bytes, and adopt a finite buffer length of 20 packets in the node. The traffic flow with higher priority exists during the whole course of learning process. The offered loading of the simulation varies between 0.6 and 1.2 corresponding to the systems' dynamics; therefore, higher loading results in heavier traffic and vice versa. For the link of 80 Mbps, the theoretical throughput is 62.5 K packets.

Four schemes of flow controlling agents were implemented individually, namely, the AIMD, the model based PID flow controller, neural network (NN) flow controller with learning ability, and the now introduced simple Q-learning flow controller in high-speed networks. The first scheme AIMD increases its sending rate by a fixed increment (0.11) if the queue length is less than the predefined threshold; otherwise the sending rate is decreased by a multiple of 0.8 of the previous sending rate to avoid congestion [15]. As for the other schemes, the sending rate has been controlled by the respective flow controller synthesized.

In order to ensure the novel controller computed and proposed to the high-speed networks to be feasible to apply, the comparisons of the simulation results among those schemes were analyzed. Four measures, throughput, PLR, buffer utilization, and packets' mean delay, are used as the performance indices. The throughput is the amount of received packets at specified nodes (switches) without retransmission. The status of the input multiplexer's buffer in node reflects the degree of congestion resulting in possible packet losses. For simplicity, packets' mean delay only takes into consideration the processing time at node plus the time needed to transmit packets.

The performance comparison of throughput, PLR, buffer utilization, and mean delay controlled by four different kinds of agents individually are shown in Figs. 10, 11, 12 and 13, respectively. The throughput for AIMD method decrease seriously at loading of 0.9. Conversely, the controller proposed remains a higher throughput even though the offered loading is over 1.0. It is obvious that PLR is high, even though we adopt AIMD scheme. However, the controller proposed can decrease the PLR enormously with high throughput and low mean delay. The controller proposed in here possesses considerably better performance over the PID flow controller and the NN flow controller in PLR, buffer utilization, and mean delay. It demonstrates once again the proposed controller possesses the ability to predict the network behaviour hence the improved performance results.

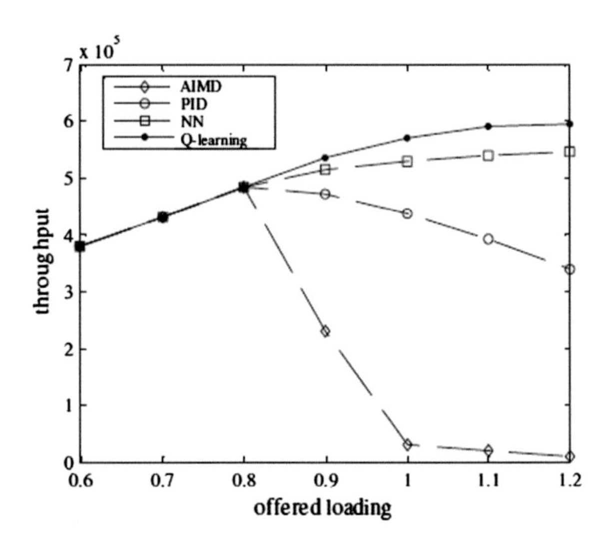


**Fig. 10** Throughput versus various offered loading

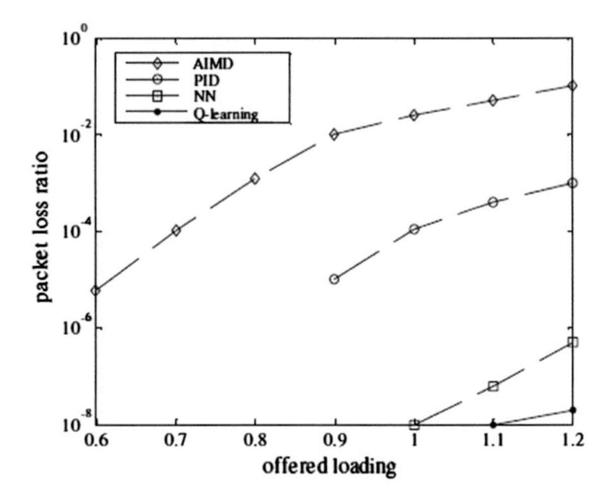


**Fig. 11** PLR versus various offered loading

## *3.3 Concluding Remarks for the Intelligent Simple Q-Learning Design*

A simple Q-learning model-independent flow controller is proposed to deal with the congestion problems in high-speed networks. Because of the interaction with the environment, given the competition of the flows with different priorities, the designed Q-learning controller does demonstrate good performance in the congestion control of high-speed networks. It can be readily seen from Figs. 10, 11, 12, and 13 that this intelligent controller indeed performs the flow control an efficiently, because it is capable of learning the network system behaviour. The simulation results show that the proposed controller is superior with respect to the performance

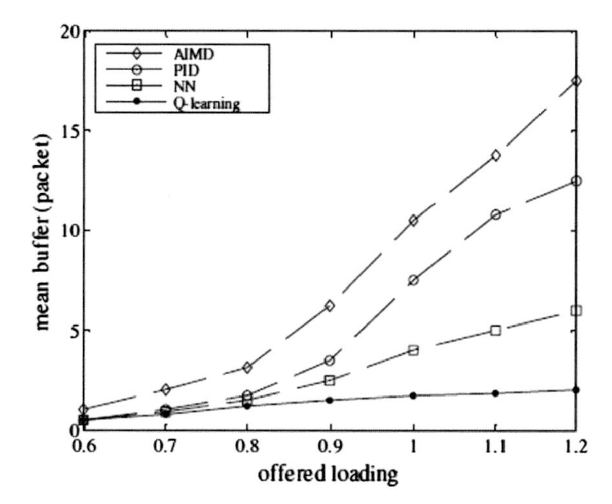


**Fig. 12** Mean buffer versus various offered loading

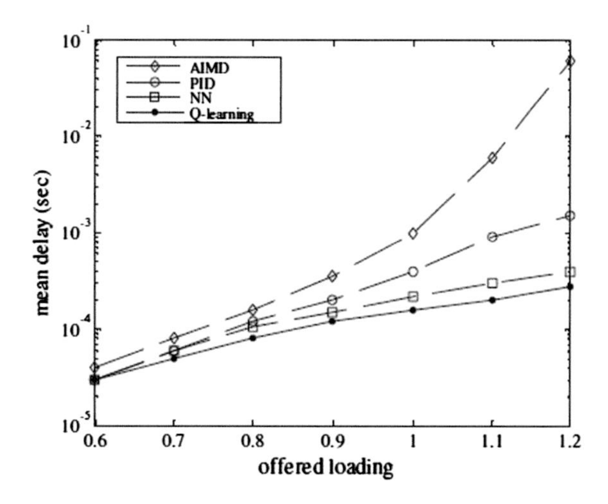


**Fig. 13** Mean delay versus various offered loading

in high-speed networks to the other flow controllers taken into comparison. However, these quality properties have been verified solely when a single bottleneck in the network is present. The case when a multiple bottleneck occurs in the network does require further investigation and possibly some modification improvement into the learning strategy. This is topic for a future research.

## 4 Conclusion

This chapter has introduced two efficient and yet feasible design solutions for learning intelligent flow controls in high-speed networks such as TCPO/ATM. One of the derived intelligent control alternatives is a multi-agent flow controller

(MAFC) based on Q-learning algorithm in conjunction with the theory of Nash equilibrium of opponents' strategies. The other one is a model-independent Q-learning control (MIQL) scheme that is also independent of prior-knowledge on communication traffic and congestion with a focus on the flow with higher priority in the network. The underlying idea in both proposed solutions is to consider the competition of the communication flows with different priorities as a two-player game. The MAFC can learn to take the best actions in order to regulate source flows that guarantee high throughput and low packet-loss ratio. The MIQL control, on the other hand, through a specific learning processing can achieve the optimum sending rate for the sources with lower priority while observing the sources with higher priority.

**Acknowledgments** The author has the honour to acknowledge and point out that this research was accomplished due to his collaboration with the young Dr. Xin Li and Dr. Yan Zheng, their mentor Prof. Yuan-Wei Jing from the Northeastern University, Shenyang, P.R. of China. Furthermore, special thanks are due to Academician Si-Ying Zhang, our common advisor and teacher, whose guidance has been instrumental. The author acknowledges their crucial merits, respectively, for fruitful carrying out this joint research endeavour.

## References

1. Aström, K.J.: Introduction: complex systems and control. In: Astöm, KJ., Albertos, P., Blanke, M., Isidori, A., Sanz, R. (eds.) Control of Complex Systems, Chapter 1, pp. 1–20. Springer, London (2001)
2. Siljak, D.D.: Decentralized Control of Complex Systems. Academic Press, Boston (1991)
3. Wiener, N.: Cybernetics or Control and Communication in the Animal and the Machine. MIT Press, Coambridge, MA (1948).
4. Tsien, H.-S.: Engineering Cybermetics. McGraw-Hill, New York, NY (1954).
5. Dimirovski, G.M., Gough, N.E., Barnett, S.: Categories in systems and control theory. Int. J. Control **8**(9), 1081–1090 (1977)
6. Gitt, W.: Information: the third fundamental quantity. Siemens Rev. **56**, 36–41 (1989)
7. Kalman, R.E.: On the general theory of control: In: Proceedings of the First International Congress on Automatic Control. Butterworth Scientific Institute, London, UK, vol. 1, pp. 481–506. Also in Russian: Trudyi I Kongressa IFAK, Izdatelystvo AN SSSR, Moskva, SSSR, vol. 2, pp. 521–547 (1961)
8. Pontryagin, L.S., Boltyanskiy, V.G., Gamkrelidze, R.V., Mishchenko, E.F.: Matematicheskaya teoriya optimalynih processov, Izdatelystvo fizichko-matematicheskoy literaturi, Moskva, SSSR. Also in English (1962): The Mathematical Theory of Optimal Processes. J. Wiley InterScience, New York (1961)
9. Savkin, A.V., Matveev, A.S.: Cyclic linear differential automata: a simple class of hybrid dynamical systems. Automatica **36**(5), 727–734 (2000)
10. Imer, O.C., Basar, T.: Control of congestion in high-speed networks. Eur. J. Control **7**(2–3), 132–144 (2001)
11. Jacobson, V.: Congestion avoidance and control. ACM Comput. Commun. Rev. **18**, 314–329 (1988)
12. Alpcan, T., Basar, T.: A game theoretical analysis of intrusion detection in access control systems. In: Proceedings of the 43rd IEEE Conference on Decision and Control, Paradise Island, Bahamas, 14–17 Dec. The IEEE Press, Piscataway, NJ, USA, pp. 1568–1573 (2004)

13. Basar, T., Olsder, G.J.: Dynamic Non-cooperative Game Theory, 2nd edn. The SIAM, Philadelphia, PA (1999)
14. Dimirovski, G.M: Network Q-Learning control prevent cyber intrusion risks: synergies of control theory and computational intelligence. In: Duman, E., Atiya, A. (eds.) Use of Risk Analysis in Computer-Aided Persuasion, vol. 88. NATO Science Sub-Series E, Chapter 20–Part III. The IOS Press, Amsterdam, Berlin Tokyo Washington DC, pp. 281–303 (2011)
15. Gevros, P., Crowcoft, J., Kirstein, P., Bhatti, S.: Congestion control mechanisms and the best effort service model. IEEE Netw. **15**(3), 16–26 (2001)
16. Imer, O.C., Compans, S., Basar, T., Srikant, R.: ABR congestion control in ATM networks. IEEE Control Syst. Mag. **21**(1), 38–56 (2001)
17. Jing, Y.-W., Li, X., Dimirovski, G.M., Zheng, Y., Zhang, S.-Y.: Nash Q-learning multi-agent flow control for high-speed networks. In: Proceedings of the 28th American Control Conference, St. Louis, MO, USA, 10–12 June, pp. 3304–3309. The IEEE Press, Piscataway, NJ, USA (2009)
18. Lestas, M., Pitsillides, A., Ioannou, P., Hadjipollas, G.: Adaptive congestion protocol: A congestion control protocol with learning capability. Comput. Netw. Int. J. Comput. Telecommun. Netw. **51**(13), 3773–3798 (2007)
19. Li, X., Dimirovski, G.M., Jing, Y.-W., Zhang, S.-Y.: A Q-learning model-independent flow controller for high-speed networks. In: Proceedings of the 28th American Control Conference, St. Louis, MO, USA, 10–12 June, pp. 1544–1548. The IEEE Press, Piscataway, NJ, USA (2009)
20. Li, X., Zhou, Y.C., Dimirovski, G.M., Jing, Y.-W.: Simulated annealing Q-learning algorithm for ABR traffic control of ATM networks. In: Proceedings of the 27th American Control Conference, Seattle, WA, USA, 11–13 June, pp. 4462–4467. The IEEE Press, Piscataway, NJ, USA (2008)
21. Li, X., Shen, X.J., Jing, Y.-W., Zhang, S.-Y.: Simulated annealing-reinforcement learning algorithm for ABR traffic control of ATM networks. In: Proceedings of the 46th IEEE Conference on Decision and Control, New Orleans, LA. USA, 12–14 Dec, pp. 5716-5721. The IEEE Press, Piscataway, NJ, USA (2007)
22. Ren, T., Zhu, Z., Yu, H., Dimirovski, G.M.: Integral sliding-mode controller for congestion problem in ATM networks. Int. J. Control **86**(3), 529–539 (2013)
23. Ren, T., Wang, C., Luo, X., Jing, Y., Dimirovski, G.M.: Robust controller design for ABR traffic control of ATM networks with time-varying multiple time delays. Int. J. Innovative Comput. Inf. Control **5**(8), 68–74 (2009)
24. Ren, T., Gao, Z., Kong, W., Jing, Y., Yang, M., Dimirovski, G.M.: Performance and robustness analysis of a fuzzy-immune flow controller in ATM networks with time-varying multiple time-delays. J. Control Theory Appl. **6**(3), 253–258 (2008)
25. Aizerman, M.A., Breverman, E.M., Rosonoer, L.I.: Theoretical foundation of the potential function method in pattern recognition learning (in Russian). Avtomatika i Telemechanika **25**, 821–837 (1964)
26. Aizerman, M.A., Breverman, E.M., Rosonoer, L.I.: The probability problem of pattern recognition learning and the potential function method (in Russian). Avtomatika i Telemechanika **25**, 821–837 (1964)
27. Chatovich, A., Okug, S., Dundar, G.: Hierarchical neuro-fuzzy call admission controller for ATM networks. Comput. Commun. **24**(11), 1031–1044 (2001)
28. Cheng, R.G., Chang, C.J., Lin, L.F.: A QoS-provisioning neural fuzzy connection admission controller for multimedia high-speed networks. IEEE ACM Trans. Netw. **7**(1), 111–121 (1999)
29. Harmer, P.K., Williams, P.D., Gunsh, G.H., Lamont, G.B.: An artificial immune system architecture for computer security applications. IEEE Trans. Evol. Comput. **6**(2), 252–280 (2002)
30. Hsiao, M.C., Tan, S.W., Hwang, K.S., Wu, C.S.: A reinforcement learning approach to congestion control of high-speed multimedia networks. Cybern. Syst. **36**(2), 181–202 (2005)

31. Hwang, K.S., Tan, S.W., Hsiao, M.C., Wu, C.S.: Cooperative multi-agent congestion control for high-speed networks. IEEE Trans. Syst. Man Cybern. Part B: Cybern. **35**(2), 255–268 (2005)
32. The MathWorks Laboratory: Matlab: Control Toolbox, Fuzzy Toolbox. LMI Toolbox, The MathWorks Inc, Natick (1991)
33. The MathWorks Laboratory: Matlab: Simulink – Dynamic System Simulation Software. The MathWorks Inc, Natick (1996)
34. Dimirovski, G.M.: Lyapunov stability in control synthesis design using fuzzy-logic and neural-networks. In: Proceedings of the 17th IMACS World Congress, Paris, 11–15 July. The IMACS and Ecole Centrale de Lille, Villeneuve d'Ascq, FR, Paper T5-I-01-0907, pp. 1–8 (2005)
35. Marinos, P.N.: Fuzzy logic and its applications to switching systems. IEEE Trans. Comput. **18** (4), 343–348 (1969)
36. Palm, R., Driankov, D.: Fuzzy switched hybrid systems – Modelling and identification. In: Proceedings of the 1998 IEEE International Conference on Fuzzy Systems, Gaithersburg, MD, pp. 130–135. The IEEE Press, Piscataway, NJ, USA (1998)
37. Tanaka, K., Masaaki, I., Wang, H.O.: Stability and smoothness conditions for switching fuzzy systems. Proceedings of the 19th American Control Conference, pp. 2474–2478. The IEEE Press, Piscataway, NJ, USA (2000)
38. Wang, H.O., Tanaka, K., Griffin, M.: An approach to fuzzy control of nonlinear systems: Stability and design issues. IEEE Trans. Fuzzy Syst. **4**(1), 14–23 (1996)
39. Ye, H., Michael, A.N.: Stability theory for hybrid dynamical systems. IEEE Trans. Autom. Control **43**(4), 464–474 (1998)
40. Zhao, J., Dimirovski, G.M.: Quadratic stability for a class of switched nonlinear systems. IEEE Trans. Autom. Control **49**(4), 574–578 (2004)
41. Yang, H., Dimirovski, G.M., Zhao, J.: Switched fuzzy systems: representation modelling, stability, and control design. In: Chountas, P., Petrounias, I., Kacprzyk, J. (eds.) Studies in Computational Intelligence Volume 109 – Intelligent Techniques and Tool for Novel Systems Architectures, Chapter 9, pp. 169–184. Springer, Berlin Heidelberg (2008)
42. Yang, M., Jing, Y., Dimirovski, G.M., Zhang, N.: Stability and performance analysis of a congestion control algorithm for networks. In: Proceedings of the 46th IEEE Conference on Decision and Control, New Orleans, LA. USA, 12–14 Dec, pp. 4453–4458. The IEEE Press, Piscataway, NJ, USA (2007)
43. Jing, Y. W.: A Study on Advanced Fuzzy and Learning Controls for Communication Netwroks and Effects of Their Applications. Private Communication Report CCN-2007-2008-Shenyang-Skopje, College of Informations Science and Engineering, Northeastern University, Shenyang, PR China (2009).
44. Watkins, C.J.C.H., Dayan, P.: Q-learning. Mach. Learn. **8**(3), 279–292 (1992)
45. Kaelbling, L.P., Littman, M.L., Moore, A.W.: Reinforcement learning: a survey. J. Artif. Intell. Res. **4**(1), 237–285 (1996)
46. Sutton, R.S., Barto, A.G.: Reinforcement Learning: An Introduction. The MIT Press, Cambridge, MA, USA (1998)
47. Littman, M.L.: Value-function reinforcement learning in Markov games. J Cogn. Syst. Res. **2** (1), 55–66 (2001)
48. Szepesvari, C., Littman, M.L.: A unified analysis of value-function-based reinforcement-learning algorithms. Neural Comput. **11**(8), 2017–2060 (1999)
49. Hu, J., Wellman, M.P.: Nash Q-learning for general-sum stochastic games. J. Mach. Learn. Res. **4**, 1039–1069 (2003)
50. Jing, Y.W., Dimirovski, G.M.: Multimedia transfer over the internet: a control based improvement of the AIMD algorithm. In: Proceedings of the 3rd IEEE International Conference on Information and Communication Technologies: From Theory to Applications, Damascus, Syria, 24–28 April. The IEEE Press, Piscataway, NJ, USA and ENST Brest, Bretagne, FR, Paper SEN06-1/1-6 (2006)

51. Dimirovski, G.M.: Complexity versus integrity solution in adaptive fuzzy-neural inference models. Int. J. Intell. Syst. **23**(5), 556–573 (2008)
52. Nash, J.F.: Non-cooperative games. Ann. Math. **54**(2), 286–295 (1951)
53. Bishop, C.M.: Pattern Recognition and Machine Learning. Springer, London, UK (2006)
54. Chiu, D.M., Jain, R.: Analysis of the increase decrease algorithms for congestion avoidance in computer networks. Comput. Netw. ISDN Syst. **17**, 1–14 (1989)
55. Bradtke, S.J., Ydstie, BE., Barto, A.G.: Adaptive linear quadratic control using policy iteration. In: Proceedings of the 13th American Control Conference, Chicago, IL, 13–15 June, pp. 3475–3479. The IEEE Press, Piscataway, NJ (1994)
56. Haykin, S.: Neural Networks and Learning Machines, 3rd edn. Pearson Prentice Hall, Upper Saddle River (2009)
57. Ljung, L.: System Identification—Theory for the User, 2nd edn. Prentice Hall PTR, Upper Saddle River (1999)

# Logical Operations and Inference in the Complex s-Logic

**Vassil Sgurev**

**Abstract** Imaginary propositional logic (*i*-logic) is being introduced through which the classical propositional logic (called in the present work real or *r*-logic) is extended to complex, summary (*s*-logic), in which the two logics above are interpreted. For this purpose a constraint (axiom) is added to the structures of the two logics—*r* and *i*, which connects heir variables and states. The *s*-logic received provides a possibility all logic equations to be solved which cannot be done in the frame of the classical propositional logic. It is proved that the *s*-logic has six states, non-equipotent between each other, i.e. it is a multi-valued logic. All possible truth tables for conjunction, disjunction, and implication between the six states and variables of the three logics—*r*, *i*, and *s* are received by the formalisms being introduced. A number of new results are discussed characteristic for the implication in the *s*-logic. On the base of the truth tables and the rules Modus Ponens and Modus Tollens a number of relations are received for the logical inference in the *s*-logic which is different in some aspect from those in the *r*-logic. Examples are given for *s*-logic application: solving problems in logical equations and the mental activity.

**Keywords** Classical propositional logic · Imaginary logic · Complex *s*-logic · Implication · Logical inference · Lattice · Boolean algebra

## 1 Introduction

One of the pillars on which the development of the modern information technologies lies is the logical approach and the wide usage of the methods of the classic propositional logic. Inference of new knowledge and intelligent information processing may be carried out through it [1–3].

---

V. Sgurev (✉)
Institute of Information and Communication Tehnologies—Bulgarian Academy of Sciences, Bulgaria, Europe
e-mail: vsgurev@gmail.com



V. Sgurev et al. (eds.), *Innovative Issues in Intelligent Systems*,
Studies in Computational Intelligence 623, DOI 10.1007/978-3-319-27267-2_5

A situation is reached in the framework of this logic, when at solving logical equations no solution of the equation can be found. Such is the case when neither of the two possible states—TRUE ($T$) or FALSE ($F$) of the set $Q_r = \{T, F\}$ of the logical variable $x \in Q_r$ does not satisfy the relation

$$F \Lambda x = T; \tag{1}$$

The usage of imaginary variables $p \in Q_i = \{i, \neg i\}$ is proposed in [3, 4], which may be in one of the states $i$, and $\neg i$ (not $i$) [5–7].

Then the solution of Eq. (1) may be found by using the state $i$, namely

$$F \wedge p = F \wedge i = T. \tag{2}$$

## 2 Defining the Real Propositional Logic (r-logic), Imaginary (i-logic), and the Complex (s-logic)

The classic propositional logic will be called *real logic* and will be denoted by index $r$, i.e. $r$-logic.

The imaginary logic will be denoted by index $i$ and will be called $i$-logic. It is defined on the base of the same algebraic structure like the classic propositional logic, namely symmetric idempotent ring, and lattice [8]. Its state $i$ is not equipotent to $T$, and in the same way $\neg i$ is not equipotent to $F$. This follows from (2) where the assumption of equipotence of $i$ with $T$ would result in a contradiction.

All results of the classic propositional logic are valid for the real and imaginary logics as they are based on one and the same algebraic structure. Both logics—real and imaginary may be considered as two parallel logics connected by the relation (2) which contains variables from one and the other logics. As both logics are based on one and the same axioms of the lattice then (2) may be considered as additional axiom which interconnects them [9, 10, 11, 12].

By analogy with the complex numbers $z = a + bi$, where $a$ and $b$ are real numbers and $i$ is the imaginary unit, complex logical variables may be introduced for the relations from (2)

$$g_1 = x_1 \vee p_1; g_2 = x_2 \wedge p_2; \tag{3}$$

$$\begin{aligned} \text{where } g_1, g_2 \in G &= \{g_1, g_1, ..g_k, ..\}; \\ x_1, x_2 \in X &= \{x_1, x_1, ..x_k, ..\}; \\ p_1, p_2 \in I &= \{p_1, p_1, ..p_k, ..\}; \end{aligned} \tag{4}$$

where $G$, $X$, and $I$ are the sets of variables in the complex, real, and imaginary logic, respectively.

Relations (3) define in essence complex (summary) logic between the real and the imaginary logic which will be further referred to as *s*-logic.

It is shown in [3, 4] that under the assumptions made, the complex *s*-logic is up to the axioms of the symmetric idempotent ring, the Boolean algebra, and the lattice, as well as to the additional constraints (2), unifying the real and imaginary logics. All requirements for associativity, idempotence of disjunction and conjunction, as well as distributivity of the disjunction with regard to conjunction, and vice versa, are also observed in the *s*-logic. At that for each complex variable $g$ the complement $\neg g$ is unique and "symmetry" of the complement exists ($\neg\neg g = g$) and the De Morgan two laws are valid, namely:

$$\neg(g_1 \vee g_2) = \neg g_1 \wedge \neg g_2; \ \neg(g_1 \wedge g_2) = \neg g_1 \vee \neg g_2. \tag{5}$$

It is expedient to define the logical operation "negation" in the *s*-logic through the relations (5).

All five logical operations of the *s*-logic—disjunction, conjunction, implication, equivalence, and negation are defined it a way, analogical to the one in *i*-logic and *r*-logic. The several following relations, proved in [1, 13] play an important role in the *s*-logic.

**Proposition 1.1** *A relation exists*:

$$T \wedge i = F \wedge i = T. \tag{6}$$

*The following chain of relations is based on Eq.* (2)

$$T \wedge i = (F \wedge i) \wedge i = F \wedge (i \wedge i) = F \wedge i = F. \tag{7}$$

*The negation of both sides of* (6) *keeping in mind* (7) *results in*

$$\neg(T \wedge i) = F \vee \neg i = T \vee \neg i = \neg T = F. \tag{8}$$

**Proposition 1.2** *A relation exists*:

$$T \vee i = F \vee i. \tag{9}$$

*If the state i in the left hand side the state i is substituted by* $i \vee \neg i$ *and* (8) *is taken in mind, then the following chain of equivalences is reached*:

$$T \vee i = T \vee (i \vee \neg i) = (T \vee \neg i) \vee i = F \vee i. \tag{10}$$

*The negation of both sides of* (9) *results in*:

$$F \vee \neg i = T \vee \neg i = F. \tag{11}$$

*It is further important to define in how many and what states the logical variables* (3) *of the s-logic, real, and imaginary logic may be.*

(a) Eight elementary states may be pointed out distributed in the following two groups corresponding to conjunction and disjunction, namely:

$$Q_k = \{(T \wedge i), (F \wedge i), (T \wedge \neg i), (F \wedge \neg i)\};$$
$$Q_d = \{(T \vee i), (F \vee i), (T \vee \neg i), (F \vee \neg i)\},$$

where $Q_s = Q_k \cup Q_d$ is the total number of states of the logical variables (3). It follows from (6) and (8) that the first two states are equivalent between them and according (2) both are equivalent to $T$. Last two states of $Q_k$ are equivalent between them and, according (7), they may be represented by $(F \wedge \neg i)$. Analogically it follows from (8) that the last two states in $Q_d$ are equivalent between them and equivalent to $T$ a well and the first two states in the same set are equivalent between them and may be represented by $(T \vee i)$. Hence

$$Q_k = \{F, (F \wedge \neg i)\}; Q_d = \{T, (T \vee i)\}; Q_s = \{T, (T \vee i), F, (F \wedge \neg i)\}.$$

In this way only four states remain from the initial eight in the s-logic after the corresponding transformations. Two of them are from the $r$-logic and two are characteristic for the $s$-logic only.

(b) The states which are characteristic of the imaginary logic are only two, namely

$$Q_i = \{i, \neg i)\}.$$

(c) There are two only states in the real $r$-logic in the same manner

$$Q_r = \{T, F\}.$$

Then only six states will be contained in general in the $s$-logic, namely $Q = Q_s \cup Q_r \cup Q_i$;

$$Q = \{(T \vee i), (F \wedge \neg i), T, F, i, \neg i\}.$$

It follows from that the $s$-logic may be considered as a six valued logic which integrates in an entity both logics—real and imaginary through the additional axiom (2)

In the algebraic structure of the $s$-logic the state $(T \vee i)$ plays the role of the unity "1" and the state $(F \wedge \neg i)$—the role of zero "0". This also follows from the relations:

$$(T \vee i) \vee (F \wedge \neg i) = ((T \vee i) \vee F) \wedge ((T \vee i) \vee \neg i) = (T \vee i) \wedge (T \vee i) = T \vee i; \tag{12}$$

$$(T \vee i) \wedge (F \wedge \neg i) = ((F \wedge \neg i) \wedge T) \vee ((F \wedge \neg i) \wedge i) = (F \wedge \neg i) \vee (F \wedge \neg i) = F \wedge \neg i. \tag{13}$$

# 3 Conjunction Between States of r, i, and s Logics

The two Tables 1 and 2 present the truth tables well known from the propositional logic of the implication in the real and imaginary logic.

The truth Table 3 in the complex *s*-logic may be checked by rows in the following way:

- for the first row: $(T \lor i) \rightarrow (T \lor i) = (F \land \neg i) \lor (T \lor i) = (F \lor (T \lor i)) \land (\neg i \lor (T \lor i)) = (T \lor i) \land (T \lor i) = (T \lor i)$;
- in the same way for the fourth row: $(F \land \neg i) \rightarrow (F \land \neg i) = (T \lor i) \lor (F \land \neg i) = = (T \lor i)$;
- for the second row: $(T \lor i) \rightarrow (F \land \neg i) = (F \land \neg i) \lor (F \land \neg i) = (F \land \neg i)$;
- for the third row: $(F \land \neg i) \rightarrow (T \lor i) = (T \lor i) \lor (T \lor i) = (T \lor i)$.

The truth of Tables 4, 5 and 6 may be checked taking in mind constraints (6) to (11) by rows respectively:

(a) Table 4

- first row: $(i \rightarrow T) = \neg i \lor T = F$;
- second row: $(i \rightarrow F) = (\neg i \lor F) = (\neg i \lor T) = F$;
- third row: $(\neg i \rightarrow T) = (\neg\neg i \lor T) = (i \lor T)$;
- fourth row: $(\neg i \rightarrow F) = (\neg\neg i \lor F) = (i \lor F) = (T \lor i)$.

**Table 1** Conjunction between states of *r*-logic

| $x_1$ | $x_2$ | $x = x_1 \land x_2$ |
|---|---|---|
| $T$ | $T$ | $T$ |
| $T$ | $F$ | $F$ |
| $F$ | $T$ | $F$ |
| $F$ | $F$ | $F$ |

**Table 2** Conjunction between states of *i*-logic

| $p_1$ | $p_2$ | $p = p_1 \land p_2$ |
|---|---|---|
| $i$ | $i$ | $i$ |
| $i$ | $\neg i$ | $\neg i$ |
| $\neg i$ | $i$ | $\neg i$ |
| $\neg i$ | $\neg i$ | $\neg i$ |

**Table 3** Conjunction between states of *s*-logic

| $g_1$ | $g_2$ | $g = g_1 \land g_2$ |
|---|---|---|
| $T \lor i$ | $T \lor i$ | $T \lor i$ |
| $T \lor i$ | $F \land \neg i$ | $F \land \neg i$ |
| $F \land \neg i$ | $T \lor i$ | $F \land \neg i$ |
| $F \lor \neg i$ | $F \land \neg i$ | $F \land \neg i$ |

**Table 4** Conjunction between states of $i$ and $r$ logics

| $p$ | $x$ | $g = p \wedge x$ |
|---|---|---|
| $i$ | $T$ | $T$ |
| $i$ | $F$ | $T$ |
| $\neg i$ | $T$ | $F \wedge \neg i$ |
| $\neg i$ | $F$ | $F \wedge \neg i$ |

**Table 5** Conjunction between states of $s$ and $r$ logics

| $g$ | $x$ | $g = g \wedge x$ |
|---|---|---|
| $T \vee i$ | $T$ | $T$ |
| $T \vee i$ | $F$ | $T$ |
| $F \wedge \neg i$ | $T$ | $F \wedge \neg i$ |
| $F \wedge \neg i$ | $F$ | $F \wedge \neg i$ |

**Table 6** Conjunction between states of $i$ and $s$ logics

| $g$ | $x$ | $g = g \wedge x$ |
|---|---|---|
| $T \vee i$ | $T$ | $T$ |
| $T \vee i$ | $F$ | $T$ |
| $F \wedge \neg i$ | $T$ | $F \wedge \neg i$ |
| $F \wedge \neg i$ | $F$ | $F \wedge \neg i$ |

(b) Table 5

- first row: $((T \vee i) \rightarrow T) = ((F \wedge \neg i) \vee T) = (T \wedge (T \vee \neg i)) = (T \wedge F) = F$;
- second row: $((T \vee i) \rightarrow F) = ((F \wedge \neg i) \vee F) = (F \wedge (F \vee \neg i)) = (F \wedge (F \vee \neg i) = F \wedge F = = F$;
- third row: $((F \wedge \neg i) \rightarrow T) = ((T \vee i) \vee T) = (T \vee i)$;
- fourth row: $((F \wedge \neg i) \rightarrow F) = ((T \vee i) \vee F) = (T \vee i)$.

(c) Table 6

- first row: $(i \rightarrow (T \vee i)) = (\neg i \vee (T \vee i)) = (T \vee i)$;
- second row: $(i \rightarrow (F \wedge \neg i)) = (\neg i \vee (F \wedge \neg i)) = ((\neg i \vee F) \wedge \neg i) = (F \wedge \neg i)$;
- third row: $(\neg i \rightarrow (T \vee i)) = (\neg\neg i \vee (T \vee i) = (T \vee i)$;
- fourth row: $(\neg i \rightarrow (F \wedge \neg i)) = (\neg\neg i \vee (F \wedge \neg i)) = ((F \vee i) \wedge i) = ((F \wedge i) \vee i) = (T \vee i)$.

## 4 Disjunction Between States of r, i, and s Logics

The results of the disjunction between the real and the imaginary logic are shown in Tables 7 and 8 respectively. The former one includes only the states $\{T, F\}$ and the latter—$\{i, \neg i\}$. The results of the disjunction in the frame of the $s$-logic only are shown in Table 9 when using its states $\{(T \vee i), (F \wedge \neg i)\}$. The first and last rows of

**Table 7** Disjunction between states of $r$ logic

| $x_1$ | $x_2$ | $x = x_1 \vee x_2$ |
|---|---|---|
| $T$ | $T$ | $T$ |
| $T$ | $F$ | $T$ |
| $F$ | $T$ | $T$ |
| $F$ | $F$ | $F$ |

**Table 8** Disjunction between states of $i$ logic

| $p_1$ | $p_2$ | $p = p_1 \vee p_2$ |
|---|---|---|
| $i$ | $i$ | $i$ |
| $i$ | $\neg i$ | $i$ |
| $\neg i$ | $i$ | $i$ |
| $\neg i$ | $\neg i$ | $\neg i$ |

**Table 9** Disjunction between states of $s$ logic

| $g_1$ | $g_2$ | $g = g_1 \vee g_2$ |
|---|---|---|
| $T \vee i$ | $T \vee i$ | $T \vee i$ |
| $T \vee i$ | $F \wedge \neg i$ | $T \vee i$ |
| $F \wedge \neg i$ | $T \vee i$ | $T \vee i$ |
| $F \wedge \neg i$ | $F \wedge \neg i$ | $F \wedge \neg i$ |

this table immediately follow from the idempotence of the disjunction. The second and third rows lead to the same results ensuing from the commutativity of disjunction.

$$\begin{aligned}(T \vee i) \vee (F \wedge \neg i) &= ((T \vee i) \vee F) \wedge ((T \vee i) \vee \neg i) = (T \vee i) \wedge (T \vee i) \\ &= (T \vee i).\end{aligned}$$

The results of the disjunction for the case when states of one only logic are used, and are shown in Tables 7, 8, and 9. It is of not less importance results to be achieved for the disjunction when states of the three different logics—$r$, $i$, and $s$ logics are used. The results of the disjunction between the different states of the real and imaginary logic are described in Table 10. The results in the first and second rows of this table follow from relations (9) and in the third and fourth rows—from (11).

**Table 10** Disjunction between states of $i$ and $r$ logics

| $p$ | $x$ | $g = p \vee x$ |
|---|---|---|
| $i$ | $T$ | $T \vee i$ |
| $i$ | $F$ | $T \vee i$ |
| $\neg i$ | $T$ | $F$ |
| $\neg i$ | $F$ | $F$ |

The disjunction between the states of the *s*-logic and the real one is reflected in the truth Table 11. The following three chains of equivalences confirm their truth. Relations from (6) to (11) are considered in them.

$$(T \vee i) \vee T = T \vee i; (T \vee i) \vee F = T \vee i;$$
$$(F \wedge \neg i) \vee T = T \wedge (T \vee \neg i) = T \wedge F = F;$$
$$(F \wedge \neg i) \vee F = F \wedge (F \vee \neg i) = F \wedge F = F.$$

The results from the disjunction between the states of the imaginary and the S-logic are demonstrated in Table 12. The following three logical chains of equivalence are deduced on the base of relations (6) to (11) that confirm the results of the truth Table 12.

$$i \vee (T \vee i) = T \vee i; \neg i \vee (F \wedge \neg i) = (F \vee \neg i) \wedge \neg i = F \vee \neg i;$$
$$i \vee (F \wedge \neg i) = (F \vee i) \wedge i = (F \wedge i) \vee i = T \vee i;$$
$$\neg i \vee (T \vee i) = T \vee i.$$

The following general conclusions may be drawn when comparing six Tables 7, 8, 9, 10, 11 and 12:

(a) The comparison of the three truth Tables 7, 8 and 9 shows that for the disjunction in them, the requirement of the classic propositional logic is observed —third column contains one “False” and three “True” states.
(b) The results shown in Table 9 for the disjunction between the states $\{(T \vee i), (F \wedge \neg i)\}$ of the S-logic coincide with the results shown in Table 12 of the disjunction of that logic and the imaginary logic. The requirements of the classic propositional logic are also observed in them—one “False” and three “True” states. This is due to the lack of specific constraints similar to those from (8).

**Table 11** Disjunction between states of *s* and *r* logics

| $g_1$ | $x$ | $g = g_1 \vee x$ |
|---|---|---|
| $T \vee i$ | $T$ | $T \vee i$ |
| $T \vee i$ | $F$ | $T \vee i$ |
| $F \wedge \neg i$ | $T$ | $F$ |
| $F \wedge \neg i$ | $F$ | $F$ |

**Table 12** Disjunction between states of *i* and *s* logics

| $p$ | $g_1$ | $g = p \vee g_1$ |
|---|---|---|
| $i$ | $T \vee i$ | $T \vee i$ |
| $i$ | $F \wedge \neg i$ | $T \vee i$ |
| $\neg i$ | $T \vee i$ | $T \vee i$ |
| $\neg i$ | $F \wedge \neg i$ | $F \wedge \neg i$ |

(c) The relation in the third row of the truth Table 10 corresponds to the constraints available in (8). It imposes specific constraints between the real and the imaginary logics due to which this result seems unnatural for the disjunction. Two states of two different logics—$T$ and $(T \vee i)$ are received in third column of Table 10. The two other states—$\{T, (F \wedge \neg i)\}$ are not used in this case of the disjunction. At that the lower "bound of False" is from the real logic and the upper "bound of True" is from the *s*-logic. "Raise" of the "Truth bound" is observed. States of the third, *s*-logic are encountered in the results of the disjunction between the real and the S-logic due to the constraint (11).

(d) The results of the disjunction between the real and S-logic, reflected in Table 11 are the same as in the previous Table 10—between the real and imaginary logics. The same considerations and results are valid like in item (c).

(e) Comparison of Tables 10, 11, and 12 leads to the conclusion that the results of the disjunction between the real logic and the other two logics are related to some "distortion" of results, described in item (c), due to the constraints (2) and (11). Such effect is not encountered at the disjunction between all states of the imaginary logic and the S-logic which remains in the frame of the classic propositional logic.

It seems expedient to check whether the relations of Tables 1, 2, 3, 4, 5 and 6 at conjunction correspond to Tables 7, 8, 9, 10, 11 and 12 at disjunction by applying the De Morgan laws. Such relations exist for the classic propositional logic. This is of great importance when logical operations are between states of different logics.

This is evident by assumption for Tables 1, 2, 7 and 8 respectively. For each of the remaining Tables 3, 4, 5 and 6 a correspondence will be sought to Tables 9, 10, 11 and 12 by checking a control row of each table:

- Negation of the relations of the first row of Table 3 through the De Morgan laws results in

$$\begin{aligned}\neg(T \vee i) &= \neg(T \vee i) \vee \neg(T \vee i) = (F \wedge \neg i) = (F \wedge \neg i) \vee (F \wedge \neg i)\\ &= F \wedge \neg i;\end{aligned}$$

  which corresponds to the relations of the fourth row of Table 9;
- The same operation for the second row of Table 4 leads to

$$\neg T = \neg(i \wedge T) = \neg i \vee F = T \vee \neg i = F;$$

  which corresponds to the third row of Table 10;
- the same operation for the third row of Table 5 results in

$$\neg(F \wedge \neg i) = \neg\,(F \wedge \neg i)\ \vee \neg T = (T \vee i)\ \vee F = T \vee i;$$

  which corresponds to the second row of Table 11;

- The same operation for the fourth row of Table 6 results in

$$\neg(F \wedge \neg i) = \neg\neg i \vee \neg(F \wedge \neg i) = i \vee (T \vee i) = T \vee i;$$

which corresponds to the first row of Table 12.

Analogic results may be respectively obtained for the other rows of the tables under consideration.

Summing up:
Results achieved for the conjunction and disjunction in the *s*-logic offer a chance some general conclusions to be made:

1. Tables 1, 2, 3, 4, 5 and 6 thoroughly define the operation "conjunction" between any pair of the six possible states of the set $Q$ of the *s*-logic. At that the constraint (2) is considered in Tables 4 and 5 between the real and the imaginary logics.
2. Disjunction between the six states of the set $Q$ of the *s*-logic is thoroughly defined in Tables 7, 8, 9, 10, 11 and 12. Constraint (9) between the real and imaginary logic is reflected in Tables 10 and 11 respectively.
3. Negation $\neg$ in the *s*-logic is defined by the relations in (5).
4. Then through the three operations thus defined—conjunction, disjunction, and negation all other relations in the *s*-logic may be defined—implication and equivalence as well as the other formulae, analogic to those of the classic propositional logic and at that keeping in mind constraint (2), equivalent to the negation (8).

A conclusion may be made from the equipollent relations received in the present work

$$F \wedge I = T \vee F \text{ and } T \vee \neg i = T \wedge F$$

that $i$ is not equivalent to $T$ neither is $\neg i$ equivalent to $F$. These relations demonstrate that the state $i$ may be considered as "truer" than $T$ and $\neg i$—as "falser" than $F$. In this sense we may accept that the imaginary logic is "senior" or "more meaningful" than the real one. And although that the real and the imaginary logics are constructed on the same logical structure, their states $i$ and $T$ which correspond to one "1" in the Boolean algebra and $\neg i$ and $F$ corresponding to zero "0" in the same algebraic structure, are different. They are united in the complex S-logic through the relations (2) and (9) and its two states $(T \vee i)$ corresponding to one and $(F \wedge \neg i)$ corresponding to zero.

## 5 Implication in s-Logic

When defining the implication the following relations from the real, imaginary and *s*-logic are of essential importance:

$$x_1 \to x_2 = \neg x_1 \vee x_2; x_1, x_2 \in \{T, F\}; \quad (14)$$

$$p_1 \to p_2 = \neg p_1 \vee p_2; p_1, p_2 \in \{i, \neg i\}; \quad (15)$$

$$g_1 \to g_2 = \neg g_1 \vee g_2; g_1, g_2 \in \{(T \vee i), (F \wedge \neg i)\}; \quad (16)$$

In the three truth Tables 13, 14 and 15, the implication's defining is carried out through relations (14)–(16) in the generally acknowledged in propositional logic manner and only between the states inside the respective logic.

Defining the implication between two different states of two different logics, united in the *s*-logic, is not of lesser interest. Three such possible cases are shown in truth Tables 16, 17 and 18 respectively, for which:

$$p \to x = \neg p \vee x; p \in \{i, \neg i\}; x, \in \{T, F\}; \quad (17)$$

$$g \to x = \neg g \vee x; g \in \{(T \vee i), (F \wedge \neg i)\}; x \in \{T, F\}; \quad (18)$$

**Table 13** Implication between states of *r* logic

| $x_1$ | $x_2$ | $x = x_1 \to x_2$ |
|---|---|---|
| $T$ | $T$ | $T$ |
| $T$ | $F$ | $F$ |
| $F$ | $T$ | $T$ |
| $F$ | $F$ | $T$ |

**Table 14** Implication between states of *i* logic

| $p_1$ | $p_2$ | $p = p_1 \to p_2$ |
|---|---|---|
| $i$ | $i$ | $i$ |
| $i$ | $\neg i$ | $\neg i$ |
| $\neg i$ | $i$ | $i$ |
| $\neg i$ | $\neg i$ | $i$ |

**Table 15** Implication between states of *s* logic

| $g_1$ | $g_2$ | $g = g_1 \to g_2$ |
|---|---|---|
| $T \vee i$ | $T \vee i$ | $T \vee i$ |
| $T \vee i$ | $F \wedge \neg i$ | $F \wedge \neg i$ |
| $F \wedge \neg i$ | $T \vee i$ | $T \vee i$ |
| $F \wedge \neg i$ | $F \wedge \neg i$ | $T \vee i$ |

**Table 16** Implication between states of *i* and *r* logics

| $p$ | $x$ | $g = p \to x$ |
|---|---|---|
| $i$ | $T$ | $F$ |
| $i$ | $F$ | $F$ |
| $\neg i$ | $T$ | $T \vee i$ |
| $\neg i$ | $F$ | $T \vee i$ |

**Table 17** Implication between states of $s$ and $r$ logics

| $g$ | $x$ | $g = g \rightarrow x$ |
|---|---|---|
| $T \vee i$ | $T$ | $F$ |
| $T \vee i$ | $F$ | $F$ |
| $F \wedge \neg i$ | $T$ | $T \vee i$ |
| $F \wedge \neg i$ | $F$ | $T \vee i$ |

**Table 18** Implication between states of $s$ and $i$ logics

| $g$ | $x$ | $g = g \rightarrow x$ |
|---|---|---|
| $T \vee i$ | $T$ | $F$ |
| $T \vee i$ | $F$ | $F$ |
| $F \wedge \neg i$ | $T$ | $T \vee i$ |
| $F \wedge \neg i$ | $F$ | $T \vee i$ |

$$p \rightarrow g = \neg p \vee g; p \in \{i, \neg i\}; g \in \{(T \vee i), (F \wedge \neg i)\}. \tag{19}$$

Due to the constraints (6) to (11) which produce some "non-symmetry" between the six states of the $s$-logic it may be expected that some peculiarities may arise in the implications.

$$x \rightarrow p = \neg x \vee p; x \in \{T, F\}; p \in \{i, \neg i\}; \tag{20}$$

$$x \rightarrow g = \neg x \vee g; g \in \{(T \vee i), (F \wedge \neg i)\}; x \in \{T, F\}; \tag{21}$$

$$g \rightarrow p = \neg g \vee p; p \in \{i, \neg i\}; g \in \{(T \vee i), (F \wedge \neg i)\}. \tag{22}$$

The truth of Tables 19, 20 and 21 may be checked through relations (6)–(11) and (17)–(19).

**Table 19** Implication between states of $r$ and $i$ logics

| $x$ | $p$ | $g = x \rightarrow p$ |
|---|---|---|
| $T$ | $i$ | $T \vee i$ |
| $T$ | $\neg i$ | $F$ |
| $F$ | $i$ | $T \vee i$ |
| $F$ | $\neg i$ | $F$ |

**Table 20** Implication between states of $r$ and $s$ logics

| $x$ | $g$ | $g = x \rightarrow g$ |
|---|---|---|
| $T$ | $T \vee i$ | $T \vee i$ |
| $T$ | $F \vee \neg i$ | $F$ |
| $F$ | $T \vee i$ | $T \vee i$ |
| $F$ | $F \wedge \neg i$ | $F$ |

(d) Table (19)

- first row: $T \rightarrow i = F \vee i = T \vee I$;
- second row: $T \rightarrow \neg i = F \vee \neg i = T \vee \neg i = F$;
- third row: $F \rightarrow i = T \vee i$;
- fourth row: $F \rightarrow \neg i = T \vee \neg i = F$.

(e) Table (20)

- first row: $(T \rightarrow (T \vee i)) = (F \vee (T \vee i)) = (T \vee i)$;
- second row: $(T \rightarrow (F \wedge \neg i)) = (F \vee (F \wedge \neg i)) = (F \wedge (F \vee \neg i)) = (F \wedge (T \vee \neg i) = F \wedge F == F$;
- third row: $(F \rightarrow (T \vee i)) = (T \vee (T \vee i) = (T \vee i)$;
- fourth row: $(F \rightarrow (F \wedge \neg i)) = (T \vee (F \wedge \neg i)) = (T \wedge (T \vee \neg i) \wedge i) = (T \vee \neg i) = F$.

(f) Table (21)

- first row: $((T \vee i) \rightarrow i) = ((F \wedge \neg i) \wedge i) = ((F \vee i) \wedge i) = ((F \wedge i) \wedge i) = (T \vee i)$;
- second row: $((T \vee i) \rightarrow \neg i) = ((F \wedge \neg i) \vee \neg i) = ((F \vee \neg i) \wedge \neg i) = ((T \vee \neg i) \wedge \neg i) = (F \wedge \neg i)$;
- third row: $((F \wedge \neg i) \rightarrow i) = ((T \vee i) \vee i) = (F \vee i)$;
- fourth row: $((F \wedge \neg i) \rightarrow \neg i) = ((T \vee i) \vee \neg i) = (T \vee i)$.

The results achieved may be checked up by one row per table and through the contraposition law observing constraints (6)–(11):

$$g_1 \rightarrow g_2 = \neg g_2 \rightarrow \neg g_1 \tag{23}$$

1. Table 15, row 1: $(T \vee i) \rightarrow (T \vee i) = (T \vee i)$;
2. Table 16, row 2: $(i \rightarrow F) = (T \rightarrow \neg i) = (F \vee \neg i) = F$;
3. Table 17, row 3: $((F \wedge \neg i) \rightarrow T) = (F \rightarrow (T \vee i)) = (T \vee (T \vee i) = (T \vee i)$;
4. Table 18, row 4: $(\neg i \rightarrow (F \wedge \neg i)) = ((T \vee i) \rightarrow i) = ((F \wedge \neg i) \vee i) = ((F \vee i) \wedge i) == ((F \wedge i) \vee i) = (T \vee i)$;
5. Table (19), row 1: $(T \rightarrow i) = (\neg i \rightarrow F) = (T \vee i)$;
6. Table (20), row 2: $(T \rightarrow (F \wedge \neg i)) = ((T \vee i) \rightarrow F) = ((F \wedge \neg i) \vee F) = (F \wedge (F \vee \neg i)) = F \wedge F = F$;
7. Table (21), row 3: $((F \wedge \neg i) \rightarrow i) = (\neg i \rightarrow (T \vee i)) = (i \vee (T \vee i)) = (T \vee i)$;

which confirms the truth of the results obtained.

**Table 21** Implication between states of $s$ and $i$ logics

| $g$ | $p$ | $g = g \rightarrow p$ |
|---|---|---|
| $T \vee i$ | $i$ | $T \vee i$ |
| $T \vee i$ | $\neg i$ | $F \wedge \neg i$ |
| $F \wedge \neg i$ | $i$ | $T \vee i$ |
| $F \wedge \neg i$ | $\neg i$ | $T \vee i$ |

The results obtained presented in Tables 15, 16, 17, 18, 19, 20 and 21 provide possibility of drawing up the following relations:

**Proposition 1** *For the imaginary and s-logic for each* $g_2 \in \{(T \vee i), (F \vee \neg i)\}$; $p$

$$\in \{i, \neg i\} \text{ and } g_1 = \begin{cases} (T \vee i) \text{ if } p = i; \\ (F \vee \neg i) \text{ if } p = \neg i; \end{cases} \tag{24}$$

*the following relation exists*:

$$(g_1 \rightarrow g_2) = (p \rightarrow g_2). \tag{25}$$

*The result above immediately follows from truth Tables* 15 *and* 21.

**Corollary 1.1** *According to the contraposition law it may be put down from* (23), (24) *and* (25):

$$(g_1 \rightarrow g_2) = (\neg g_2 \rightarrow \neg g_1) = (p \rightarrow g_2) = (\neg g_2 \rightarrow \neg p). \tag{26}$$

**Proposition 2** *For the real, imaginary, and complex s-logic for each* $x \in \{T, F\}$; $p \in \{i, \neg i\}$; $g_1 \in \{(T \vee i), (F \wedge \neg i)\}$ *if* (24) *is observed, then*

$$(p \rightarrow x) = (g_1 \rightarrow x). \tag{27}$$

*This relation immediately follows from the truth Tables* 16 *and* 17.

**Corollary 2.1** *According to the contraposition law, relations* (24) *and* (27), *the following chain of equivalences follows*:

$$(p \rightarrow x) = (\neg x_2 \rightarrow \neg p) = (g_1 \rightarrow x) = \neg x \rightarrow \neg g_1. \tag{28}$$

**Corollary 2.2** *The results in the two Truth Tables* 16 *and* 19 *obey the contraposition law* (23). *The same may be also stated about the results of Tables* 5 *and* 8 *as well as those from Tables* 6 *and* 9. *This also follows from the two chains of equivalences* (27) *and* (28).

The results shown in Tables 13, 14, 15, 16, 17, 18, 19, 20 and 21 are bound to the following features of the implication:

1. In truth Tables 13, 14, 15, 18 and 21 the general requirements of the logical operation "implication" are observed—the false state of the implication is from the set

   $$I_f = \{F, \neg i, (F \vee \neg i)\}; \tag{29}$$

   respectively and it is reached from a true premise of the set

   $$I_t = \{T, i, (T \vee i)\}; \tag{30}$$

   and conclusion from $I_f$ from (29). For all other values of the premise and the conclusion from $I_f$ and $I_t$ the operation "implication" takes true state from the set $I_t$.

2. The general rule of the previous item is not observed in Tables 16, 17, 19 and 20 only in the cases when the premise is in a state of the real logic and the conclusion —in a state of the imaginary or *s*-logic, and vice versa. This is due to the constraints (6) or (8) which connect the real logic with the imaginary and *s*-logic. Due to this the truth Tables 16, 17, 19, and 20 have unusual look, alien to implication.

The results obtained from the truth Tables 13, 14, 15, 16,16, 17, 18, 19, 20 and 21 exhaust all possible combinations between the real, imaginary, and *s*-logic and the premises, conclusions, and implications of these logics.

## 6 Inference Rules Through Implication in the s-Logic

Some of the rules for logical inference exposed in the present section are well known from the classic propositional logic and others immediately follow from the results of the truth tables of implication 3.3 to 3.9.

1. Inference rule Modus Ponens (MP)

1.1. For the real logic (Table 13)

$$\frac{(x_1 \to x_2), x_1}{x_2}; \tag{31}$$

where if $(x_1 \to x_2) = T$ and $x_1 = T$, then $x_2 = T$.

1.2. For the imaginary logic (Table 14)

$$\frac{(p_1 \to p_2), p_1}{p_2}; \tag{32}$$

where if $(p_1 \to p_2) = i$ and $p_1 = i$, then $p_2 = i$.

1.3. For the complex *s*-logic (Table 15)

$$\frac{(g_1 \to g_2), g_1}{g_2}; \tag{33}$$

where if $(g_1 \to g_2) = (T \vee i)$ and $g_1 = (T \vee i)$, then $g_2 = (T \vee i)$.
The conclusion immediately follows from the results presented in Table 15.

1.4. For the results immediately presented in Table 16 for the implication between the states of the imaginary and real logic:

$$\frac{(p \to x)}{p}; \tag{34}$$

where if $(p \to x) = F$ and $x \in \{T, F\}$, then $p = i$.

1.5. Logical inference between the states of the $s$-logic and the real logic (Table 17):

$$\frac{(g \rightarrow x)}{g}; \tag{35}$$

where if $(g \rightarrow x) = F$ and $x \in \{T, F\}$, then $g = (T \vee i)$.

1.6. Logical inference between the states of the $s$-logic and the imaginary logic (Table 18)

$$\frac{(p \rightarrow g), p}{g}; \tag{36}$$

where if $(p \rightarrow g) = (T \vee i)$ and $p = i$, then $g = (T \vee i)$.

1.7. Logical inference between the states of the real and imaginary logic (Table 19)

$$\frac{(x \rightarrow p)}{p}; \tag{37}$$

where if $(x \rightarrow p) = (T \vee i)$ and $x \in \{T, F\}$, then $p = i$.

1.8. Logical inference between the states of the real and the complex $s$-logic (Table 20)

$$\frac{(x \rightarrow g)}{g}; \tag{38}$$

where if $(x \rightarrow g) = (T \vee i)$ and $x \in \{T, F\}$, then $g = (T \vee i)$.

1.9. Logical inference between the states of the complex s-logic and the imaginary logic (Table 21)

$$\frac{(g \rightarrow p), g}{p}; \tag{39}$$

where if $(g \rightarrow p) = (T \vee i)$ and $g = (T \vee i)$, then $p = i$.

The results received (31)–(39) directly follow from the logical relations presented in the truth Tables 13, 14, 15, 16, 17, 18, 19, 20 and 21. These results ensue from the application of the rule Modus Ponens applied to the special features of the $s$-logic.

The following nine logical relations (40)–(48) may be received from the same Tables 13, 14, 15, 16, 17, 18, 19, 20 and 21 but by applying the logical inference rule Modus Tollens (MT).

2. Inference rule Modus Tollens (MT)

2.1. For the real logic (Table 13):

$$\frac{(x_1 \to x_2), \neg x_2}{\neg x_1}; \tag{40}$$

where if $(x_1 \to x_2) = T$ and $\neg x_2 = F$, then $\neg x_1 = F$.

2.2. For the imaginary logic (Table 14)

$$\frac{(p_1 \to p_2), \neg p_2}{\neg p_2}; \tag{41}$$

where if $(p_1 \to p_2) = i$ and $\neg p_2 = \neg i$, then $\neg p_1 = \neg i$.

2.3. For the complex *s*-logic (Table 15)

$$\frac{(g_1 \to g_2), \neg g_2}{\neg g_2}; \tag{42}$$

where if $(g_1 \to g_2) = (T \vee i)$ and $\neg g_2 = \neg(T \vee i) = (F \wedge \neg i)$, then $\neg g_1 = \neg (T \vee i) = (F \wedge \neg i)$.

2.4. For the logical inference between the states of the imaginary and the real logic (Table 16)

$$\frac{(p \to x)}{\neg p}; \tag{43}$$

where if $(p \to x) = (T \vee i)$ and $x \in \{T, F\}$, then $\neg p = \neg i$.

2.5. Logical inference between the states of the *s*-logic and the real logic (Table 17) through the Modus Tollens rule

$$\frac{(g \to x)}{\neg g}; \tag{44}$$

where if $(g \to x) = (T \vee i)$ and $x \in \{T, F\}$, then $\neg g = \neg(T \vee i) = (F \wedge \neg i)$.

2.6. Logical inference between the states of the imaginary logic and the *s*-logic (Table 18)

$$\frac{(p \to g), \neg g}{\neg g}; \tag{45}$$

where if $(p \to g) = (T \vee i)$ and $\neg g = \neg(T \vee i)$, then $\neg g = (F \wedge \neg i)$.

2.7. Logical inference between the states of the real and the imaginary logic (Table 19) through the Modus Tollens rule

$$\frac{(x \rightarrow p)}{\neg p}; \tag{46}$$

where if $(x \rightarrow p) = F$ and $x \in \{T, F\}$, then $\neg p = \neg i$.

2.8. Logical inference between the states of the real and the complex $s$-logic (Table 20)

$$\frac{(x \rightarrow g)}{\neg g}; \tag{47}$$

where if $(x \rightarrow g) = F$ and $x \in \{T, F\}$, then $\neg g = \neg(T \vee i)$, then $\neg g = (F \wedge \neg i)$.

2.9. Logical inference between the states of the complex $s$-logic and the imaginary logic (Table 21)

$$\frac{(g \rightarrow p), \neg p}{\neg g}; \tag{48}$$

where if $(g \rightarrow p) = (T \vee i)$ and $\neg p = \neg i$, then $\neg g = \neg(T \vee i) = (F \wedge \neg i)$.

The results achieved in the present work provide a possibility some most general conclusions to be made:

1. Having in mind that the relations the relations

$$F \vee T = T \text{ and } T \wedge F = F$$

exist in the $r$-logic it follows from (7) and (8), that

$$F \wedge i = F \vee T = T \text{ and } T \vee \neg i = T \wedge F = F. \tag{49}$$

Hence a conclusion can be made that $i$ not equipotent to $T$ and $\neg i$ to $F$. In this sense the sate $i$ may be considered as "truer" than $T$ and $\neg i$—"falser" than $F$, i.e. the states of the imaginary logic demonstrate greater seniority than those of the real logic. These states are united in the $s$-logic considering the axiom-requirement (2). Two complex states—$(T \vee i)$ and $(F \wedge \neg i)$ are distinguished in it.

2. The results received for the $s$-logic on the base of Modus Ponens and Modus Tollens offer chance knowledge bases to be constructed in the complex $s$-logic, from which new complex knowledge can be retrieved through inference machines. New instruments and specialized languages are necessary for carrying out this activity.

3. The problem for possible applications is essential for the *s*-logic as a whole and for the logical inference in particular:
   Three possible directions exist in this trend:
   3.1 Solving of logical equations of type (2) for which there is no solution real propositional logic. It is shown in the present work how this could be realized
   3.2 Formation of new models of the mental activity with richer potentialities and instruments then those used up to now.

As an example for this the following propositions may be pointed out connected with relation (2) of the *r*-logic and *i*-logic:

- Proposition *A* which is in the state "False" of the real logic:

  "My neighbor states I make noise and disturb him";

- Proposition *B* which is in state *i* of the imaginary logic:

  "My neighbor is Member of Parliament".

- Complex proposition *C*:

  "My neighbor states I make noise and disturb him and my neighbor is a Member of Parliament".

No matter that "I don't make noise and I don't disturb the Member of Parliament", the complex proposition *C* is true and my proposition *A* from *False* is not as FALSE but as TRUE.

It follows from this that the propositions in the real and imaginary logic are different in power; they are interpreted in a different manner and are in non-equipotent groups of true and false states. They are united through the *s*-logic.

## References

1. Kleene, S.C.: Mathematical Logic. John Wiley & Sons, N.Y. (1967)
2. Kowalski, R.: Logic for Problem Solving, Elsevier North Holland Inc. (1979)
3. Sgurev, V.: An essay on complex valued propositional logic. Artif. intell. Decis. Making, Institute of System Analysis RAS, **2**, 95–101 (2014)(□□c□□□y□ c□c□e□□o□o a□a□□□a PAH), http://www.aidt.ru, ISSN 2071-8594
4. Sgurev, V.: An approach to constructing a complex propositional logic. C.R. Acad. Bulg. Sci., Tome 66, **11**, 1623–1632, ISSN 1310_1331, "Marin Drinov" Academic Publishing House, Sofia (2013)
5. Ionov, A.S., Petrov, G.A.: Quaternion Logic as a Base Of New Computational Logic. http://zhurnal.ape.relarn.ru/articles/2007/047.pdf (in Russian)
6. Nguyen, H.T., Kreinovich, V., Shekhter, V.: On the Possibility of Using Complex Values in Fuzzy Logic For Representing Inconsistencies. http://www.cs.utep.edu/vladik/1996/tr96–7b.pdf
7. Louis H.K.: Virtual logic—the flagg resolution. Cybern. Hum. Knowing, 6(1), 87–96 (1999)
8. Johnson, B.: Topics in Universal Algebra. Springer-Verlag, N.Y. (1972)

9. Tamir, D.E., Kandel, A.: Axiomatic theory of complex fuzzy logic and complex fuzzy classes. Int. J. Comput. Commun. Control, ISSN 1841–9836, E-ISSN 1841-9844, **6**(3), 562–576 (2011)
10. Richard, G., Shoup, A.: Complex logic for computation with simple interpretations for physics. Interval Res. Palo Alto, CA 94304, http://www.boundarymath.org/papers/CompLogic.pdf
11. Aizenberg, I.: Complex-Valued Neural Networks with Multi-Valued Neurons, Springer (2011)
12. Lucas, C.: A Logic of complex values. In: Proceedings of the 1st international conference on neutrosophy, neutrosophic logic, set, probability and statistics, University of New Mexico, 1–3, 121–138 (2001). (ISBN 1-931233-55)
13. Mendelson, E.: Introduction to Mathematical Logic. D.von Nostrand Com. Inc., Princeton (1975)
14. Sgurev, V.: On way to the complex propositional logic. Int. Conf. "Automatics and Informatics—2013", Sofia, Bulgaria, 3–7 (2013)

# Generalized Nets as a Tool for the Modelling of Data Mining Processes

**Krassimir Atanassov**

**Abstract** Short remarks on Generalized net theory are given. Some possible applications of the generalized net apparatus as means for modelling of data-mining processes are discussed.

**Keywords** Data mining · Generalized net · Modelling

## 1 Introduction

This paper discusses the origin, the current state of research and the applications in the area of Data Mining (DM) of one extension of Petri Net (PN, [176]), called "*Generalized Net (GN)*".

Since 1983, about one thousand papers related to the concept of GNs have been published. A large part of them is included in the bibliographies [2, 180].

GNs are defined as extensions of the ordinary PNs and their modifications, but in a way that is different in principle from the ways of defining the other types of PNs. The additional components in the GN-definition give more and better modelling capabilities and determine the place of GNs among the separate types of PNs, similar to the place of the Turing machine among the finite automata.

In 1990 Prof. Mesnard motivated the author to collect the results in a book [18]. In 2007, the author published a new book with his results—[29]. For the last 25 years GNs have been used for modelling of processes in the areas of systems theory [22] and of Artificial Intelligence (AI) [24, 27, 32, 69, 90, 126, 193, 212].

K. Atanassov (✉)
Department of Bioinformatics and Mathematical Modelling,
Institute of Biophysics and Biomedical Engineering - Bulgarian Academy of Sciences,
Acad. G. Bonchev Street Bl. 105, 1113 Sofia, Bulgaria
e-mail: krat@bas.bg

K. Atanassov
Intelligent Systems Laboratory, Asen Zlatarov University, 8000 Bourgas, Bulgaria



V. Sgurev et al. (eds.), *Innovative Issues in Intelligent Systems*,
Studies in Computational Intelligence 623, DOI 10.1007/978-3-319-27267-2_6

Now they have a lot of applications in medicine [35, 200, 201], chemical industry, public transportation, etc.

Here we give the basic notions from the GN theory, describe some applications of GNs in AI and the benefits of this, and discuss the possibilities for application of GNs for modelling of Data Mining-tools.

# 2 Short Remarks on Generalized Nets

## 2.1 *Definition of the Concept of a Generalized Net*

The concept of a GN is described in details in [18, 29].

The GNs are defined in a way that is different in principle from the ways of defining the other types of PNs.

The first basic difference between GNs and the ordinary PNs is the "place—transition" relation [231]. Here, the transitions are objects of more complex nature. A transition may contain $m$ input places and $n$ output places, where the integers $m$, $n \geq 1$.

Formally, every transition is described by a seven-tuple (Fig. 1):

$$Z = \langle L', L'', t_1, t_2, r, M, \square \rangle,$$

where:

(a) $L'$ and $L''$ are finite, non-empty sets of places (the transition's input and output places, respectively); for the transition in Fig. 1 these are

$$L' = \{l'_1, l'_2, \ldots, l'_m\}$$

and

$$L'' = \{l''_1, l''_2, , \ldots, l''_n\};$$

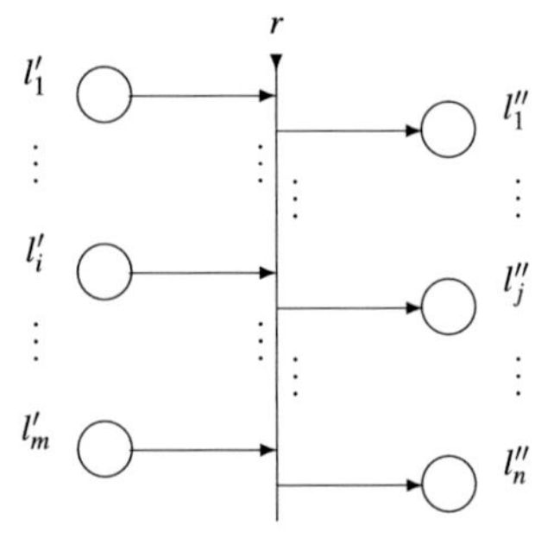


**Fig. 1** The form of transition

(b) $t_1$ is the current time-moment of the transition's firing;
(c) $t_2$ is the current value of the duration of its active state;
(d) $r$ is the transition's *condition* determining which tokens will pass (or *transfer*) from the transition's inputs to its outputs; it has the form of an Index Matrix (IM; see [14, 31]):

$$r = \begin{array}{c|c} & l''_1 \ldots l''_j \ldots l''_n \\ \hline l'_1 & \\ \vdots & r_{i,j} \\ l'_i & (r_{i,j} \; - \text{ predicate }) \\ \vdots & (1 \le i \le m, 1 \le j \le n) \\ l'_m & \end{array};$$

$r_{i,j}$ is the predicate that corresponds to the $i$-th input and $j$-th output place. When its truth value is "*true*", a token from the $i$-th input place transfers to the $j$-th output place; otherwise, this is not possible;

$M$ is an IM of the capacities of transition's arcs:

$$M = \begin{array}{c|c} & l''_1 \ldots l''_j \ldots l''_n \\ \hline l'_1 & \\ \vdots & m_{i,j} \\ l'_i & (m_{i,j} \ge 0 - \text{ natural number }) \\ \vdots & (1 \le i \le m, 1 \le j \le n) \\ l'_m & \end{array};$$

$\square$ is an object of a form similar to a Boolean expression. It may contain as variables the symbols that serve as labels for a transition's input places, and $\square$ is an expression built up from variables and the Boolean connectives $\wedge$ and $\vee$ . Its semantics is defined as follows:

$\wedge(l_{i_1}, l_{i_2}, \ldots, l_{i_u})$ – every place $l_{i_1}, l_{i_2}, \ldots, l_{i_u}$ must contain at least one token,
$\vee(l_{i_1}, l_{i_2}, \ldots, l_{i_u})$ – there must be at least one token in all places $l_{i_1}, l_{i_2}, \ldots, l_{i_u}$, where $\{l_{i_1}, l_{i_2}, \ldots, l_{i_u}\} \subset L'$.

When the value of a type (calculated as a Boolean expression) is "*true*", the transition can become active, otherwise it cannot.

The ordered four-tuple

$$E = \langle\langle A, \boldsymbol{\pi_A}, \boldsymbol{\pi_L}, \boldsymbol{c}, \boldsymbol{f}, \boldsymbol{\theta_1}, \boldsymbol{\theta_2}\rangle, \langle \boldsymbol{K}, \boldsymbol{\pi_K}, \boldsymbol{\theta_K}\rangle, \langle \boldsymbol{T}, \boldsymbol{t^o}, \boldsymbol{t^*}\rangle, \langle \boldsymbol{X}, \boldsymbol{\Phi}, \boldsymbol{b}\rangle\rangle$$

is called a GN if:

(a) $A$ is a set of transitions;
(b) $\pi_A$ is a function giving the priorities of the transitions, i.e., $\pi_A : A \rightarrow N$, where $N = \{0, 1, 2, \ldots\} \cup \{\infty\}$;
(c) $\pi_L$ is a function giving the priorities of the places, i.e., $\pi_L : L \rightarrow N$, where $L = pr_1 A \cup pr_2 A$, and $pr_i X$ is the $i$-th projection of the $n$-dimensional set, where $n \in N$, $n \geq 1$ and $1 \leq k \leq n$ (obviously, $L$ is the set of all GN–places);
(d) $c$ is a function giving the capacities of the places, i.e., $c : L \rightarrow N$;
(e) $f$ is a function that calculates the truth values of the predicates of the transition's conditions (for the GN described here, let the function $f$ have the value "*false*" or "*true*", that is, a value from the set $\{0, 1\}$;
(f) $\boldsymbol{\theta_1}$ is a function which indicates the next time-moment when a certain transition $Z$ can be activated, that is, $\theta_1(t) = t'$, where $pr_3 Z = t$, $t' \in [T, T + t^*]$ and $t \leq t'$. The value of this function is calculated at the moment when the transition ceases to function;
(g) $\boldsymbol{\theta_2}$ is a function which gives the duration of the active state of a certain transition $Z$, i.e., $\boldsymbol{\theta_2}(t) = t'$, where $pr_4 Z = t \in [T, T + t^*]$ and $t' \geq 0$. The value of this function is calculated at the moment when the transition starts to function;
(h) $K$ is the set of the GN's tokens. In some cases, it is convenient to consider it as a set of the form

$$K = \underset{l \in Q^I}{\cup} K_l,$$

where $K_l$ is the set of tokens that enter the net from place $l$, and $Q^I$ is the set of all input places of the net;
(i) $\pi_K$ is a function which gives the priorities of the tokens, that is, $\pi_K : K \rightarrow N$;
(j) $\theta_K$ is a function which gives the time-moment when a given token can enter the net, that is, $\theta_K(\alpha) = t$, where $\alpha \in K$ and $t \in [T, T + t^*]$;
(k) $T$ is the time-moment when the GN starts to function. This moment is determined with respect to a fixed (global) time-scale;
(l) $t^o$ is an elementary time-step, related to the fixed (global) time-scale;
(m) $t^*$ is the duration of the functioning of the GN;
(n) $X$ is the set of all initial characteristics which the tokens can obtain on entering the net;
(o) $\Phi$ is the characteristic function that assigns new characteristics to every token when it makes the transfer from an input to an output place of a given transition;
(p) $b$ is a function which gives the maximum number of characteristics a given token can obtain, that is, $b : K \rightarrow N$.

For example, if $b(\alpha) = 1$ for any token $\alpha$, then this token will enter the net with some initial characteristic (marked as its zero-characteristic) and subsequently it will keep only its current characteristic.

When $b(\alpha) = \infty$, token $\alpha$ will keep all its characteristics. When $b(\alpha) = k < \infty$, except its zero-characteristic, token $\alpha$ will keep its last $k$ characteristics (characteristics older than the last $k$-th one will be "forgotten"). Hence, in general, every token $\alpha$ has $b(\alpha) + 1$ characteristics on leaving the net.

It is also convenient to assume that functions $f$, $\theta_1$, $\theta_2$ and $\Phi$ have other forms. For example, function $\Phi$ can be represented in the form:

$$\Phi = \bigcup_{i=1}^{|L-Q^I|} \Phi_i,$$

where $\Phi_i$ calculates the characteristics that the tokens will receive in the $i$-th GN place and $|X|$ is the cardinality of the set $X$.

A given GN may not have some of the above components. In these cases, any redundant component will be omitted. The GNs of this kind form a special class of GNs called "*reduced GNs*".

The static part of a given GN is specified by the elements of the set $pr_{1,2,6,7}A$, where for a given $n$-dimensional set $X (n \geq 2)$

$$pr_{i_1,i_2,\ldots i_k}X = \prod_{j=1}^{k} pr_{i_j}X$$

$(1 \leq i_j \leq n, 1 \leq j \leq k, i_{j'} \neq i_{j''}$ for $j' \neq j'')$, that is, the static part of a GN is specified by the collection of the following elements for each transition: the input and output places, the index matrix of the arcs and the transition type. The dynamic character of the net is due to the GN's tokens and the transitions' conditions ($pr_5A$), the temporal character comes from components $T$, $t^o$, $t^*$ and from the elements of the set $pr_{3,4}A$. Finally, components $\Phi$, $X$ and $b$ play the role of the memory of the GN.

Various functions are also related to these four GN components: functions $\pi_A$, $\pi_L$, $c$ to the static structure; $f$, $\pi_K$ to the dynamical elements; $\theta_1$, $\theta_2$ and $\theta_K$ to the temporal components.

The definition of a GN is more complex than the definition of a PN and so the algorithms of the tokens' transfer in GNs are also more complex. On the other hand, as GNs are more general, the algorithms for the tokens' movements in the GN are more general than those for PNs. In a PN implementation, parallelism is reduced to a sequential firing of its transitions and the order of their activation in the general case is probabilistic or dependent on the transitions' priorities, if such exist. GN's algorithms provide means for a more detailed modelling of the described process. The algorithms for the tokens transfers take into account the priorities of the places, transitions and tokens, i.e., they are more precise.

## 2.2 Remarks on the Theory of GNs

Following [18, 29] we shall discuss some elements of the theory of GNs.

First, we note that GNs have more than 20 conservative extensions (i.e., extensions for which their functioning and the results of their work can be represented by an ordinary GN). The most important of them, described in [29], are:

- **Intuitionistic fuzzy GNs of type 1**: their transition condition predicates are evaluated into the set $[0, 1]^2$ with a degree of truth ($\mu$) and a degree of falsity ($\nu$), so that $\mu + \nu \leq 1$ (see [12]);
- **Intuitionistic fuzzy GNs of type 2** are intuitionistic fuzzy GNs of type 1, which have "quantities" instead of tokens, which "flow" from input to output GN places;
- **Intuitionistic fuzzy GNs of type 3** are extensions of the intuitionistic fuzzy GNs of type 1, for which the tokens characteristics are also estimated with a degree of truth ($\mu^*$) and a degree of falsity ($\nu^*$), so that $\mu^* + \nu^* \leq 1$;
- **Intuitionistic fuzzy GNs of type 4** are extensions of the intuitionistic fuzzy GNs of types 1 and 3;
- **Colour GNs** are GNs whose tokens and transition arcs are coloured in different colours and the tokens transfer depends on these colours;
- **GNs with global memory** have a global tool (that is common for all GN-components) for storing data (during the functioning of the GN) or determining the values of various parameters related to the modelled processes;
- **GNs with optimization components** use optimization procedures for determining the way tokens transfer from transition input to output places;
- **GNs with additional clocks** comprise tools for recording the duration of the predicates' truth-values evaluations. Should this process take more time than a specified limit, it stops and the predicate is assigned its previous value determining the future functioning of the net;
- **GNs with complex transition types** have additional conditions for activation of the GN transition types;
- **GNs with 3-dimensional structure**;
- **GNs with stop-conditions** have additional conditions for termination of the GN functioning;
- In the **Opposite GNs**, tokens can transfer (within a separate GN area) in the opposite direction, i.e., from output to input places. This allows us to determine the necessary conditions for the desired flow of a given process (see also [177, 178]);
- **GNs with volumetric tokens** [53];
- **GNs where tokens can obtain variables as characteristics** can be used for solving optimization problems related to the processes modelled by GNs;
- **GNs with characteristic functions of the places** [3].

The different extensions of the GNs have a lot of real applications. For example, the intuitionistic fuzzy GNs of different types are used for models in chemical and, specifically, in petrochemical industry. Some of these models contain additional clock and/or stop-conditions. They are suitable, e.g., for the models of rectification columns in a petrochemical refinery. The GNs with optimization components can be used for describing of decision making control processes and for some more theoretical GN-models, e.g., for ant-colony optimization algorithms. The GNs with volumetric tokens are used for models of Internet processes, while GNs with place characteristics in some medical and ecological models. The GNs with 3-dimensional structure, the colour GNs and the GNs with complex transition type provide better visualization of the modeled processes.

Operations and relations are defined both over the transitions, and over GNs in general. The operations, defined over GNs—"union", "intersection", "composition" and "iteration" (see [18])—do not exist anywhere else in the PN theory. They can be transferred to virtually all other types of PNs (obviously with some modifications concerning the structure of the corresponding nets). These operations are useful for constructing GN models of real processes.

In [18] different properties of the operations over transitions and GNs are formulated and proved. Certain relations over transitions and GNs are also introduced there; these are, however, beyond the scope of the present book.

The idea of defining operators over the set of GNs in the form suggested below dates back to 1982 year (see [18]). It is a proper extension of the Valk's idea from [237].

Now, the operator aspect has an important place in the theory of GNs. Six types of operators are defined in its framework. Every operator assigns to a given GN a new GN with some desired properties. The comprised groups of operators are:

- global ($G$ –) operators,
- local ($P$ –) operators,
- hierarchical ($H$ –) operators,
- reducing ($R$ –) operators,
- extending ($O$ –) operators,
- dynamic ($D$ –) operators.

The *global operators* transform, according to a definite procedure, a given entire net or all its components of a certain type. There are operators that alter the form and structure of the transitions ($G_1$, $G_2$, $G_3$, $G_4$, $G_6$), the temporal components of the net ($G_7$, $G_8$); the duration of its functioning ($G_9$), the set of tokens ($G_{10}$), the set of the initial characteristics ($G_{11}$); the characteristic function of the net ($G_{12}$) (this is a function which is the union of all places' characteristic functions); the evaluation function ($G_{13}$), or other functions of the net ($G_5$, $G_{14}$, …, $G_{20}$).

The second type of operators are *local operators*. They transform single components of some of the transitions of a given GN. There are three types of local operators: temporal local operators ($L_1$, $L_2$, $L_3$, $L_4$), which change the temporal components of a given transition, matrix local operators ($L_5$, $L_6$), which change some of the index matrices of a given transition, and different additional operators: one that alters the transition's type ($L_7$), one that alters the capacity of some of the places in the net ($L_8$), one that alters the characteristic function of an output place

($L_9$), or the evaluation function associated to the transition condition predicates of the given transition ($L_{10}$).

Any of these operators ($L_i$, $1 \leq i \leq 10$) can be extended to a global one ($\overline{L}_i$, $1 \leq i \leq 10$) by defining the corresponding operator in such a way that it would transform all components of a specified type in every transition of the net.

The third type of operators are the *hierarchical operators*. These are of 6 different types and fall into two groups according to their way of action: extending a given GN ($H_1$, $H_3$, $H_5$ $H_6$) and shrinking a given GN ($H_2$, $H_4$ and $H_5$). The $H_5$ operator can be extending as well as shrinking, depending on its form. According to their object of action the operators enter again into three groups: acting upon or giving as a result a place ($H_1$ and $H_2$), acting upon or giving as a result a transition ($H_3$, $H_4$ and $H_5$) and acting upon or giving as a result a subnet ($H_5$ and $H_6$).

The hierarchical operators $H_1$ and $H_3$ replace a given place or transition, respectively, of a given GN with a whole new net. Conversely, operators $H_2$ and $H_4$ replace a part of a given GN with a single place ($H_2$) or transition ($H_4$). Operator $H_5$ changes a subnet of a given GN with another subnet. Finally, operator $H_6$ changes a GN-token to a subnet. Extending operators can be viewed as tools for magnifying the modelled process' structure; while shrinking operators—as a means of integration and ignoring the irrelevant details of the process.

The next (fourth) group of operators defined over GNs produces a new, reduced GN from a given net. They allow the construction of elements of the classes of reduced GNs. To find the place of a given PN modification among the classes of reduced GNs, it must be compared to a reduced GN obtained by an operator of this type. These operators are called *reducing operators*.

Operators from the fifth group extend a given GN. These operators are called *extending operators*. The extending operators are respectively associated with each of the GN extensions.

Finally, the operators from the last—sixth—group are related to the ways the GN functions, so that they are called *dynamic operators*. They determine the algorithms for tokens' transfer or the process of calculating of transition condition predicates and also allow or forbid splitting or uniting of tokens.

The operators of different types, as well as the others that can be defined, have a major theoretical and practical value. On the one hand, they help us study the properties and the behaviour of GNs. On the other hand, they also facilitate the modelling of many real processes. The basic properties of the operators are discussed in [18, 22].

"*A Self-Modifying GN*" (SMGN) is constructed and described in detail in [18]. A SMGN has the property of changing its structure (number of transitions, places, tokens, transition condition predicates, token characteristics, place and arc capacities, etc.) and the token transfer strategy during the time of the GN-functioning. These changes are realized by operators which can be defined over the GN. Of course, not all operators can be applied over a GN during the time of its functioning. Some of them are applicable only before, and others—only after the GN-functioning. The conditions satisfied by the operators applicable during the time of the GN-functioning are discussed in [18].

# 3 Generalized Nets and the Others Types of Petri Nets

As it can be seen from above, the GN definition is essentially more complex than the definitions of the other types of PNs. On the other hand, it is valid that the functioning and the result of the work of the ordinary PNs [176], E-nets [156], time PNs (versions of [76, 149, 181]), PRO-nets [155], coloured PNs (versions of [120, 121, 197, 241]), regular PNs [127], self-modifying nets [237], predicate/transition nets [94, 95], stochastic PNs [202] and [151], generalized E-nets [10, 15], M-nets [104], super nets [84], generalized modification PNs [182], fuzzy PNs [145] and others can be represented by GNs (see, for example, [11, 13, 18, 36–38, 80, 98, 99]).

The relations between GNs and the most important classes of PN-modifications are shown on Fig. 2, where $\Sigma_{PN}$, $\Sigma_{EN}$, $\Sigma_{TPN}$, $\Sigma_{SPN}$, $\Sigma_{CPN}$, $\Sigma_{SMN}$, $\Sigma_{PRON}$, $\Sigma_{PTN}$, $\Sigma_{FPN}$, $\Sigma_{SN}$, $\Sigma_{MN}$, $\Sigma_{GEN}$, $\Sigma_{GMPN}$ and $\Sigma_{GN}$ are the respective classes of all ordinary PNs, all E-nets, all time PNs, all stochastic PNs, all colour PNs, all self-modifying PNs, all PRO-nets, all predicative-transition nets, all fuzzy PNs, all super nets, all M-nets, all generalized E-nets, all generalized modified PNs and all GNs.

For each of the above-mentioned classes of PNs, there exists a GN that is universal for the elements of this class (see, for example, [16, 18, 106]).

Therefore, the models of real processes, which can be described by some of the above-mentioned types of nets, can be described by a GN, too, while GNs also give the possibility of generating, using, storing and analyzing more detailed information in comparison to the other types of nets. For example, if there are two separate processes described by some kind of PNs, then the fact that one of the processes should influence the other one is represented by a token that enters a certain

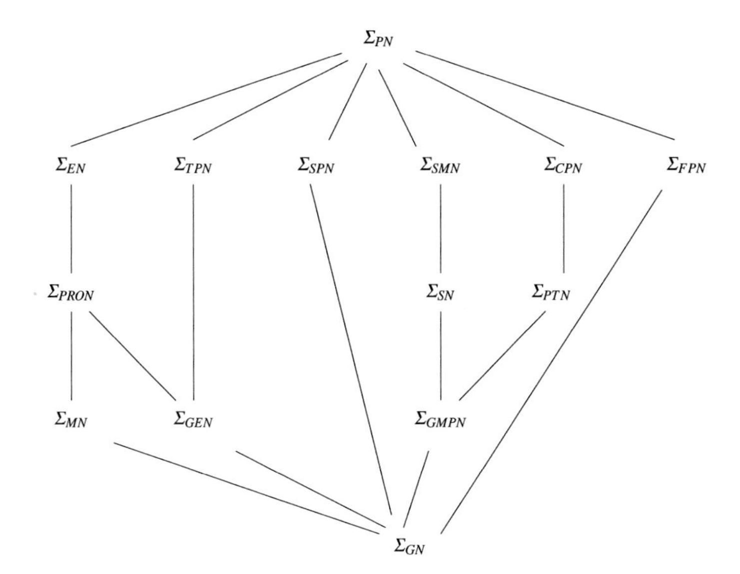


**Fig. 2** GNs and the most important classes of PN-modifications

(previously determined) place. Hence, there are a special token and a place for the aim (a special transition is possible, too). In the case of GNs, this situation can be modelled, for example, by the predicate of a transition and/or by the characteristic of a token which is of functional meaning for the model, i.e., which is not included in the model only for the above-mentioned aim.

Using the token characteristics, we can store all the necessary information about the process and after this, we can retrieve it with respect to some of the previously formulated aims.

On the other hand, GNs can modify their structure and/or behaviour during the time of their functioning essentially better than the nets of Valk, and of Etzion and Yoeli. These modifications can represent the modifications in the real modelled process.

For instance, compared to the other types of PNs, GNs can be *non-invariant* in time. They can start their functioning at a certain time-moment accounted by a real time-scale, as mentioned in [18]:

The aims of introducing GNs are:

1. To provide a possibility to compare the different types of nets both as mathematical objects and as a means of modelling parallel processes.
2. To investigate the most general properties of GNs and transfer them to other nets.
3. To give net-means of the most detailed modelling of real processes.

In a sense, GNs can play the role of the Turing machine for the different types of finite automata. In this relation it must be noted that for every type of finite automaton and for the Turing machine there is a GN that is universal for the elements of the corresponding type of objects (see, for example, [18]).

## 4 Generalized Nets and Artificial Intelligence up to Now

First, following [9, 23, 24], we give a short list of the different areas which various authors consider to lie in the foundation of the artificial intelligence framework:

(a) different kinds of automatic theorem proving;
(b) methods for searching paths in labyrinths;
(c) intellectual games;
(d) pattern recognition;
(e) scene analysis;
(f) natural languages and semantic networks;
(g) robotics;
(h) expert systems;
(i) (artificial) neural networks;
(j) genetic algorithms;
(k) computer learning;

(l) Turing machines and different types of finite automata;
(m) planning systems;
(n) multiagent systems;
(o) heuristics based problems related to scheduling, assignment, optimization, decision making, etc.;
(p) LISP, PROLOG, and their modifications;
and others.

Of course, this classification is not complete. In addition, it is old—it was prepared in 1991, when the author formulated the hypothesis that if there exists a unique mathematical tool that can describe all areas of AI, then this area of knowledge will be developed essentially. His thesis from 1991 was that the apparatus of GNs is such tool.

Now, the author is prone to think that Data Mining—as an essentially new area of sciences—can be interpreted as a natural extension and union of some of AI-areas. Therefore, we discuss how to represent each of the above areas by GNs and also how both sides (AI and GN theory) will benefit from that. As far as the latter is concerned, imagine a mathematical tool that can play the role of a relatively universal language for describing all scientific areas listed above (and even more). Moreover, we shall see that the individual areas of AI have a common background, although at the moment it is not clear whether all or only part of them. Therefore, this will facilitate the transfer of ideas from one area to another, as well as their stronger formalization and further development.

There exists, however, another problem—how to generalize and extend this description, within the framework of the GN-description of each event, adding new (perhaps not yet existing but theoretically possible) elements in a way that allows the newly obtained process (object) be described by a GN, as well. If we can achieve this, then it will be clear that GNs are not only capable of describing processes (objects), but they serve to construct new, even not yet existing, processes (objects).

Finally, a third problem arises, namely to search for the possible directions of a further development of GN-methodology and new objects to be described by it.

Now, we discuss shortly the already existing results. They are collected in the books [24, 32, 40, 41, 69, 90, 126, 131, 144, 193, 199, 212], which are based on a lot of papers in journals and conference proceedings.

In book [24], nine GN-models of Expert Systems (ESs) are described. Here, we discuss the results from this book in more detail, because some ideas from there can be transfered directly for the DM-case.

The first of the models from [24] shows how a GN-model of a given ES can be constructed. This model contains information about the separate ES components (Data Base (DB), Knowledge Base (KB) hypotheses), which are represented by GN–transitions, places, tokens and token characteristics. It shows that we can construct one GN-model for *any* ES. In the second and third GN-models from [24], a part of the GN-components that correspond to the ES-components are changed only by token characteristics. In the second GN-model the DB is represented by a characteristic of a specially constructed token, while in the third GN-model the same is done for the KB.

Therefore, the two GN-models (the second one and, particularly, the third one) are already independent of the concrete DB and the concrete KB of a given ES, being modelled by the GN. Hence, the third GN-model is *universal* for the class of *all* ESs.

The fourth GN-model is an extension of the third one, but with the possibility to represent KB rules containing the operation "negation" in their antecedents. The remaining five GN-models, described in [24], are devoted to extensions of the concept of an ES. The new ESs contain new components, which can be represented by the GN-tools. Of course, these new components are not all of those necessary for modelling concrete expert processes, but they illustrate the possibility for extending the ES-structures, and in separate cases they can be useful.

The fifth GN-model represents the functioning and the results of the work of an ES, with priorities of the data within its DB. The data have now a specific priority. At the time of the ES functioning (within the context of a GN-model), new data (represented by GN-tokens) can enter the DB changing the existing information in it (if the new data are in contradiction with the data already existing, and if the priority of the new data is larger than the priority of the old), or confirm it. There are special tools, described by GN-subnets, which can check the correctness of the new data and this information will enter the DB only if it is not in contradiction with the existing rules of the ES's KB.

The seventh GN-model is devoted to an analogous extension of the ES, but now related to its KB. The rules there have priorities and they can be changed or confirmed as in the previous case. Now, there are GN-subnets that check the correctness of the new rules and they will enter the DB only if they are not in contradiction with the existing rules of the ES's KB and with the existing data in the ES's DB.

The sixth GN describes an ES that contains "metafacts". This new concept is similar to DB facts and simultaneously to the KB rules. In practice, the metafacts can be interpreted as facts about the DB-facts, as well as, as simple KB-rules.

The concept of an Intuitionistic Fuzzy ES (IFES) was introduced in [21] as an extension of fuzzy ESs. The estimations of the truth-values of the facts there have the form $\langle m, n \rangle$, where $m, n \in [0, 1]$ and $m + n \leq 1$. Numbers $m$ and $n$ correspond to the degree of validity and the degree of non-validity of the fact. In this case, there possibly exists a degree of uncertainty $p$, for which $p = 1 - m - n \geq 0$. The three components of the Intuitionistic Fuzzy Set (IFS) [25] and its derivatives (intuitionistic fuzzy logics, intuitionistic fuzzy graphs, intuitionistic fuzzy abstract systems, intuitionistic fuzzy ESs, and so on), give greater possibilities for the real processes modelling than ordinary fuzzy objects. The eighth GN-model describes the functioning and the results of the work of an IFES.

As discussed in [25], modal types of operators can be defined over the IFS, and in an IFS-model the axioms of some modal logic systems are valid. Therefore, the eighth GN-model can represent the functioning of an ES using modal logic operators.

Finally, the ninth GN-model, described in [24], represents the functioning and the results of the work of an ES with rules, related to a fixed time-scale, which is given as characteristics of a special GN-token. In this way, we can construct a temporal ES using temporal logic operators.

A continuation of these ideas is given in [33, 34, 42, 45, 48, 49, 69 71, 126, 172, 229, 232], where the functioning and the results of the work of other types of ESs, relation databases, uncertain data and knowledge engineering processes are described by GNs.

The book [32] contains GN-models of process of machine learning of: neural networks, genetic algorithms, intellectual games, GNs and intuitionistic fuzzy GNs, abstract systems, and others. The process of machine learning described by a GN essentially extends the standard machine learning process, but we can now also take into account many facts generated at the time of the GN-functioning which are related to the trading process, as well as to the object of learning.

The processes of pattern recognition and decision making are represented by GN-models in the books [40, 41] and papers [96, 97, 154, 157, 158, 160, 233–235]. Multi-criteria decision making procedures are described by GNs in [46, 161, 162], while in near future the procedures of intercriteria analysis (see [44, 55, 57] will be described by GNs, too.

The GN-models of a part of the existing types of neural networks are described in books [131, 144, 212] and in papers [1, 4, 7, 8, 43, 50–52, 81, 100, 102, 132–143, 148, 153, 173, 174, 205–211, 213– 228, 233–235].

In book [90] and papers [54, 56], the Ant Colony Optimization procedures, introduced by M. Dorigo [82, 83] are described by GNs, while in book [193] and papers [47, 165–171, 185–187, 189–192, 194, 195], the functioning and the results of the work of separate types of genetic algorithms are represented by GNs.

Existing research shows that there is the possibility for GNs to be used as tools for modelling of practically all AI paradigms.

## 5 Generalized Nets and Data Mining—Possibilities for the Future

What is Data Mining? The answer of this question is so unclear, as well as the answer of the question for the areas of AI. Again, there are different answers with respect to the opinions of the specialists, giving answers. For example:

> The aim of DM is to make sense of large amounts of mostly unsupervised data, in some domain [73];
>
> The aim of DM is to extract implicit, previously unknown and potentially useful (or actionable) patterns from data. DM consists of many up-to-date techniques such as classification (decision trees, naive Bayes classifier, *k*-nearest neighbor, NNs), clustering (*k*-means, hierarchical clustering, density-based clustering), association (one-dimensional, multi-dimensional, multilevel association, constraint-based association) [236];
>
> DM stands at the confluence of the fields of statistics and machine learning [203];

DM is a tool, not a magic wand [119];

DM is a term that covers a broad range of techniques being used in a variety of industries [196];

DM is the core of the knowledge discovery in databases process, involving the inferring of algorithms that explore the data, develop the model and discover previously unknown patterns [146].

Data Mining is a process of finding reasonable correlations, repeating patterns and trends in large DBs. There are a lot of papers and books, devoted to DM. As a basis of our research, we use the publications [6, 58, 59, 60, 67, 72, 73, 75, 77–79, 101, 103, 105, 113–119, 122–125, 146, 147, 159, 163, 164, 175, 179, 183, 196, 198, 203, 204, 230, 236, 238–240]. In the literature, different areas of AI are determined as components of DM, for example, the algorithms of decision making, pattern recognition, neural networks, genetic algorithms, etc.

Extending the text from [24], here we make a review of some of the problems related to the above ones—those already existing, and those planned for future research within the framework of the theory of GNs. Everywhere we emphasize on:

- the way of the GN-description of the process (object) up to now (if such already exists);
- other ways for GN-realization of this;
- possible extensions or generalizations of already existing GN-models of the corresponding processes (objects) and ways for their modifications;
- possibilities for construction of new GN-models.

## 5.1 GN-Descriptions of Pattern and Speaker Recognition, Scene Analysis, Planning Systems, Problems Related to Scheduling, Assignment, Optimization, Decision Making, etc., Based on Heuristics

Books [40, 41] contain the results in the areas of pattern and speaker recognition. GN-models of processes of face, signature, handwriting and typewriting texts are described. On this basis, the processes of decision making for obtaining classified access are represented by GNs. The most interesting in this case is the possibility for automatic generation of tests for the persons who do not satisfy all but just a part of the access conditions. In this case, the control tool determines other tests (for which information exists in the respective DBs) and preserves/stores the changes in the information about the persons who must obtain access. So, the next time, these persons are checked against the new and older data existing for them.

In the areas of planning systems, problems related to scheduling, assignment and others based on heuristics, there are only a few attempts to model processes by GNs. We hope that their number will increase in the future. At the moment the only research in the above areas is related to the problems of the decision making [41, 161, 162], the travelling salesman [17, 19, 20] and to the transportation problem [18]. For the last problem it is shown that on one hand the process of its solution is described by a GN, and on the other hand, this GN can be extended in such a way that in the process of GN-functioning (other) transportational problems can be solved with the aim of optimizing the GN-tokens' transfers.

After constructing a GN for the movement of one travelling salesman, the constructed GN was extended. Its transition condition predicates and its tokens' characteristic functions were modified, and the new GN is capable of describing the following problem: determine the best move of each of a fixed set of travelling salesmen, collected in $n$ groups, if:

- every two salesmen, from different groups compete;
- for each travelling salesman, the types of the products which he sells, are known;
- for each product of a given salesman, its price, its quantity, the date of its production and its expiry date are known (i.e., a part of the salesman production may be discarded);
- for each graph vertex (a built-up area), the necessary quantity of different types of products is known;
- for each two adjacent vertices (built-up areas), the distances between them, the cost of the path and the possibility for connection between them at a given time-moment are known.

The following question is very interesting (and its answer can be obtained within the framework of a corresponding GN-model).

Can a GN-model, similar to the above one, be constructed so that the "travelling salesmen" search for concrete information in a DB or a DataWarehouse (DW)? Now, the "salesmen" will collect information instead of selling things and the attributes of this information (quality, quantity, time-moments (or periods) of validity, etc.) correspond to product attributes. Since the behaviour of the "salesmen" is similar to the agents from multi-agents systems (see, e.g. [74, 107]), below, we denote the objects from the "salesmen"-type as "agents".

Some GN-models of decision making processes are given in books [41, 161, 162]. In near future, they can be essentially extended, and they can find applications as solutions of decision-making agents that must collect suitable information from a DB/DW/KB and to represent it in a predefined form.

Another important question is related to the development of algorithms for automatic learning of the already constructed GN-models.

In [128–130], it is shown that the basic UML-components can be represented by GNs.

## *5.2 GN-Descriptions of the Process of Natural Language Analysis, Semantic Networks and Translation*

At the moment, there exist only two attempts for research in this direction and they are related to the GN-description of semantic networks in [152] and of a translation process in [39]. After the first step, the question arises about extending the semantic networks in the direction of, for instance, their (intuitionistic) fuzziness and the representation of the new nets by GNs. In future, there will be constructed GNs, which describe the translation process of information from a DB/DW from one language to another.

## *5.3 Expert Systems, Data Bases and Data Warehouses*

A lot of colleagues already assert that the ESs are dying. The author supports the idea that they will live their "Renaissance", obtaining a special place in the instrumentation of DM. Preserving their basic purpose to generate new knowledge by answering hypotheses, we can essentially extend the area of their possibilities. When some unclear situation arises in a process controlled by DM-tools, and when some hypotheses for its future development are generated, then the new type of ESs can help.

As we noted above, in a series of papers collected in [24], we described the basic steps of the process of the functioning and the results of work of ESs.

Briefly, GN-models of ESs:

- contain different logical operations (conjunction, disjunction, implication, negation, etc.), quantifiers ("for existence" and "for all") and modal operators in the hypotheses and in the antecedents of the ES-rules,
- can be self-modifying at the time of their functioning,
- can have priorities of the facts and rules,
- can have metafacts and metarules,
- can have intuitionistic fuzzy values of the aspects of estimations of facts and rules (see [21, 30]),
- can answer questions related to the time ("at the moment", "once", "sometimes", "for a long/short time", "often", "rarely", etc.),
- can be a combination of all of the above types.

Similar directions for extensions of DBs, data warehouses, OLAP-structures, etc. can be realized. The first steps for GN-modelling of these structures have been discussed in [69, 126].

We assume that solving each of the above problems or, of course, all of them, will promote not only the theory and application of GNs, but also the research in the area of DM, too.

Bearing in mind all of the above, we think that it is clear that GNs can really make a claim for a place within DM. How central is this place? That will depend on how successfully the problems above can be solved.

How shall we benefit from having the possibility of describing all the above areas by GNs? First, this will mean that these DM-tools have a common basis, a common language by means of which all separate tools can be described, and second, we shall be able to transform methods already developed (like GN-models) from one DM-tool to another. This can be achieved by GN operations and operators, as we will discuss shortly, below.

Finally, we must note that the thesis formulated in the introduction may be an overstatement, but nevertheless it could turn out to be true, at least partially.

## 5.4 GNs for Clusterisation and Classification of Data, GN-Models of the Processes of Rule Extraction, of Solution/Decision-Tree Generation and of Associative Rule Construction

The first attempts of GN models of clusterisation and classification procedures are discussed in [61–66], but until now none of the processes of rule extraction, of solution-tree generation and of associative rule construction exist. It will be important that in future GN-models for all these procedures and processes be developed. For example, it will be interesting to construct a GN that contains as subnets the GNs that describe the functioning and the results of the work of the separate clusterisation procedures.

## 5.5 GN-Models of Knowledge Discovery Processes and of Processes for Imputation (Filling in) of Missing Data

> The Knowledge Discovery Process (KDP), also called knowledge discovery in databases, seeks new knowledge in some application domain. It is defined as the nontrivial process of identifying valid, novel, potentially useful, and ultimately understandable patterns in data. [73].

Two missing data imputation methods, the mean and the hot deck, are described in the same book. Both types of processes can be described by GNs and this will be an aim of the author for the future.

## 5.6 GN-Models of Processes in OLAP-Structures

The OLAP-structures (see, e.g. [73]) can be well represented by the means of IMs [31] and especially by multi-dimensional IMs. IMs were developed as tools for formal mathematical description of the GN-transitions and logic of the modelled process. On the other hand, it is easily seen that the processes of realization of the operations, relations and operators over IMs can be described by GNs. So, GNs can be used as a tool for description of the operations in OLAP-structures [68, 70].

## 5.7 GN-Models of Neural Networks and Evolutionary Algorithms

The first results related to the GN-interpretation of neural networks date back to 1990 [108–110]. A possible extension of the concept “neural network” was shown (where the neurons can either die, or be born), and the way of the GN-interpretation of these nets was described. A GN-representation of a class of intuitionistic fuzzy neural networks (see [111]) was described in [112].

During the last 10 years, universal GNs have been constructed for half of the different types of neural networks. Each of these GNs describes the functioning and the results of the work of any neural network from the respective type. The results obtained by the moment are collected in [131, 144, 212] and the authors of this series plan to construct universal GNs for the rest types of neural networks.

In [93], it is mentioned that

> The paradigm of Evolutionary Algorithms (EAs) consists of stochastic search algorithms inspired by the process of neo-Darwinian evolution. … There are several kinds of EAs, such as Genetic Algorithms, Genetic Programming, Classifier Systems, Evolution Strategies, Evolutionary Programming, Estimation of Distribution Algorithms, etc.

In this direction of research, up to now there have been some initial results, but the serious work is still the future.

In [85], the first GN-model of Ant Colony Optimization (ACO) procedure is described. The model was used as a basis for generating new ACO-procedures and in the book [90], a comparison between the results of the work of the traditional and the new procedures is given. This is a good illustration of the fact, that the GN-models can be used as a methodological tool for generating new effective procedures for the area of data mining, as well.

The results of a series of papers devoted to GN-models of different types of ACO procedures, e.g., [86–89, 91, 92], were collected in the book [90].

Similarly, as we mentioned above, in the last years the series of papers, related to GN-models of genetic algorithms were collected in the book [193]. The ACO-and GA-procedures can be used for optimization of the search-procedure of the agents that have to collect suitable information from a DB/DW/KB and represent it in a predefined form. In [184, 188], the GN-models of the Bat algorithm and of the Firefly algorithm are described, respectively.

## 5.8 GN-Models of a Procedure for Inductive Reasoning

As is mentioned in [105], *the rule induction is one of the fundamental tools of DM. Usually rules are expressions of the form*

$$\begin{gathered} if\,(attribute_1, value_1)\&(attribute_2, value_2)\&\ldots\&(attribute_n, value_n) \\ then\,(decision, value). \end{gathered}$$

Obviously, the process of evaluation of the truth values of the members of the antecedent can be represented by the sequential characteristic of a fixed token in the GN and when it has obtained all $n$ characteristics, it will obtain in addition the characteristic "(decision, value)". Therefore, the discussed procedure is representable by GNs. But now, we can use, e.g., an intuitionistic fuzzy GN from the third type. In this case, we obtain sequentially the following $n$ characteristics

$$\begin{gathered} (attribute_1, value_1, \mu_1, \nu_1), \\ (attribute_2, value_2, \mu_2, \nu_2), \\ \ldots \\ (attribute_n, value_n, \mu_n, \nu_n), \end{gathered}$$

where, in the simplest case

$$\mu_i = \begin{cases} 1, & \text{if value}_\text{i} \text{ is anticipated (expected, correct, etc..)} \\ 0, & \text{if value}_\text{i} \text{ is not anticipated (not expected, incorrect, etc..)} \\ *, & \text{if there is not an information for attribute}_i \end{cases}$$

and $\nu_i = 1 - \mu_i$. The $(n + 1)$-th characteristic can have the form

$$(decision, value, \mu, \nu),$$

where

$$\mu = \frac{p}{n}, \quad \nu = \frac{q}{n}$$

and $p$ is the number of degrees $\mu_i$ that are equal to 1, $q$ is the number of degrees $\nu_i$ that are equal to 1. Obviously, $p + q \le n$. Therefore, the evaluation $\langle \mu, \nu \rangle$ is an intuitionistic fuzzy pair. Hence, we obtain more precise estimation for the validity of the procedure for inductive reasoning. If in the beginning we determine some threshold of validity $t_\nu$, then we can assert that a *decision* is sufficiently valid if $\mu > t_\nu$. On the other hand, if we determine some threshold of non-validity $t_n$ then we can assert that a *decision* is sufficiently valid if $\nu < t_n$.

## 5.9 GN-Models of the Statistical Procedures for Data Mining

Up to now, there have been no GN-models of the statistical procedures for Data Mining, but their development is realisable in terms of GNs. Indeed, in some subset of a given GN-model of a real process we can collect the necessary information in the form of characteristics of especially determined GN-tokens. Other GN-tokens will collect the results of the statistical data processing. So, all statistical methods can obtain GN-realizations.

## 5.10 GN-Models of the Collaborative Data Mining Processes

Following [150], we mention that "collaborative DM aims to combine the results generated by isolated experts, by enabling the collaboration of geographically dispersed laboratories and companies." The description of GN in Sect. 2 is a guarantee that the separate components of the collaborative DM processes can be represented by synchronized subnets of a larger GN.

## 5.11 Some Other Trends for Future GN-Interpretations

In [146], it is mentioned that there are several trends for future research and implementation, including:

- Active DM—closing the loop, as in control theory, where changes to the system are made according to the knowledge discovery in databases (KDD) results and the full cycle starts again. Stability and controllability, which will be significantly different in these types of systems, need to be well-defined.
- Full taxonomy—for all the nine steps of the KDD process. We have shown a taxonomy for the DM methods, but a taxonomy is needed for each of the nine steps. Such a taxonomy will contain methods appropriate for each step (even the first one), and for the whole process as well.
- Meta-algorithms—algorithms that examine the characteristics of the data in order to determine the best methods, and parameters (including decompositions).
- Benefit analysis—to understand the effect of the potential KDD or DM results on the enterprise.
- Problem characteristics—analysis of the problem itself for its suitability to the KDD process.
- Mining complex objects of arbitrary type—Expanding Data Mining inference to include also data from pictures, voice, video, audio, etc. This will require adapting and developing new methods (for example, for comparing pictures using clustering and compression analysis).
- Temporal aspects—many data mining methods assume that discovered patterns are static. However, in practice patterns in the database evolve over time. This poses two important challenges. The first challenge is to detect when concept drift occurs. The second challenge is to keep the patterns up-to-date without inducing the patterns from scratch.

- Distributed Data Mining—The ability to seamlessly and effectively employ Data Mining methods on databases that are located in various sites. This problem is especially challenging when the data structures are heterogeneous rather than homogeneous.
- Expanding the knowledge base for the KDD process, including not only data but also extraction from known facts to principles (for example, extracting from a machine its principle, and thus being able to apply it in other situations).
- Expanding Data Mining reasoning to include creative solutions, not just the ones that appear in the data, but being able to combine solutions and generate another approach.

Elements of all these trends can be represented by GN-tools and this will be one of our future research aims.

## 6 Simple Generalized Net Models of Data Mining Tools

The PNs and their extensions are often used for simulation of processes. The GN-models are also used for simulation, but they give possibility for other applications as well.

Following [28], we will note that GNs can be used for optimization of the modelled processes. Up to now, there are two GN-models of optimization procedures. In [18], there is constructed a GN describing the process of transportational problem solving, while in [17] GN-models of some travelling salesman problems are discussed. Practically, the process of solving all other optimization problems can be described in terms of GNs.

If there is a suitable interface between a computer with a program tool for GN-realization from one side and from another—the environment, and other program products, realizing DM-tools, then the GNs can be used for control of real processes flowing in the frameworks of the environment in real time. The information from the environment enters the GN in the form of the initial or current tokens' characteristics or as parameters for the transition condition predicates. Below, we will discuss this possibility in more details.

Now, let us assume that $DM_1$, $DM_2$, …, $DM_n$ are GN-models of $n$ different DM-tools. Let a GN $G$ describe some real process that uses one or more DM-tools. Then, we can construct a GN $E$ containing the GN $G$ as a subnet and simultaneously, $E$ can contain as a subnet the net $F$, so that it describes the process of some DM-tool-functioning. The GN $F$ will contain at least three transitions (see Fig. 3). In the GN-place $x$, for the whole time of the GN-functioning, there is a token with an initial and permanent characteristic

"*criteria for selection of a suitable DM-tool*".

The transitions of the GN $F$ have the following forms.

**Fig. 3** GN $F$

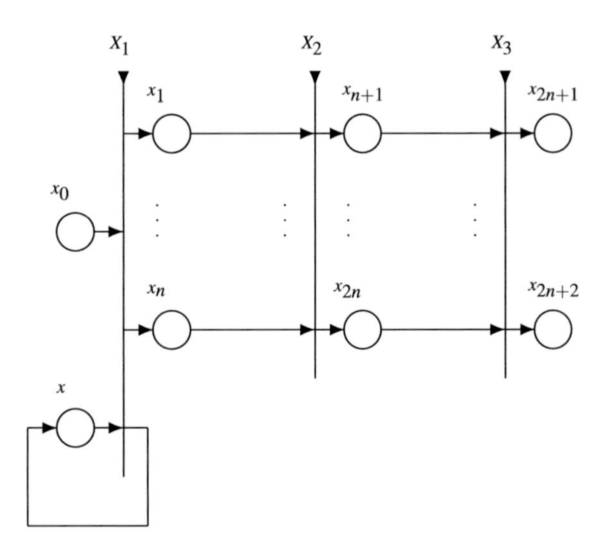


$$X_1 = \langle \{x_0, x\}, \{x_1, ..., x_n, x\}, \begin{array}{c|cccc} & x_1 & \ldots & x_n & x_0 \\ \hline x & false & \ldots & false & true \\ x_0 & V_{0,1} & \ldots & V_{0,n} & false \end{array}, \vee(x_0, x)\rangle,$$

where for $1 \leq j \leq n$:

$V_{0,j} =$ "there is a need for the $j$-th DM-tool".

The token in place $x_0$ splits to some tokens, with respect to the number of the transition condition predicates with truth-value "true". The token that enters place $x_j(1 \leq j \leq n)$ obtains as a characteristic the necessary initial information for using the $j$-th DM-tool. This information must be included in the initial characteristic of the current token that enters place $x_0$.

In some cases, only one DM-tool will be selected, and the process in $G$ will be controlled on the basis of the results of the execution of this DM-tool. More interesting is the case when at least two DM-tools are selected.

The second transition is related to the application of operator $H_6$ over some places of subnet $H$. It has the form

$$X_2 = \langle \{x_1, \ldots, x_n\}, \{x_{n+1}, \ldots, x_{2n}\},$$

$$\begin{array}{c|ccc} & x_{n+1} & \ldots & x_{2n} \\ \hline x_1 & V_{1,n+1} & \ldots & V_{1,2n} \\ \vdots & \vdots & \vdots & \vdots \\ x_n & V_{n,n+1} & \ldots & V_{n,2n} \end{array}, \vee(x_1, ..., x_n)\rangle,$$

where:

$$V_{i,n+j} = \begin{cases} true, & \text{if} \quad i = j \\ false, & otherwise \end{cases}$$

$(1 \le i, j \le n)$.

The token entering place $x_{n+i}(1 \le i \le n)$, obtains as a characteristic the result of the execution of the $j$-th DM-tool.

$$X_3 = \langle \{x_{n+1}, \ldots, x_{2n}\}, \{x_{2n+1}, x_{2n+2}\},$$

$$\begin{array}{c|cc} & x_{2n+1} & x_{2n+2} \\ \hline x_{n+1} & V_{n+1,2n+1} & V_{n+1,2n+2} \\ \vdots & \vdots & \vdots \\ x_{2n} & V_{2n,2n+1} & V_{2n,2n+2} \end{array}, \vee(x_{n+1}, \ldots, x_{2n})\rangle,$$

where:

$V_{n+i,2n+1}$ = "the results of the DM-tool are useful",

$$V_{n+i,2n+2} = \neg V_{n+i,2n+1},$$

$(1 \le i \le n)$, where $\neg P$ is the negation of predicate $P$.

Now we can add a GN $H$ (with one or more transitions) that contains tokens $\varepsilon_1$, $\varepsilon_2$, …, $\varepsilon_n$ which correspond to the GN-models of the DM-tools $E_1$, $E_2$, …, $E_n$. The last GN can have, e.g., the form from Fig. 4. In this case, tokens $\varepsilon_1$, $\varepsilon_2$, …, $\varepsilon_n$ stay in places $y_{n+1}$, $y_{n+2}$, …, $y_{2n}$, respectively, in the initial time-moment of the GN, whose two transitions are the following.

$$Y_1 = \langle \{y_{n+1}, \ldots, y_{2n}, y_{2n+1}, \ldots, y_{3n}\}, \{y_1, \ldots, y_n, y_{n+1}, \ldots, y_{2n}\},$$

$$\begin{array}{c|cccccc} & y_1 & \ldots & y_n & y_{n+1} & \ldots & y_{2n} \\ \hline y_{n+1} & W_{n+1,1} & \ldots & W_{n+1,n} & false & \ldots & false \\ \vdots & \vdots & \ddots & \vdots & \vdots & \ddots & \vdots \\ y_{2n} & W_{2n,1} & \ldots & W_{2n,n} & false & \ldots & false \\ y_{2n+1} & false & \ldots & false & W_{2n+1,n+1} & \ldots & W_{2n+1,2n} \\ \vdots & \vdots & \ddots & \vdots & \vdots & \ddots & \vdots \\ y_{3n} & false & \ldots & false & W_{3n,n+1} & \ldots & W_{3n,2n} \end{array},$$

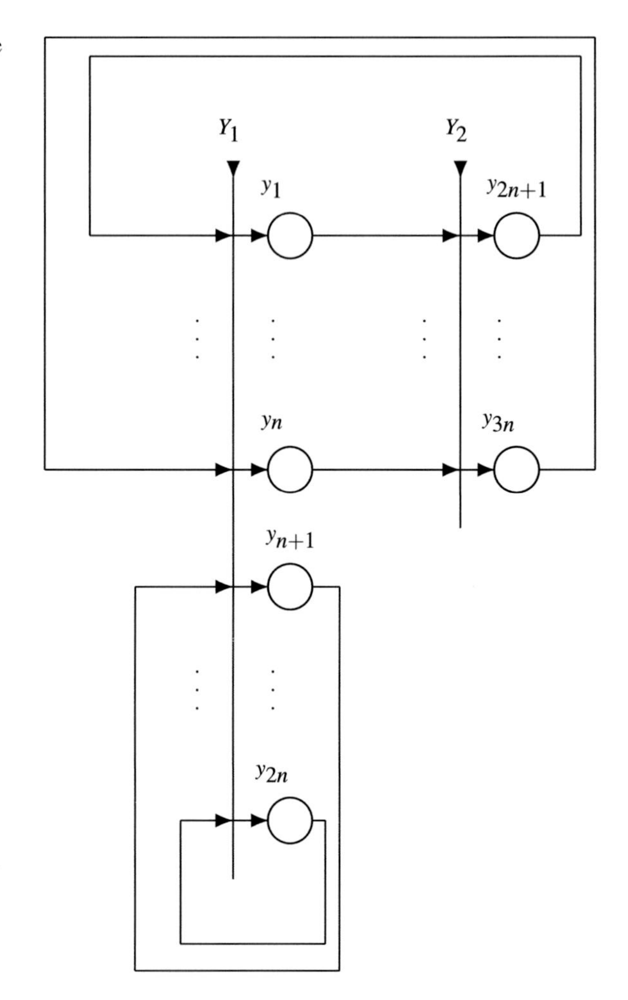


**Fig. 4** A possible form of the obtained GN

Where:

$$W_{n+i,j} = \begin{cases} \text{"execution of the j-th DM-tool is necessary"}, & \text{if } \; i = j \\ \mathit{false}, & \text{otherwise} \end{cases}$$

$(1 \leq i, j \leq n)$;

$$W_{2n+i,n+j} = \begin{cases} \mathit{true}, & \text{if } \; i = j \\ \mathit{false}, & \text{otherwise} \end{cases}$$

$(1 \leq i, j \leq n)$.

$$Y_2 = \langle \{y_1, \ldots, y_n\}, \{y_{2n+1}, \ldots, y_{3n}\}, \begin{array}{c|ccc} & y_{2n+1} & \ldots & y_{3n} \\ \hline y_1 & W_{1,2n+1} & \ldots & W_{1,3n} \\ \vdots & \vdots & \vdots & \vdots \\ y_n & W_{n,2n+1} & \ldots & W_{n,3n} \end{array}, \vee(y_1, \ldots, y_n)\rangle,$$

where:

$$W_{i,2n+j} = \begin{cases} true, & \text{if} \quad i = j \\ false, & \text{otherwise} \end{cases}$$

$(1 \leq i, j \leq n)$.

Each of the $\varepsilon_i$-tokens is a $H_6$-token, i.e., these tokens can be changed by the hierarchical operator $H_6$. Entering the respective place $y_i$, it invokes the procedure of applying operator $H_6$ over the GN and changing place $y_i$ with the GN representing $E_i$.

In GN *F* we can have a subnet that is a combination of both nets *F* and *H*, e.g., in the form given in Fig. 5.

In the frameworks of this new subnet we can describe both the process of generating and of using of the concrete DM-tool. This subnet corresponds to the case when we must realize one or more DM-procedures and comparing their results we must find the best results, i.e., the ones for which there is a consensus, or those with the highest estimations. As we mentioned in the beginning, in some cases these activities can be repeated several times. For example, we can apply optimization procedures over the results of the used DM-tools. Therefore, the subnet from Fig. 5 must have cycles, e.g., like the net from Fig. 6.

The subnet from Fig. 3 can be modified to the net from Fig. 7. Now, in place $y$ there is a token with an initial and current characteristic

"*criteria for stopping the long-lasting DM-tool-functioning*".

If the time for execution of the DM-tools in places $x_{n+1}, \ldots, x_{2n}$ is too long, the processes can be stopped and only the already ready results will be used for determining the best solution.

Now, transition $X_2$ will be changed, as follows:

$$X_2 = \langle \{x_1, \ldots, x_n, y\}, \{x_{n+1}, \ldots, x_{2n}, y\}, \begin{array}{c|cccc} & x_{n+1} & \ldots & x_{2n} & y \\ \hline x_1 & V_{1,n+1} & \ldots & V_{1,2n} & false \\ \vdots & \vdots & \vdots & \vdots & \vdots \\ x_n & V_{n,n+1} & \ldots & V_{n,2n} & false \\ y & false & \ldots & false & true \end{array}, \wedge(\vee(x_1, \ldots, x_n), y)\rangle,$$

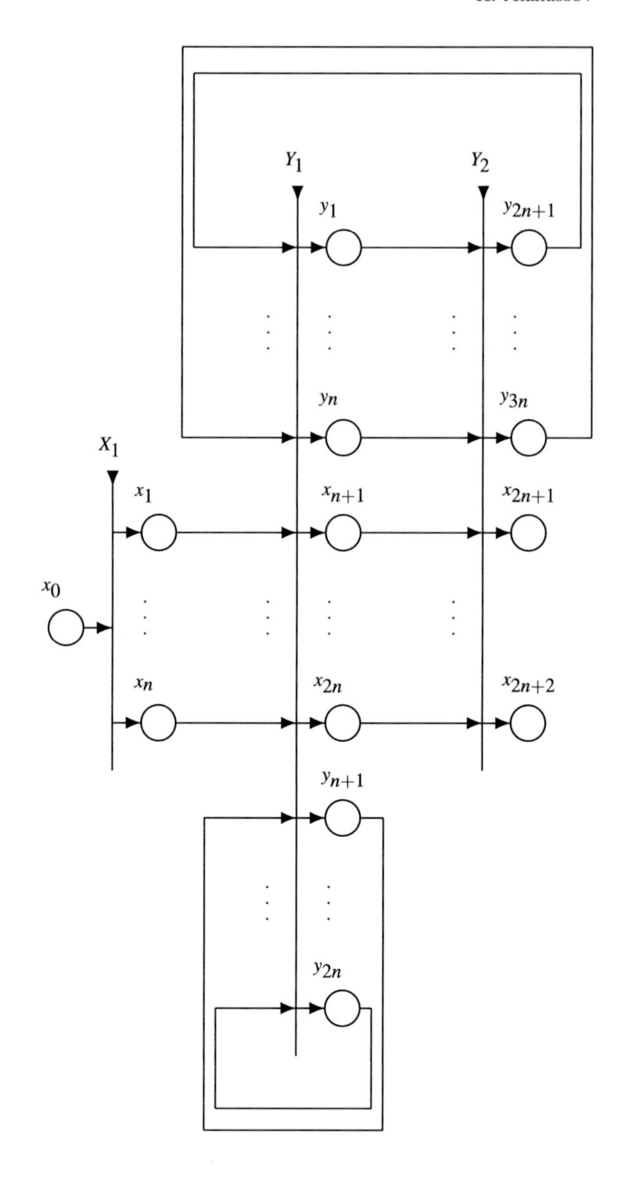


**Fig. 5** A subnet that is a combination of both nets *F* and *H*

where:

$$V_{i,n+j} = \begin{cases} true, & if \quad i = j \\ false, & \text{otherwise} \end{cases}$$

$(1 \leq i, j \leq n)$.

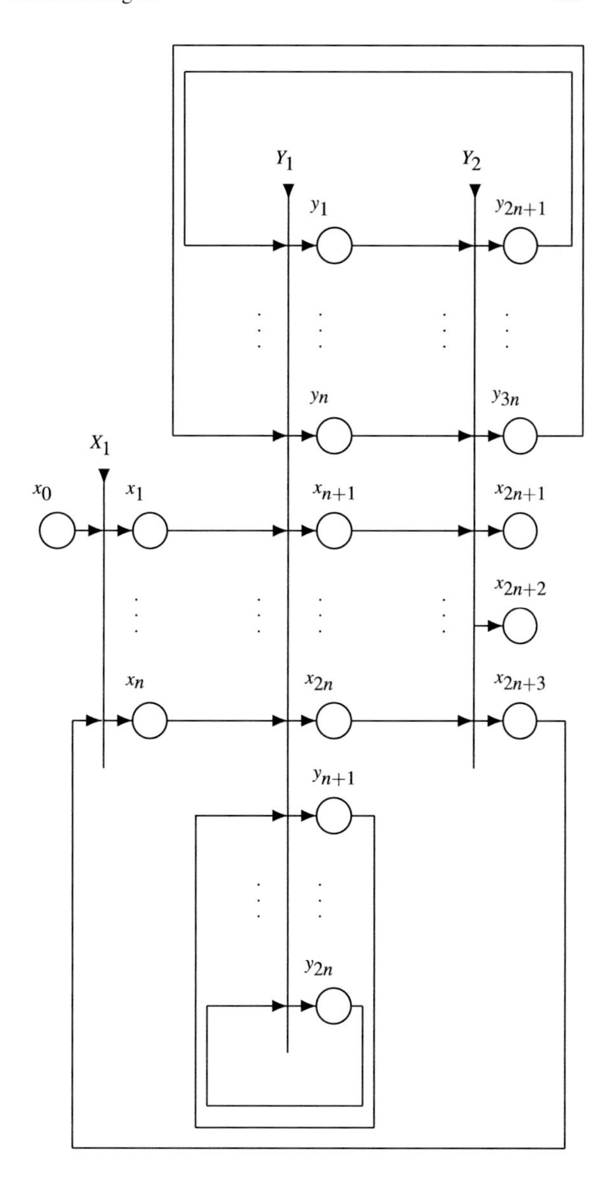


**Fig. 6** Possible cycles in the subnet

In [28], the authors discussed the idea that in the frames of a given GN an optimization procedure can be realized and its results can influence the tokens' transfer, the tokens' characteristics and/or the truth values of the transition condition predicates.

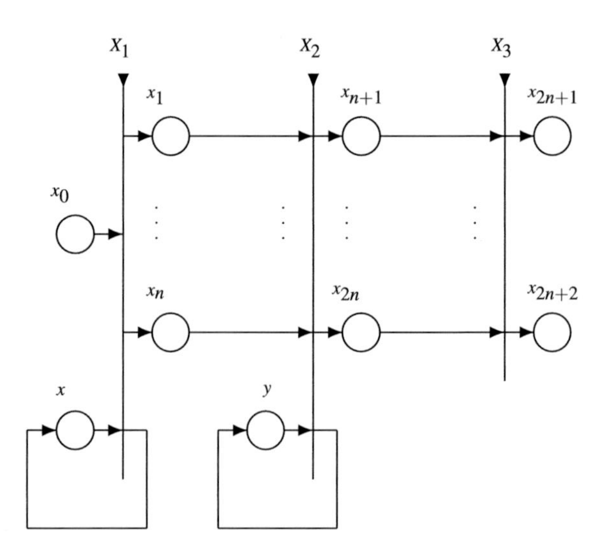


**Fig. 7** The GN $F$ transformed

# 7 Two Other Generalized Net Models

Each model of control process referring to Human Thinking (HT), as well as to the management of a firm, or of any flexible manufacturing system, is a matter of research both in cybernetics, as well as in artificial intelligence. These processes are modelled by different means and with variable success.

The apparatus of the GNs offers us the possibility of representing the dynamics of the modelled process in space and time (within the frameworks of the human body or in the manufacture), as well as its development as a set of parallelly flowing subprocesses.

The main model of the discussed process of HT and/or of Control and Decision Making of Manufactures (CDMM) is shown in Fig. 8. There we can envisage the places of all areas of AI (see Fig. 9) and their relations.

In [26, 32] we described two GN-models. The second one has two forms. Here, following an idea from P. Angelov's book [5], we extend the characteristic of the $\rho$-token in the first GN-model, keeping the second one with its two forms.

The basic aim of the constructed GN-model below is to show the relations between the separate components of the modelled process. In this way the common components of the processes of HT and/or CDMM are more clearly distinguished.

The GN-model contains some types of tokens that represent different basic activities (processes) of HT and in the CDMM.

$\alpha$-tokens enter place $l_1$ with an initial characteristic

*"current external effect, type, power, time-moment, etc."*

(we shall use the words "effect" and "stimulus" as synonyms). They will pass through places $l_1$, $l_2$ and $l_4$.

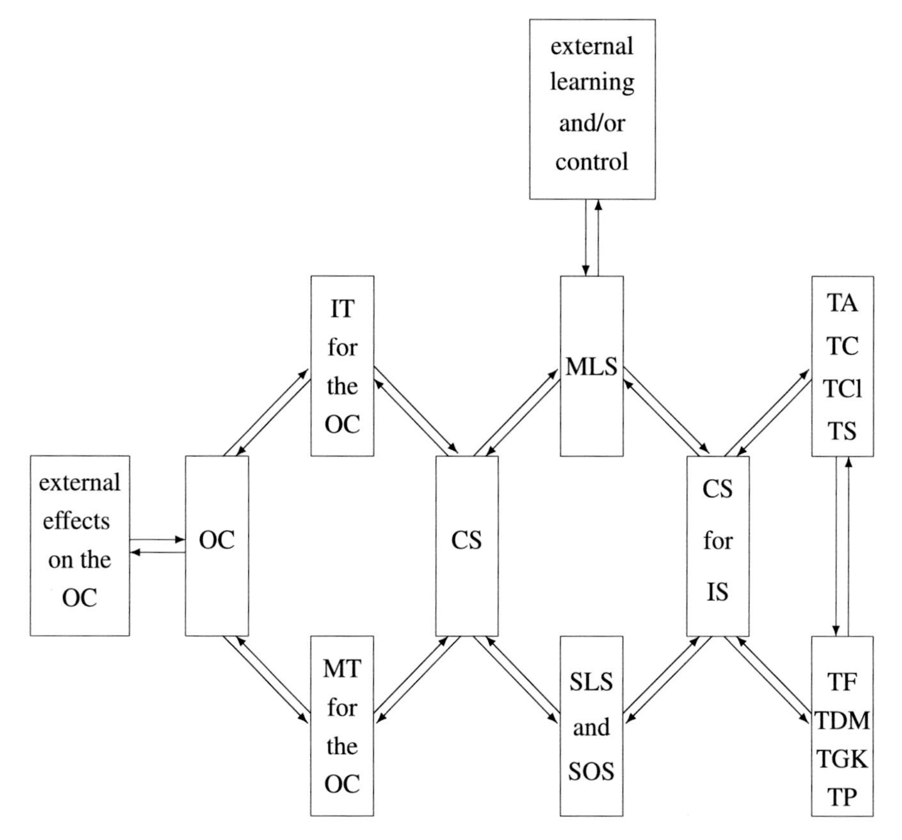


**Fig. 8** *OC* object of control, *IT* information tools, *MT* manipulating tools, *CS* control system, *MLS* machine learning system, *SLS* self–learning system, *SOS* self–organizing system, *TA* tools for analysis, *TC* tools for clusterisation, *TCl* tools for classification, *TDM* tools for decision making, *TF* tools for formalization, *TGK* tools for generating of new knowledge, *TP* tools for prognoses, *TS* tools for sorting

Only $\alpha$- and $\zeta$-tokens are external tokens. All the other tokens are internal ones. They are situated in the initial time-moment, as follows.

$\beta$-tokens are situated in place $l_7$ with an initial characteristics

"*current status of the object of control*",

if this object is only one, or

"*current status of a specific part of the object of control*",

if there is a set of tokens, corresponding to the separate parts of this object, which (in this case) is interpreted as a complex system. They will pass through places $l_7$ and $l_3$.

$\gamma$-tokens are situated in place $l_{12}$ with an initial, and afterwards current characteristic

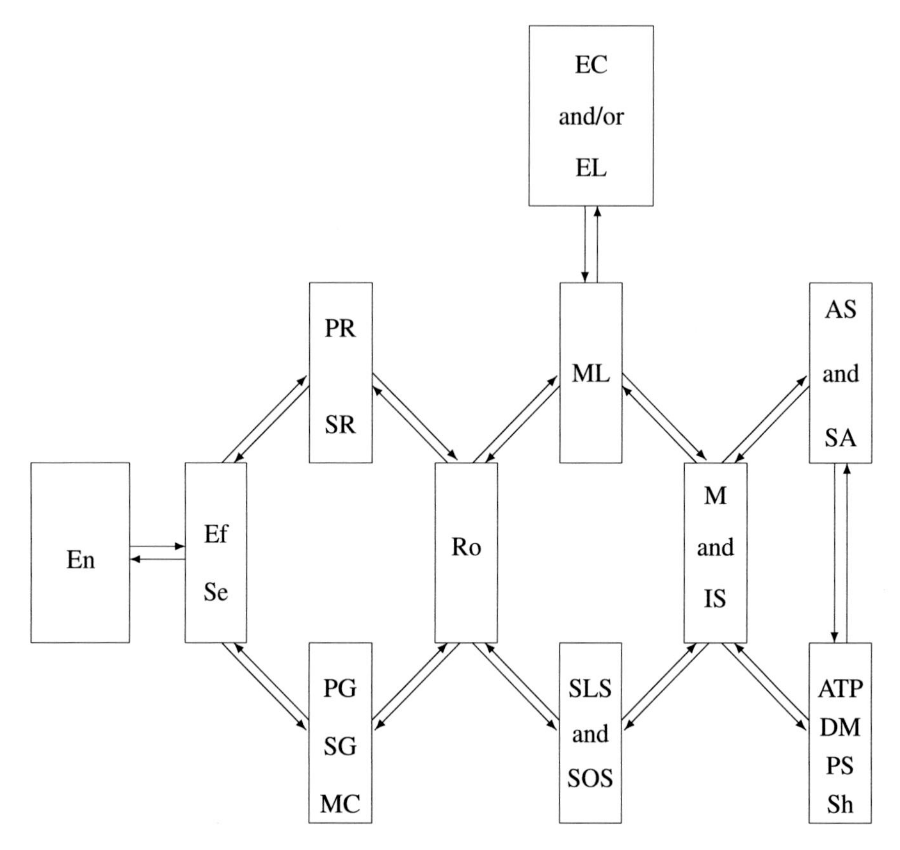


**Fig. 9** *En* environment, *Ef* effectors, *Se* sensors, *PR* pattern recognition, *SR* speak recognition, *PG* pattern generation, *SG* speak generation, *MC* motion control, *Ro* robotics, *EC* expert control, *EL* expert learning, *ML* machine learning, *SLS* self–learning system, *SOS* self–organizing system, *M* memory, *IS* information system, *AS* analysis of situations, *SA* scene analysis, *ATP* automatic theorem proving, *DM* decision making, *PS* planning system, *Sch* scheduling

*"current status of the information tools for tracing the object of control"*

They can be either one or more. They will pass through places $l_5$, $l_8$, $l_9$ and $l_{12}$.

$\delta$-tokens are situated in place $l_{15}$ with an initial and afterwards current characteristic

*"current status of the tools for control".*

They can be either one or more. They will pass through places $l_6$, $l_{10}$, $l_{11}$ and $l_{15}$.

The only $\varepsilon$-token is situated only in place $l_{16}$ with an initial and afterwards current characteristic

*"current status of the management system"*

and it will loop only in this place.

$\zeta$-tokens enter the GN through place $l_{17}$ with an initial characteristic

"*current external learning or control for the modelled process in general; time-moment, etc.*".

They will transfer in place $l_{19}$ with a final characteristic

"*information about the state of the modelled process in general, time-moment, etc.*".

$\eta$-tokens are situated in places $l_{13}$ and $l_{18}$ with no initial characteristics. They can be either one or more tokens. They will pass through places $l_{18}$ and $l_{13}$, respectively.

$\theta$-tokens are situated in places $l_{20}$ and $l_{25}$ with no initial characteristic. They can be either one or more tokens. They will pass through places $l_{25}$ and $l_{20}$, respectively.

The only $\iota$-token is situated only in place $l_{21}$ with an initial and afterwards current characteristic

"*current status of the machine learning system*"

and it will loop only in this place.

$\kappa$-tokens are situated in places $l_{22}$ and $l_{28}$ without any initial characteristics. They can be either one or more tokens. They will pass through places $l_{28}$ and $l_{22}$, respectively.

$\lambda$-tokens are situated in places $l_{14}$ and $l_{23}$ without any initial characteristics. They can be either one or more tokens. They will pass through places $l_{23}$ and $l_{14}$, respectively.

The only $\mu$-token is situated only in place $l_{24}$ with an initial and afterwards current characteristic

"*current status of the self-learning and self-organizing system*"

and it will loop only in this place.

The only $\nu$-token is situated only in place $l_{29}$ with an initial and afterwards current characteristic

"*current status of the information system of the management system*"

and it will loop only in this place.

$\xi$-tokens are situated in places $l_{26}$ and $l_{30}$ without any initial characteristics. They can be either one or more tokens. They will pass through places $l_{30}$ and $l_{26}$, respectively.

$o$-tokens are situated in places $l_{28}$ and $l_{34}$ without any initial characteristics. They can be either one or more tokens. They will pass through places $l_{34}$ and $l_{28}$, respectively.

$\pi$-tokens are situated in places $l_{31}$ and $l_{33}$ without any initial characteristics. They can be either one or more tokens. They will pass through places $l_{33}$ and $l_{31}$, respectively.

The only $\rho$-token is situated only in place $l_{32}$ with an initial and afterwards current characteristic *"current status of the tools for analysis, clusterisation and classification,"*

and it will loop only in this place.

The only $\sigma$-token is situated only in place $l_{35}$ with an initial and afterwards current characteristic

*"current status of the tools for decision making, prognosis, etc."*

and it will loop only in this place.

The GN-transitions are the following.

$$Z_1 = \langle \{l_1, l_7\}, \{l_2, l_3\}, \begin{array}{c|cc} & l_2 & l_3 \\ \hline l_1 & true & false \\ l_7 & false & true \end{array} \rangle.$$

The $\alpha$-tokens obtain the characteristic

*"changes of the form, structure, status, ... of the external effects"*

in place $l_7$ and the $\beta$-tokens obtain the characteristic

*"changes of the form, structure, status, ... of the object of the control"*

in place $l_3$.

We shall discuss two examples.

1. Let us have a flexible manufacturing system producing various details. We can detect the changes in the details status as a result of the process in place $l_2$ and we can account the changes in the machines status (either no changes, or there is a congestion, or there is a damage, etc.) in place $l_3$.
2. Let us discuss the following situation: a needle is pricking a person. It keeps its form or it breaks (the characteristic of the $\alpha$-token), while the person feels pain and/or has a haemorrhage (the characteristic of the $\beta$-token). We can take into account the status of the receptors as a result of this incident.

In a more complex model we can assume that the capacities of places $l_2$ and $l_3$ are more than 1, e.g., the natural numbers $c_2$ and $c_3$, and in this case we can trace a lot of the external effects and the internal reactions that can flow simultaneously.

$$Z_2 = \langle \{l_2, l_3, l_8, l_{11}\}, \{l_4, l_5, l_6, l_7\},$$

$$\begin{array}{c|cccc} & l_4 & l_5 & l_6 & l_7 \\ \hline l_2 & W_{2,4} & false & false & false \\ l_3 & false & false & false & W_{3,7} \\ l_8 & false & W_{8,5} & false & false \\ l_{11} & false & false & W_{11,6} & false \end{array} \rangle,$$

where

$W_{2,4}$ "the action of the external effect is over",

$W_{3,7}$ "the action of the object of control related to the current external effect, is over",

$W_{8,5}$ "it is necessary to obtain additional information about the object of control (its way of functioning, the reaction to the external effects, etc.) with respect to the current $\varepsilon$-token characteristic",

$W_{11,6}$ "it is necessary to use a tool for control with respect to the current $\varepsilon$-token characteristic"

The $\alpha$-token obtains the characteristic

*"the final expression of the form, structure, etc. of the current external effect, after finishing its contact with the object of control"*

in place $l_4$. The $\beta$-token's characteristic in place $l_7$ is shown above. The $\gamma$-token obtains the characteristic

*"information, obtained from the object of control, related to the current external effect"*

in place $l_5$. The $\delta$-token obtains the characteristic

*"information about the use of the tool for control"*

in place $l_6$.

For the two examples above we have now the following interpretations.

1. In the case of flexible manufacturing systems, we can obtain the necessary information about the processed details and about the corresponding machines by suitable pickups/sensors (this information is represented as a current characteristic of a $\gamma$-token) and we can trace the status of the machine (which is represented as a current characteristic of a $\delta$-token).
2. In the case of a living organism, the $\gamma$- and $\delta$-tokens obtain information about the status of the sensors (as synonym of *sense organs, sensorium/sensory apparatus)* and about the status of the organs and systems that react to the external effects. For example, a person eats some food and his perceptions of it (in terms of taste, flavour, temperature, etc.) are interpreted as characteristics of the $\gamma$-tokens, and the information about the status of the teeth, tongue, oesophagus, etc. are interpreted as characteristics of the $\delta$-tokens.

$$Z_3 = \langle \{l_5, l_{12}\}, \{l_8, l_9\}, \begin{array}{c|cc} & l_8 & l_9 \\ \hline l_5 & false & W_{5,9} \\ l_{12} & W_{12,8} & false \end{array} \rangle,$$

where

$W_{5,9}$ = "the current $\gamma$-token's characteristic satisfies some criterion, determined preliminary"

(this criterion can be e.g.: $x^{\gamma}_{cu} \geq L_{\gamma}$, where $L_{\gamma}$ is a fixed number),

$W_{12,8}$ = "a new piece of information about the object of control is necessary".

The $\gamma$-tokens' characteristics in place $l_8$ is discussed above, while these tokens keep their previous characteristics or obtain as a characteristic some parts of their previous ones (e.g., the parts that satisfy the above mentioned criterion) changed in a suitable way in place $l_9$, for example: the nerve signal for pleasant/unpleasant perceptions, reaching the central nerve system of the person.

In the context of the examples above, we can say that transition $Z_3$ can be interpreted, respectively, as:

(1) an informational channel for information from the object of control (e.g., about the details of a given flexible manufacturing system), or
(2) neural paths for sending information from some receptors to the cerebrum.

In the first case we must have suitable technical tools (pickups/sensors, video-cameras, etc.), while in the second case nature has taken care for that.

$$Z_4 = \langle \{l_6, l_{16}\}, \{l_{10}, l_{11}\}, \begin{array}{c|cc} & l_{10} & l_{11} \\ \hline l_6 & W_{6,10} & false \\ l_{16} & false & W_{16,11} \end{array} \rangle,$$

where

$W_{6,10} =$ "the current $\delta$-token's characteristic satisfies some criterion, determined preliminary"

(this criterion can be e.g.: $x^{\gamma}_{cu} \geq L_{\delta}$, where $L_{\delta}$ is a fixed number),

$W_{16,11} =$ "new information about the status of the tools for control is necessary".

The $\delta$-tokens characteristic in place $l_{11}$ is discussed above, while these tokens keep their previous characteristics or obtain as a characteristic some parts of their previous ones (e.g., these parts that satisfy the above mentioned criterion) changed in a suitable way in place $l_{10}$, for example: the nerve signal for pain in the tooth as a result of hot soup, going to the central nerve system of the person.

Transition $Z_4$ has the following interpretations within the frameworks of the two examples above:

(1) a system for detail processing,
(2) tools for reaction to external effects.

For example, if one's hand touches a sharp piece, the information about this reaches the cerebrum and afterwards it reaches by means of a command the corresponding muscles that pull the hand away from the dangerous place. Similar is the situation with the reaction set up by the information that either a given detail is already ready and must be replaced on the production line, or that it has broken and must be rejected, or that there is a failure of the processing machine. After any of these pieces of information, there is a reaction of the control system which leads to correction of a current behaviour of the controlled object.

$$Z_5 = \langle \{l_9, l_{10}, l_{16}, l_{18}, l_{23}\}, \{l_{12}, l_{13}, l_{14}, l_{15}, l_{16}\},$$

$$\begin{array}{c|ccccc} & l_{12} & l_{13} & l_{14} & l_{15} & l_{16} \\ \hline l_9 & true & false & false & false & true \\ l_{10} & false & false & false & true & true \\ l_{16} & false & false & false & false & true \\ l_{18} & false & W_{18,13} & false & false & true \\ l_{23} & false & false & false & W_{23,15} & true \end{array} \rangle,$$

where

$W_{18,13} = W_{23,15} =$ "there is a necessity for information about the control system".

The $\gamma$- and $\delta$-tokens do not obtain any characteristic in places $l_{12}$ and $l_{15}$, respectively, the $\varepsilon$-token characteristic was discussed above. The $\eta$-tokens obtain the characteristic

$$\begin{cases} \text{“}\textit{information about the control system,} & \text{if its previous} \\ \textit{which is necessary for the learning} & \text{characteristic is empty} \\ \textit{and self}-\text{organized system}'', & \\ & \\ \text{“}*'', & \text{otherwise} \end{cases},$$

in place $l_{13}$. The $\lambda$-tokens obtain the characteristic

$$\begin{cases} \text{“}\textit{information about the control system,} & \text{if its previous} \\ \textit{which is necessary for the self}-\text{learning} & \text{characteristic is empty} \\ \textit{and self}-\text{organizing system}'', & \\ & \\ \text{“}*'', & \text{otherwise} \end{cases}$$

in place $l_{14}$.

Transition $Z_5$ has a central role in our GN-model. In the example with the flexible manufacturing system it corresponds to the control system that realizes the process of decision making related to the detail processing. This decision making can be realized:

- only by a computer (for example, if a suitable expert system exists),
- only by a dispatcher (operator), or
- both by a dispatcher aided by a computer (in dialogue).

Within the framework of the example of a human body, transition $Z_5$ can be interpreted as a representative of the control functions of the cerebrum.

The $\varepsilon$-token, as well as the $\iota$-, $\lambda$-, $\mu$-, $o$- and $\rho$-tokens plays the role of local memory where the information related to the respective transition is stored. Together with these tokens the $\mu$-token has the most special place, because it represents the global memory of the process, described by the GN-model.

The next transitions also have analogues in human activity as well as in manufacturing processes, but they are obvious enough and special examples are hardly necessary for them.

$$Z_6 = \langle \{l_{13}, l_{17}, l_{21}, l_{25}\}, \{l_{18}, l_{19}, l_{20}, l_{21}\},$$

$$\begin{array}{c|cccc} & l_{18} & l_{19} & l_{20} & l_{21} \\ \hline l_{13} & W_{13,18} & false & false & true \\ l_{17} & false & true & false & true \\ l_{21} & false & false & false & true \\ l_{25} & false & false & W_{25,20} & true \end{array} \rangle,$$

where

$W_{13,18} = W_{25,20} =$ "there is a necessity for information about the learning system".

The $\eta$- and $\theta$-tokens obtain the characteristics

$$\begin{cases} \text{"}\textit{information about the learning system,} & \text{if its previous} \\ \textit{which is necessary for the control} & \text{characteristic is empty} \\ \textit{system}\text{"}, & \\ \text{"}*\text{"}, & \text{otherwise} \end{cases}$$

in place $l_{18}$, and

$$\begin{cases} \text{"}\textit{information about the learning system,} & \text{if its previous} \\ \textit{which is necessary for the information} & \text{characteristic is empty} \\ \textit{system of the control system}\text{"}, & \\ \text{"}*\text{"}, & \text{otherwise} \end{cases}$$

in place $l_{20}$, respectively. The $\iota$- and $\zeta$-token characteristics were discussed above.

This transition represents the process of learning and control, when the control system or a person is being trained by a teacher. On the other hand, the next transition represents the process of self-learning and self-organization. They are realized on the basis of knowledge already collected during the time of the functioning of the GN-model.

We must note that the self-organization does not coincide with the self-modifications that some GN can realize, using suitable operators (see Appendix 1). The self-learning and self-organizations do not influence the structure of the present GN-model and/or the way of its tokens transfers.

$$Z_7 = \langle \{l_{14}, l_{24}, l_{28}\}, \{l_{22}, l_{23}, l_{24}\}, \begin{array}{c|ccc} & l_{22} & l_{23} & l_{24} \\ \hline l_{14} & false & W_{14,23} & true \\ l_{24} & false & false & true \\ l_{28} & W_{28,22} & false & true \end{array} \rangle,$$

where

$W_{14,23} = W_{28,22} =$ "there is a necessity for information about the self-learning and self-organizing system".

The $\lambda$- and $\kappa$-tokens obtain the characteristics

$$\begin{cases} \text{"}\textit{information about the self}-\text{learning} & \text{if its previous} \\ \textit{and self}-\text{organizing system, which is} & \text{characteristic is empty} \\ \textit{necessary for the control system}\text{"}, & \\ \text{"}*\text{"}, & \text{otherwise} \end{cases}$$

in place $l_{22}$, and

$$\begin{cases} \text{"}\textit{information about the learning system, which is necessary for the information system of the control system}\text{"}, & \text{if its previous characteristic is empty} \\ \text{"}*\text{"}, & \text{otherwise} \end{cases}$$

in place $l_{23}$, respectively.

The $\mu$-token characteristics were discussed above.

As we noted above, the role of the $\mu$-token is more special. It circulates only in place $l_{24}$, obtaining as a characteristic all the information that enters in the form of tokens characteristics in transition $Z_8$, which corresponds to the information system of the control system. If transition $Z_5$ represents the cerebrum, then transition $Z_8$, in a combination with the two previous and the two next transitions, corresponds to the high nerve, particularly human, activity. All data, generated in each of these five transitions, is collected by the $\mu$-token when these data are a correct characteristic of the $\eta$-, $\kappa$-, $\nu$- or $\pi$-token, entering a corresponding place of transition $Z_8$.

$$Z_8 = \langle \{l_{20}, l_{22}, l_{29}, l_{30}, l_{34}\}, \{l_{25}, l_{26}, l_{27}, l_{28}, l_{29}\},$$

$$\begin{array}{c|ccccc} & l_{25} & l_{26} & l_{27} & l_{28} & l_{29} \\ \hline l_{20} & W_{20,25} & false & false & false & true \\ l_{22} & false & false & false & W_{22,28} & true \\ l_{29} & false & false & false & false & true \\ l_{30} & false & W_{30,26} & false & false & true \\ l_{34} & false & false & W_{34,27} & false & true \end{array} \rangle,$$

where

$W_{20,25} = W_{22,28} = W_{30,26} = W_{34,27} =$ "there is a necessity for information about the control system".

The $\theta$-, $\kappa$-, $\xi$-, and $o$-tokens obtain the characteristics

$$\begin{cases} \text{"}\textit{information about the information system of the control system, which is necessary for the learning system}\text{"}, & \text{if its previous characteristic is empty} \\ \text{"}*\text{"}, & \text{otherwise} \end{cases}$$

in places $l_{25}$,

$$\begin{cases} \text{"}\textit{information about the information system of the control system, which is necessary for the analysis, clusterisation and classification tools}\text{"}, & \text{if its previous characteristic is empty} \\ \text{"}*\text{"}, & \text{otherwise} \end{cases}$$

in places $l_{26}$,

$$\begin{cases} \text{“\textit{information about the information system of the control system, which is necessary for the tools for formalization, decision making and prognostics}''}, & \text{if its previous characteristic is empty} \\ \text{“}*\text{''}, & \text{otherwise} \end{cases}$$

in places $l_{27}$,

$$\begin{cases} \text{“\textit{information about the information system of the control system, which is necessary for the self}–learning and \textit{self}–organizing system''}, & \text{if its previous characteristic is empty} \\ \text{“}*\text{''}, & \text{otherwise} \end{cases}$$

in places $l_{28}$, respectively, as the $\nu$-token's characteristic was discussed above.

$$Z_9 = \langle \{l_{26}, l_{32}, l_{33}\}, \{l_{30}, l_{31}, l_{32}\}, \begin{array}{c|ccc} & l_{30} & l_{31} & l_{32} \\ \hline l_{26} & W_{26,30} & false & true \\ l_{32} & false & false & true \\ l_{33} & false & W_{33,31} & true \end{array} \rangle,$$

where

$W_{26,30} = W_{33,31} =$ “there is a necessity for information about the tools for analysis, clusterisation and classification”.

The $\xi$- and $\pi$-tokens obtain the characteristics

$$\begin{cases} \text{“\textit{information about the analysis, clusterisation and classification tools, which is necessary for the information system of the control system}''}, & \text{if its previous characteristic is empty} \\ \text{“}*\text{''}, & \text{otherwise} \end{cases}$$

in place $l_{30}$ and

$$\begin{cases} \text{“\textit{information about the information system of the control system, which is necessary for the tools for formalization, decision making and prognostics}''}, & \text{if its previous characteristic is empty} \\ \text{“}*\text{''}, & \text{otherwise} \end{cases}$$

in place $l_{31}$, respectively. The $\rho$-token's characteristic was discussed above.

$$Z_{10} = \langle \{l_{27}, l_{31}, l_{35}\}, \{l_{33}, l_{34}, l_{35}\}, \begin{array}{c|ccc} & l_{33} & l_{34} & l_{35} \\ \hline l_{27} & false & W_{27,34} & true \\ l_{31} & W_{31,33} & false & true \\ l_{35} & false & false & true \end{array} \rangle,$$

where

$W_{27,34} = W_{31,33} =$ "there is a necessity for information about the tools for formalization, decision making and prognostics".

The $\pi$- and $o$-tokens obtain the characteristics

$$\begin{cases} \text{"}\textit{information about the tools for formalization, decision making and prognostics which is necessary for the analysis, clusterisation and classification tools}\text{"}, & \text{if its previous characteristic is empty} \\ \text{"}*\text{"}, & \text{otherwise} \end{cases}$$

in place $l_{33}$ and

$$\begin{cases} \text{"}\textit{information about the tools for formalization, decision making and prognostics which is necessary for the information system of the control system}\text{"}, & \text{if its previous characteristic is empty} \\ \text{"}*\text{"}, & \text{otherwise} \end{cases}$$

in place $l_{34}$, respectively. The $\sigma$-token's characteristic was discussed above.

Finally, we expand a little more on the GN-model. In its present form it shows in general lines the ways of functioning as a system of the human body or a flexible manufacturing system, using elements of AI. We can elaborate in details on the separate components of the GN-model. One of the most interesting directions of detailization is related to the process of memorizing. It must be represented by a new GN being a sub-net of the present net. What can be represented in its framework is not only the fact that we (people) remember information, and when necessary we can reproduce it, but also the fact that we forget information. If (when) we construct such a GN-model, we shall describe processes essentially nearer to real human thinking. In the case of the manufacturing process controlled by a computer (the computer does not "forget"!), this process will correspond to the situation when information is destroyed or lost intentionally or non-intentionally from the computer memory (e.g., as a result of acasual failure). In a detailed GN-model of the processes in the memory, these reasons can be represented, too, and the conditions when they appear can be simulated and studied.

We should also note that this model can be based on intuitionistic fuzzy (see [30]) estimations of the degrees of memorizing and forgetting. Thus this way we shall describe in more details such situations for which we know a given things in certain boundaries. We can also represent the situation when our knowledge of some situations is based on other (contextual) relations. For example, we can interpret some information only in relation with other information which we have from the past. The former or the latter information enters and we can recognize the moment when it is obtained as a result of decision making or planning activity.

Now, we shall study the possibility for machine learning of the GN-model described above. There are at least two different ways of realizing so. Let us call them, respectively, a short and an extended GN-model of learning.

When we study human behaviour, we can use some external effects on one's sensors to record one's reactions and/or we can submit to one some information, related to extending one's knowledge or give one some orders to follow to correct one's behaviour. But we do not know one's interior activities, i.e., the solutions that one obtains on the basis of our (external) effects. Therefore, the object of our studying for us is a kind of a "black box".

The process of learning of the process can be represented by the GN from Fig. 11, where $E$ is the GN from Fig. 10 (now it is a sub-net) and transition $Z_0$ has the form:

$$Z_0 = \langle \{l_a, l_c, l_4, l_{19}\}, \{l_1, l_6, l_7, l_c\}, \begin{array}{c|cccc} & l_1 & l_{17} & l_b & l_c \\ \hline l_a & W_{a,1} & W_{a,17} & false & false \\ l_4 & false & false & true & false \\ l_{19} & false & false & true & false \\ l_c & false & false & false & true \end{array} \rangle,$$

where

$W_{a,1}$ "the token is from $\alpha$-type";
$W_{a,17}$ "the token is from $\zeta$-type".

Tokens from $\alpha$- or $\zeta$-types enter place $l_a$ with their initial characteristics, discussed above. These tokens do not obtain next characteristics in places $l_1$, $l_{17}$, and $l_b$.

Token $\tau$ stays only in place $l_c$ obtaining as a current characteristic the information, obtained from sub-net $E$.

Essentially more complex is the possibility in which the modelled process is not a "black box". In this case, the separate information components of the net (represented by places $l_{16}$, $l_{21}$, $l_{24}$, $l_{29}$, $l_{32}$, and $l_{35}$) can send it to the additional part of the net, which interprets the learning process of the constructed model. We can change the net from Fig. 10 to a new net. To each of places $l_{16}$, $l_{21}$, $l_{24}$, $l_{29}$, $l_{32}$, $l_{35}$ in

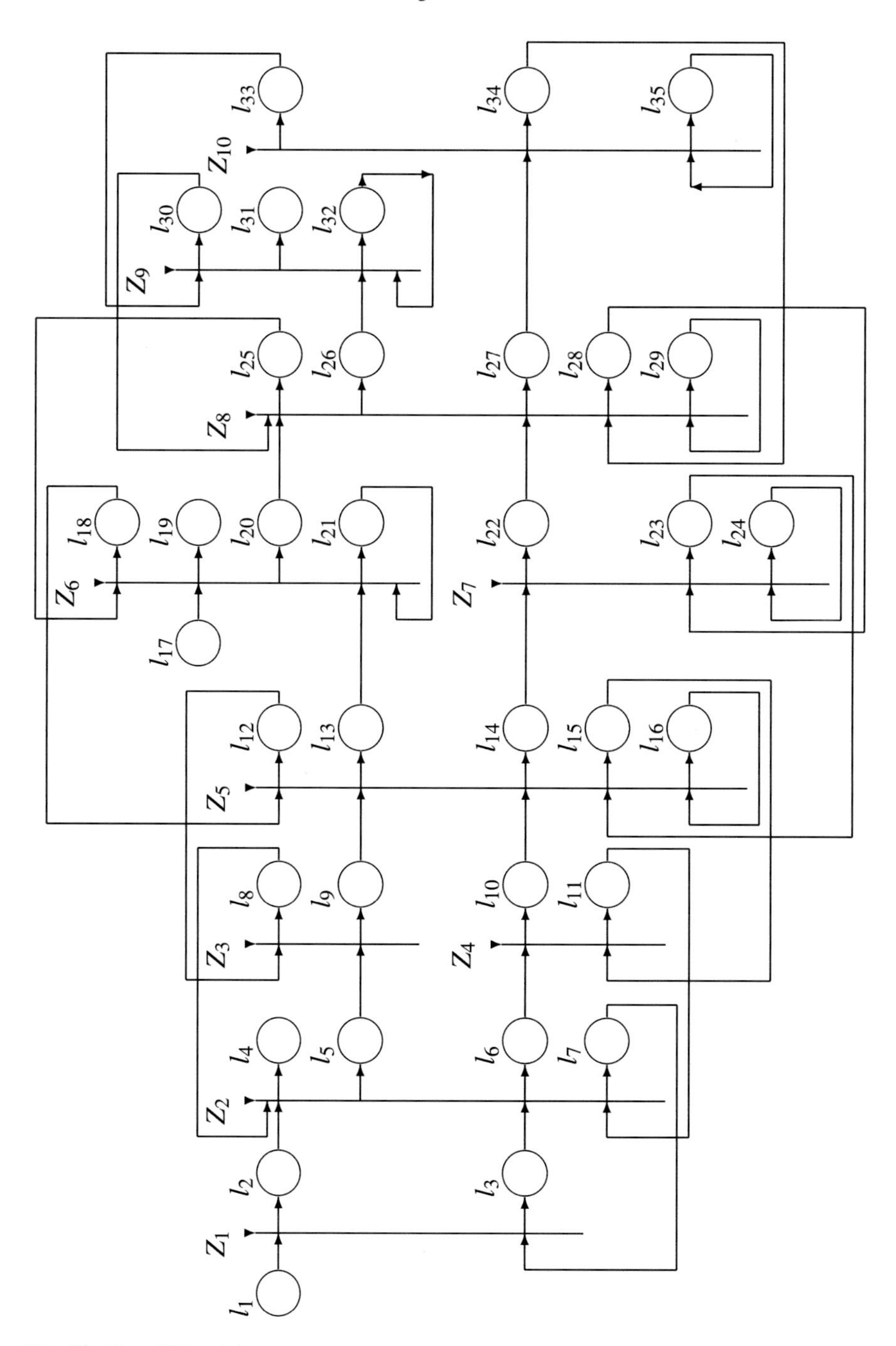


**Fig. 10** First GN-model

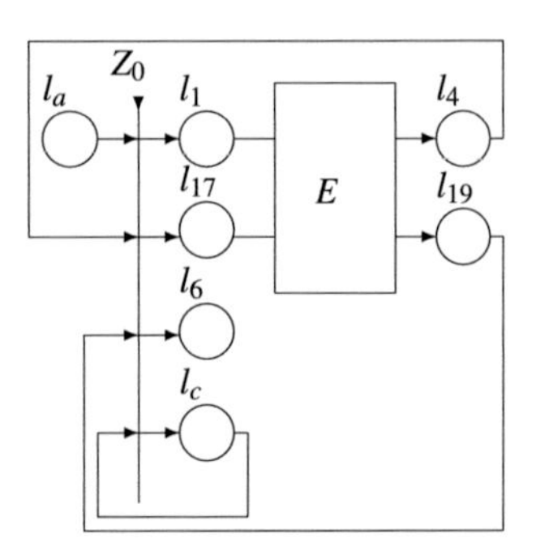


**Fig. 11** Second GN-model (first form)

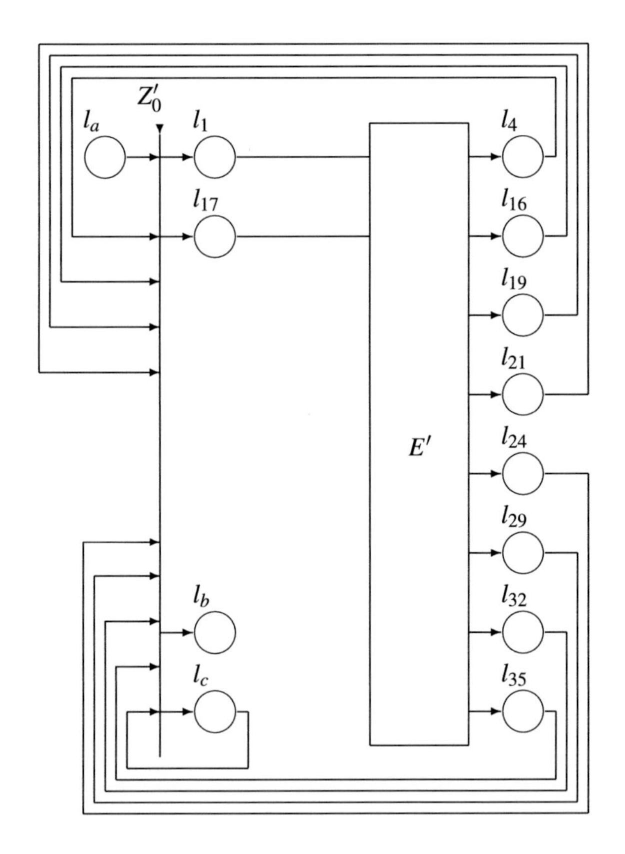


**Fig. 12** Second GN-model (second form)

the new net an additional place $l'_{16}$, $l'_{21}$, $l'_{24}$, $l'_{29}$, $l'_{32}$, $l'_{35}$, respectively, is added, the last ones being outputs for the GN. Now, we can construct the GN from Fig. 12, similar to Fig. 11, where $E'$ is the new form of the sub-net.

Now the GN contains the additional transition $Z_0'$ with the form:

$$Z_0' = \langle \{l_a, l_4, l_{16}', l_{19}, l_{21}', l_{24}', l_{29}', l_{32}', l_{35}', l_c\}, \{l_1, l_{17}, l_b, l_c\},$$

$$\begin{array}{c|cccc} & l_1 & l_{17} & l_b & l_c \\ \hline l_a & W_{a,1} & W_{a,17} & false & false \\ l_4 & false & false & true & false \\ l_{16} & false & false & true & false \\ l_{19} & false & false & true & false \\ l_{21} & false & false & true & false \\ l_{24} & false & false & true & false \\ l_{29} & false & false & true & false \\ l_{32} & false & false & true & false \\ l_{35} & false & false & true & false \\ l_c & false & false & false & true \end{array} \rangle,$$

where predicates $W_{a,1}$ and $W_{a,17}$ have the above forms.

Tokens from $\alpha$-, $\varepsilon$-, $\zeta$-, $\iota$-, $\mu$-, $\nu$-, $\rho$-, $\sigma$-, and $\tau$-types have the discussed above characteristics.

## 8 Conclusion

Here for the first time we have launched the idea for execution of one or more DM-tools in the frames of a given GN, and GN models of DM-tools added to it. This idea will be a basis of new GN-applications, related to modelling, simulation, optimization and control of real processes.

## References


1. Aladjov, H., Atanassov, K., Shannon, A.: Generalized net model of temporal learning algorithm for artificial neural networks. In: Proceedings of First International IEEE Conference on Intelligent Systems, Vol. 1, pp. 190–193 (2002)
2. Alexieva, J., Choy, E., Koycheva, E.: Review and bibloigraphy on generalized nets theory and applications. In: Choy, E., Krawczak, M., Shannon, A., Szmidt, E. (eds.) A Survey of Generalized Nets , Raffles KvB Monograph No. 10, pp. 207–301 (2007)
3. Andonov, V., Atanassov, K.: Generalized nets with characteristics of the places. Comptes Rendus de l'Academie bulgare des Sciences, Tome **66**(12), 1673–1680 (2013)
4. Andreev, S., Atanassov, K., Sotirov, S.: Generalized net model of a social network with intuitionistic fuzzy estimation. Notes on Intuitionistic Fuzzy Sets **20**(3), 72–83 (2014)
5. Angelov, P.: Autonomous Learning Systems. John Wiley and Sons, Chichester (2013)
6. Angelov, P., Filev, D., Kasabov, N.: Evolving Intelligent Systems. John Wiley and Sons, Hoboken (2010)
7. Antonov, A.: Presentation of neuron by generalized net, issues in the representation and processing of uncertain imprecise information: fuzzy sets, intuitionistic fuzzy sets, generalized nets, and related topics In: Akademicka Oficyna Wydawnictwo EXIT, Warsaw, pp. 1–10 (2005)

8. Antonov, A., Hadjitodorov, S.: Concurrent algorithm for learning of neural networks. In: IEEE 6th International Conference Intelligent Systems, pp. 225–228 (2012)
9. Atanassov, K.: Introduction in Generalized Nets Theory. Pontica-Print, Bourgas (1992). (in Bulgarian)
10. Atanassov, K.: Algebraic aspect of E–nets. In: Proceedings of International Symposium on the Automation of Science Research, pp. 143–148. Varna (1981) (in Russian)
11. Atanassov, K.: The generalized nets and the other graphical means for modelling, AMSE Rev. **2**(1), 59–64 (1985)
12. Atanassov, K.: Intuitionistic fuzzy sets. Fuzzy sets Syst. **20**(1), 87–96 (1986)
13. Atanassov, K.: Generalized nets and the classes of reduced generalized nets resulting from them. AMSE Rev. **3**(4), 1–6 (1986)
14. Atanassov, K.: Generalized index matrices. Compt. Rend. de l'A-ca-de-mie Bulgare des Sci. **40**(11), 15–18 (1987)
15. Atanassov, K.: The generalized E–nets—predecessors of the generalized nets. AMSE Rev. **5** (3), 5–9 (1987)
16. Atanassov, K.: The generalized net which represents all Petri nets. AMSE Rev. **12**(3), 33–37 (1990)
17. Atanassov, K.: A generalized net, representing the travelling salesman problem. AMSE Rev. **14**(4), 61–64 (1990)
18. Atanassov, K.: Generalized Nets. World Scientific, New Jersey (1991)
19. Atanassov, K.: Generalized nets and extensions of the travelling salesman problem. AMSE Rev, **21**(2), 16–26 (1992)
20. Atanassov, K. Generalized nets and some travelling salesman problems. In: Atanassov, K. (ed.) Applications of Generalized Nets, pp. 68–81. World Scientific, Singapore (1993)
21. Atanassov, K.: Remark on intuitionistic fuzzy expert systems. BUSEFAL **59**, 71–76 (1994)
22. Atanassov, K.: Generalized Nets and Systems Theory. "Prof. M. Drinov". Academic Publishing House, Sofia (1997)
23. Atanassov, K.: Generalized nets and artificial intelligence. Adv Model. Anal. B **37**(1–2), 37–51 (1997)
24. Atanassov, K.: Generalized Nets in Artificial Intelligence. Vol. 1: Generalized nets and Expert Systems, "Prof. M. Drinov". Academic Publishing House, Sofia (1998)
25. Atanassov, K.: Intuitionistic Fuzzy Sets. Springer, Heidelberg (1999)
26. Atanassov, K.: Generalized net models of special abstract processes. In: Proceedings of the Conference "Bioprocess systems' 2000", pp. I.4–I.10, Sofia, 11–13 Sept 2000
27. Atanassov, K.: Generalized nets as tools for modelling in the area of the artificial intelligence. Adv Stud Contemp Math. **3**(1), 21–42 ( 2001)
28. Atanassov, K.: Generalized nets as tools for modelling, optimization and simulation in the area of the artificial intelligence. In: Atanassov, K., Hryniewicz, O., Kacprzyk, J. (eds.) Soft Computing Foundations and Theoretical Aspects, pp. 19–51. Academicka Oficyna Wydawnicza EXIT, Warszawa (2004)
29. Atanassov, K.: On Generalized Nets Theory, "Prof. M. Drinov". Academic Publishing House, Sofia (2007)
30. Atanassov, K.: On Intuitionistic Fuzzy Sets Theory. Springer, Berlin (2012)
31. Atanassov, K.: Index Matrices: towards an Augmented Matrix Calculus. Springer, Cham (2014)
32. Atanassov, K., Aladjov, H.: Generalized Nets in Artificial Intelligence. Vol. 2: Generalized nets and Machine Learning. "Prof. M. Drinov". Academic Publishing House, Sofia (2000)
33. Atanassov, K., Chountas, P., Kolev, B., Sotirova, E.: Generalized net model of an expert system with temporal components. Adv. Stud. Contemp. Math **12**(2), 255–289 (2006)
34. Atanassov, K., Chountas, P., Kolev, B., Sotirova, E.: Generalized net model of a self–developing expert system In: Proceedings of the 10th International Conference on Intuitionistic Fuzzy Sets, pp. 35–40, Sofia, 28–29 Oct 2006

35. Atanassov, K., Daskalov, M. Georgiev, P., Kim, S., Kim, Y., Nikolov, N., Shannon, A., Sorsich, J.: Generalized Nets in Neurology "Prof. M. Drinov". Sofia, Academic Publishing House (1997)
36. Atanassov, K., Dimitrov, E.: On the representation of M–nets by generalized nets In: Proceedings of the XIV Spring Conference of the Union of Bulg. Math., Sunny Beach, pp. 317–322 (1985)
37. Atanassov, K., Dimitrov, E.: Theorem for representation of the Self–modification nets by generalized nets. AMSE Rev. **5**(3), 1–4 (1987)
38. Atanassov, K., Dimitrov, E.: Theorem for representation of the generalized modification of Petri nets by generalized nets. AMSE Rev. **5**(1), 8–13 (1987)
39. Atanassov, K., Gluhchev, G.: Generalized net model of the automatic natural language translation. In: Proceedings of the Eighth International Workshop on Generalized Nets, Sofia, pp. 32–37, 26 June 2007
40. Atanassov, K., Gluhchev, G., Hadjitodorov, S., Shannon, A., Vassilev, V.: Generalized Nets and Pattern Recognition. KvB Visual Concepts Pty Ltd, Monograph No. 6, Sydney (2003)
41. Atanassov, K., Gluhchev, G., Hadjitodorov, S., Kacprzyk, J., Shannon, A., Szmidt, E., Vassilev, V.: Generalized Nets Decision Making and Pattern Recognition. Warsaw School of Information Technology, Warszawa (2006)
42. Atanassov, K., Hadjiski, M.: Generalized nets and intelligent systems. Int. J. Gen Syst. **39**(5), 457–470 (2010)
43. Atanassov, K., Krawczak, M., Sotirov, S.: Generalized net model for parallel optimization of feed–forward neural network with variable learning rate backpropagation algorithm, Advanced Intelligent systems from theory to practice. Springer, pp. 361–372 (2010)
44. Atanassov, K., Mavrov, D., Atanassova, V.: Intercriteria decision making: A new approach for multicriteria decision making, based on index matrices and intuitionistic fuzzy sets. Issues in Intuitionistic Fuzzy Sets and Generalized Nets **11**, 1–8 (2014)
45. Atanassov, K., Orozova, D., Sotirova, E., Chountas, P., Tasseva, V.: Generalized net model of expert system validity testing process. In: Proceedings of the international conference of BFU, pp. 165–173 (2007)
46. Atanassov, K., Pasi, G., Yager, R.: Intuitionistic fuzzy interpretations of multi–criteria multi–person and multi–measurement tool decision making. Int. J. Syst. Sci. **36**(14), 859–868 (2005)
47. Atanassov, K., Pencheva, T.: Generalized net model of simple genetic algorithm modifications. Issues in Intuitionistic Fuzzy Sets and Generalized Nets **10**, 97–106 (2013)
48. Atanassov, K., Peneva, D., Taseva, V., Sotirova, E., Orozova, D.: Generalized net model of expert systems with frame—Type Data Base with intuitionistic fuzzy estimations. In: First International Workshop on Intuitionistic fuzzy sets, Generalized nets and Knowledge Engineering, University of Westminster, London, UK, pp. 111–116, 2–6 September (2006)
49. Atanassov, K., Sotirova, E., Orozova, D.: Generalized net model of expert systems with frame-type data base. Jangjeon Math. Soc. **9**(1), 91–101 (2006)
50. Atanassov, K., Sotirov, S.: Optimization of a neural network of self–organizing maps type with time–limits by a generalized net. Adv. Stud. Contemp Math. **13**(2), 213–220 (2006)
51. Atanassov, K., Sotirov, S., Antonov, A.: Generalized net model for parallel optimization of feed–forward neural network. Adv. Stud. Contemp. Math. **15**(1), 109–119 (2007)
52. Atanassov, K., Sotirov, S., Shannon, A.: Generalized net model of the hierarchical neural networks. In: Proceedings of the 13th International Workshop on Generalized Nets, pp. 8–14, London, UK, 29 Oct 2012
53. Atanassova, V.: Generalized nets with volumetric tokens. Compt. rend. Acad. bulg. des Sci. **65**(11), 1489–1498 (2012)
54. Atanassova, V., Atanassov, K.: Ant colony optimization approach to tokens movement within generalized nets. Lecture Notes in Computer Science, vol. 6046, pp. 240–247. Springer, Berlin (2011)

55. Atanassova, V., Doukovska, L., Atanassov, K., Mavrov, D.: InterCriteria Decision Making approach to EU Member States Competitive Analysis. In: Proceedings of 4th International Symposium on Business Modeling and Software Design, Luxembourg, Grand Duchy of Luxembourg, pp. 289–294, 24–26 June 2014
56. Atanassova, V., Fidanova, S., Chountas, P., Atanassov, K.: A generalized net with an ACO–algorithm optimization component. Lect. Notes Comput. Sci. **7116**, 190–197 (2012)
57. Atanassova, V., Mavrov, D., Doukovska, L., Atanassov, K.: Discussion on the threshold values in the intercriteria decision making approach. Notes on Intuitionistic Fuzzy Sets, **20** (2), 94–99 (2014)
58. Berti–Equille, L.: Measuring and modelling data quality for quality–awareness in data mining. In: Guillet, F., Hamilton, H. (eds.) Quality Measures in Data Mining, pp. 101–126. Springer, Berlin (2007)
59. Bramer, M.: Principles of Data Mining. Springer, London (2013)
60. Bull, L., Ester, B.–M., Holmes, J.: Learning classifier systems in data mining: An introduction, In: Bull, L., Ester, B.–M., Holmes, J. (eds.) Learning Classifier Systems in Data Mining, pp. 1–15. Springer, Berlin (2008)
61. Bureva, V., Sotirova, E., Atanassov, K.: Hierarchical generalized net model of the process of clustering. Issues in Intuitionistic Fuzzy Sets and Generalized Nets **11**, 73–80 (2014)
62. Bureva, V., Sotirova, E., Atanassov, K.: Hierarchical generalized net model of the process of selecting a method for clustering. In: Proceedings of the 15th International Workshop on Generalized Nets, pp. 39–48 (2014)
63. Bureva, V., Chountas, P., Atanassov, K.: A generalized net model of the process of decision tree construction. In: Proceedings of the 13th International Workshop on Generalized Nets London, pp. 1–729 (2012)
64. Bureva, V.: Generlized net model of the creating of associative rules, annual of informatics section. Union Scientists Bul. **5**, 73–83 (2012). (in Bulgarian)
65. Bureva, V., Sotirova, E.: Generalized net of the process of association rules discovery by Eclat algorithm using weather databases. In: Proceedings of the 14th International Workshop on Generalized Nets, IWGN2013, Burgas, pp. 1–10, 29–30 November 2013
66. Bureva, V., Sotirova, E., Chountas, P.: Generalized Net of the Process of Sequential Pattern Mining by Generalized Sequential Pattern Algorithm (GSP), Intelligent Systems'2014, pp. 831–838. Springer, Cham (2015)
67. Büyüközkan, G., Feyaioglu, O.: Accelerating the new product introduction with intelligent data mining, In: Ruan, D., Chen, G., Kerre, E., Wets, G. (eds.). Intelligent Data Mining: Techniques and Applications, pp. 337–354. Springer, Berlin (2005)
68. Chountas, P., Kolev, B., Tasseva, V., Atanassov, K.: Generalized net model for binary operations over intuitionistic fuzzy OLAP cubes. In: Proceedings of the Eighth International Workshop on Generalized Nets, Sofia, pp. 66–72, 26 June 2007
69. Chountas, P., Kolev, B., Rogova, E., Tasseva, V., Atanassov, K.: Generalized Nets in Artificial Intelligence. Vol. 4: Generalized nets, Uncertain Data and Knowledge Engineering. "Prof. M. Drinov" Academic Publishing House, Sofia (2007)
70. Chountas, P., Rogova, E., Atanassov, K.: The notion of H-IFS: An approach for enhancing the OLAP capabilities in oracle10g. Int. J. Intell. Syst. **26**(3), 262–283 (2011)
71. Chountas, P., Sotirova, E., Kolev, B., Atanassov, K.: On intuitinistic fuzzy expert system with temporal parameters, using intuitionistic fuzzy estimations, Computational Intelligence, Theory and Applications, pp. 241–249. Springer, Berlin (2006)
72. Cios, K.J., Pedrycz, W., Swiniarski, R.: Data Mining Methods for Knowledge Discovery, Kluwer (1998)
73. Cios, K., Pedrycz, W., Swiniarski, R., Kurgan, L.: Data Mining. A Knowledge Discovery Approach, Springer, New York (2007)

74. Cohen, P., Cheyer, A., Wang, M., Baeg, S. C.: OAA: an Open Agent Architecture, AAAI Spring Symposium (1994)
75. Cox, E.: Fuzzy Modeling and Genetic Algorithms for Data Mining and Exploration. Elsevier, Amsterdam (2005)
76. Crespi–Reghizzi, S., Mandrioli, D.: Some algebraic properties of petri nets, Alma Erequeza, **XLV**(2) 1976
77. Dahan, H., Cohen, Sh, Rokach, L., Maimon, O.: Proactive Data Mining with Decision Trees. Springer, New York (2014)
78. Data mining: foundations and intelligent paradigms, vol. 2: statistical, bayesian, time series and other theoretical aspects. In: Holmes, D., Jain, L.(eds.), Springer, Berlin (2012)
79. Data mining: foundations and intelligent paradigms, vol. 3: medical, health, social, biological and other applications. In: Holmes, D., Jain, L.(eds.), Springer, Berlin (2012)
80. Dimitrov, E., Atanassov, K.: Theorem for representation of M–nets by generalized nets. AMSE Rev. **6**, 5–12 (1988)
81. Dimitrova, M., Vasilev, K., Sotirov, S.: Generalized net model of the process of the prognosis biomass accumulation with neural network, developments in fuzzy sets, intuitionistic fuzzy sets, generalized nets and related topics. vol II: Applications, System Research Institute, Polish Academy of Science. Warsaw, pp. 79–90 (2010)
82. Dorigo, M.: Optimization, Learning and Natural Algorithms. Ph.D. thesis. Politecnico di Milano, Italy (1992) (in Italian)
83. Dorigo, M., Stutzle, T.: Ant Colony Optimization. MIT Press (2004)
84. Etzion, T., Yoeli, M.: Super–nets and their hierarchy. Theor. Comput. Sci. **23**, 243–272 (1983)
85. Fidanova, S., Atanassov, K.: Generalized net models of the process of ant colony optimization with intuitionistic fuzzy estimations. In: Atanassov, K., Shannon, A. (eds.) Proceedings of the Ninth International Workshop on Generalized Nets, vol. 1, pp. 41–48, Sofia, 4 July 2008
86. Fidanova, S., Atanassov, K.: Generalized net models for the process of hybrid ant colony optimization. Comptes Rendus de l'Academie bulgare des Sciences, Tome **61**(12), 1535–1540 (2008)
87. Fidanova, S., Atanassov, K.: Generalized net models of the process of ant colony optimization. Issues in Intuitionistic Fuzzy Sets and Generalized Nets **7**, 108–114 (2008)
88. Fidanova, S., Atanassov, K.: Generalized net models and intuitionistic fuzzy estimations of the process of ant colony optimization. Issues in Intuitionistic Fuzzy Sets and Generalized Nets **8**, 109–124 (2010)
89. Fidanova, S., Atanassov, K.: Generalized nets as tools for modelling of the ant colony optimization algorithms. Lect. Notes Comput. Sci. **5910**, 326–333 (2010)
90. Fidanova, S., Atanassov, K., Marinov, P.: Generalized nets in artificial intelligence. vol. 5: generalized nets and ant colony optimization. "Prof. M. Drinov" Academic Publishing House, Sofia (2011)
91. Fidanova, S., Marinov, P., Atanassov, K.: Generalized net models of the process of ant colony optimization with different strategies and intuitionistic fuzzy estimations. In: Proceedings of the Jangjeon Mathematical Society, vol. 13(1), 1–12 (2010)
92. Fidanova, S., Marinov, P., Atanassov, K.: New evaluations of ant colony optimization start nodes. Control Cybern. **43**(3), 471–485 (2014)
93. Freitas, A.A.: A review of evolutionary algorithms for data mining. In: Maimon, O., Rokach, L. (eds.)Data Mining and Knowledge Discovery Handbook 2nd edn. Springer, New York, pp. 371–400 (2010)
94. Genrich, H.: Predicate/transition nets. Lect. Notes Comput. Sci. **254**, 207–247 (1986)
95. Genrich, H., Lautenbach, K.: The analysis of distributed systems by means of predicate/transition nets. Lect. Notes Comput. Sci. **70**, 123–146 (1979)

96. Gluhchev, G., Atanassov, K., Hadjitodorov, S., Shannon, A.: A generalized net model of the process of scene analysis. Cybern. Inf. Technol. **9**(1), 13–17 (2009)
97. Gluhchev, G., Atanassov, K., Vassilev, V.: A generalized net model of scene analysis process. Advances in Fuzzy Sets, Intuitionistic Fuzzy Sets, Generalized Nets and Related Topics. Vol. II: Applications, Academic Publishing House EXIT, Warszawa, pp. 63–68 (2008)
98. Gochev V., Atanassov, K.: Over generalized net reprersentation of one type of fuzzy Petri nets. Adv. Stud. Contemp. Math. **8**(1), 59–64 (2004)
99. Gochev, V., Atanassov, K., Chountas, P.: A generalized net representing the functioning and the results of the work of fuzzy Petri nets. In: Proceedings of the 10th ISPE International Conference on Concurrent Engineering "Advanced Design, Production and Management Systems", pp. 1009–1012, Madeira, 26–30 Jul 2003
100. Gocheva, P., Sotirov, S., Gochev, V.: Implementation of generalized nets models of feedforward neural networks. Issues in Intuitionistic Fuzzy Sets and Generalized Nets **10**, 125–135 (2013)
101. Granichin, O., Volkovich, Z., Toledano-Kitai, D.: Randomized Algorithms in Automatic Control and Data Mining. Springer, Berlin (2015)
102. Grigorova, G., Vassilev, K., Sotirov, S.: Generalized net model of the process of the prognosis biomass accumulation using TEMPO–amine metal complexes with neural network, new trend in Fuzzy Sets, intuitionistic Fuzzy Sets, generalized nets and related topics. Applications. Vol. II, System Research Institute, Polish Academy of Science, Warsaw, pp. 57–66 (2013)
103. Grosan, V., Abraham, A.: Intelligent Systems—A Modern Approach. Springer, Berlin (2011)
104. Grutzner, R.: Entwurtsbegleitende Verhaltensanalyse von Rechnersystemen auf der Basis von M-Netzen, Analyse und Synthese von Rechnersystemen, Problemseminar, Nassau. Teil **1**, 84–88 (1982)
105. Grzymala–Busse, J.W.: Rule induction. In: Maimon, O., Rokach, L. (eds.) Data Mining and Knowledge Discovery Handbook, 2nd edn, Springer, New York, 249–265 (2010)
106. Gyurov, P.: A generalized net representing the functioning of all coloured Petri nets and results of this functioning. Adv Model. Anal. A, AMSE Press, **26**(1), 1–9 (1995)
107. Haddadi, A.: Communication and Cooperation in Agent Systems. Springer, Berlin (1995)
108. Hadjyisky, L., Atanassov, K.: Theorem for representation of the neuronal networks by generalized nets. AMSE Rev. **12**(3), 47–54 (1990)
109. Hadjyisky, L., Atanassov, K.: A generalized net, representing the elements of one neuron network set. AMSE Rev. **14**(4), 55–59 (1990)
110. Hadjyisky, L., Atanassov, K.: Generalized nets representing the elements of neuron networks. In: Atanassov, K. (ed.) Applications of Generalized Nets, pp. 49–67. World Scientific, Singapore, (1993)
111. Hadjyisky, L., Atanassov, K.: Intuitionistic fuzzy model of a neural network. BUSEFAL **54**, 36–39 (1993)
112. Hadjyisky, L., Atanassov, K.: Generalized net model of the intuitionistic fuzzy neural networks. Adv. Model Anal. AMSE Press, **23**(2), 59–64 (1995)
113. Han, J., Kamber, M.: Data Mining: concepts and Techniques, Morgan Kaufmann (2006)
114. Hand, D., Mannila, H., Smyth, P.: Principles of Data Mining, MIT Press, Chap 1 Introduction 7 (2001)
115. Hastie, T., Tibshirani, R., Friedman, J.: The Elements of Statistical Learning—Data Mining, Inference and Prediction. Springer, New York (2001)
116. Hilderman, R., Peckham, T.: Statistical methodologies from mining potentially interesting contrast sets. In: Guillet, F., Hamilton, H. (eds.) Quality Measures in Data Mining, pp. 153–177. Springer, Berlin (2007)
117. Holmes, D., Tweedale, J., Jain, L.: Data mining techniques in clustering, association and classification. In: Holmes, D., Jain, L. (eds.) Data Mining: foundations and Intelligent Paradigms, vol. 1: Clustering, Association and Classification, pp. 1–6. Springer, Berlin (2012)

118. Hong, T.–P., Chen, C.–H., Wu, Y.–L., Tseng, V. S.: Fining active membership functions in fuzzy data mining. In: Lin, T.Y., Xie, Y., Wasilewska, A. Liau, C.–J.(eds.) Data Mining: foundations and Practice, pp. 179–196. Springer, Berlin (2008)
119. Introduction to Data Mining and Knowledge Discovery 3rd edn. Two Crows Corporation, Potomac (1999)
120. Jensen, K.: Coloured Petri nets and the invariant–method. Theor Comput Sci. **14**(3), 317–336 (1981)
121. Jensen, K.: Coloured Petri nets. EATCS Monographs on Theoretical Computer Science, vol. 1. Berlin, Springer (1992)
122. Kasabov, N.: Evolving Connectionist Systems. Springer, London (2007)
123. Kecman, V.: Learning and Soft Computing, MIT Press (2001)
124. Klose, A., Nrnberger, A., Nauck, D., Kruse, R.: Data Mining with neuro-fuzzy models. In: Data Mining and Computational Intelligence, pp. 1–35. Springer, Berlin (2001)
125. Klosgen, W., Zytkow, J. (eds.): Handbook of Data Mining and Knowledge Discovery. Oxford University Press, New York (2002)
126. Kolev, B., El–Darzi, E., Sotirova, E., Petronias, I., Atanassov, K., Chountas, P., Kodogianis, V.: Generalized Nets in Artificial Intelligence. vol. 3: generalized nets, Relational Data Bases and Expert Systems. "Prof. M. Drinov" Academic Publishing House, Sofia (2006)
127. Kotov, V.: An algebra for parallelism based on Petri nets. Lect. Notes Comp. Sci. **64**, 39–55 (1978)
128. Koycheva, E.: Examples of basic components in the UML, represented by generalized nets. In: Atanassov, K., Kacprzyk, J., Krawczak, M. (eds.) Issues in Intuitionistic Fuzzy Sets and Generalized Nets, pp. 79–87. Wydawnictwo WSISiZ, Warszawa (2004)
129. Koycheva, E.: Representability of extended UML sequence diagrams with loops by generalized nets. In: Atanassov, K., Kacprzyk, J., Krawczak, M. (eds.) Issues in Intuitionistic Fuzzy Sets and Generalized Nets, vol. 2, pp. 93–96. Wydawnictwo WSISiZ, Warszawa (2004)
130. Koycheva, E.: Entwurfsbegleitende Leistungsanalyse mit UML, MARTE und Generalisierten Netzen. Oldenbourg Velag, Munchen (2013)
131. Krawczak, M.: Multilayer Neural Systems and Generalized Net Models. Akademicka Oficyna Wydawnicza EXIT, Warszawa (2003)
132. Krawczak, M.: Modelling of adjoint neural networks by generalized nets. In: Proceedings of the 9th IEEE International Conference on Methods and Models in Automation and Robotics MMAR 2003, Miedzyzdroje (Poland), pp. 33–42, Technology University of Szczecin, 25–28 August 2003
133. Krawczak, M.: Generalized net models of multilayer neural networks. Adv. Stud. Contemp. Math. **7**(1), 69–86 (2003)
134. Krawczak, M.: Generalized net modelling concept–neural networks models, computer aiding of social, economical and environment development. Systems Research Institute, Polish Academy of Sciences, pp. 203–216. Warsaw (2004)
135. Krawczak, M.: On a generalized net model of MLNN simulation. Soft Computing Tools, Techniques and Applications, pp. 157–171. Academicka Oficyna Wydawnicza EXIT, Warszawa (2004)
136. Krawczak, M.: An example of generalized nets applications to modelling of neural networks simulation, current issues in data and knowledge engineering, pp. 297–308. Academicka Oficyna Wydawnicza EXIT, Warszawa (2004)
137. Krawczak, M.: Generalized net models of MLNN learning algorithms. Lect. Notes Comput. Sci. **2**(3697), 25–30 (2005)
138. Krawczak, M.: Modelling of adjoint neural networks by generalized nets. Issues in the Representation and Processing of Uncertain Imprecise Information: Fuzzy Sets, Intuitionistic Fuzzy Sets, Generalized Nets, and Related Topics Akademicka Oficyna Wydawnictwo EXIT, Warszawa, pp. 217–227 (2005)

139. Krawczak, M., Sotirov, S., Sotirova, E.: Generalized net model for parallel optimization of multilayer neural network with time limit. IEEE Intell. Syst. **IS12**, 173–177 (2012)
140. Krawczak, M., Aladjov, H.: Generalized net model of backpropagation learning algorithm In: Proceedings of the Third International Workshop on Generalized Nets, pp. 32–36, Sofia, 1 October 2002
141. Krawczak, M., Aladjov, H.: Generalized net model of adjoint neural networks. Adv. Stud. Contemp. Math. **7**(1), 19–32 (2003)
142. Krawczak, M., Sotirov, S., Atanassov, K.: Multilayer Neural Network Modelling by Generalized Nets. Warsaw School of Information Technologies (2010)
143. Krawczak, M., El–Darzi, E., Atanassov, K., Tasseva, V.: Generalized net for control and optimization of real processes through neural networks using intuitionistic fuzzy estimations. Notes on Intuitionistic Fuzzy Sets **12**(2), 54–60 (2007)
144. Krawczak, M., Sotirov, S., Atanassov, K.: Multilayer Neural Networks and Generalized Nets. Warsaw School of Information Technology, Warsaw (2010)
145. Looney, C.G.: Fuzzy Petri nets for rule–based decisionmaking. IEEE Trans. Syst. Man Cybern **18**(1), 178–183 (1988)
146. Maimon, O., Rokach, L.: Introduction to knowledge discovery and data mining. In: Maimon, O., Rokach, L. (eds.) Data Mining and Knowledge Discovery Handbook, 2nd edn, pp. 1–15. Springer, New York (2010)
147. Meyer–Nieberg, S., Beyer, H.–G.: Self–adaptation in evolutionary algorithms. In: Lobo, F., Lima, C., Michalewicz, Z. (eds.) Parameter Setting in Evolutionary Algorithms, Studies in Computational Intelligence, no. 54, pp. 47–75. Springer, Berlin (2007)
148. Mengov G., Pulov, S., Atanassov, K., Georgiev, K., Trifonov, T.: Modeling neural signals with a generalized net. Adv. Stud. Contemp. Math. **7**(2), 155–166 (2003)
149. Merlin, P.: A Study of the Recoverability of Computer Systems. Dissertation. University of California (1974)
150. Moyle, S.: Collaborative Data Mining, In: Maimon, O., Rokach, L. (eds.) Data Mining and Knowledge Discovery Handbook, 2nd edn, pp. 1029–1039. Springer, New York (2010)
151. Natkin, S.: Les reseaux de Petri stochastiques et leur application a l'evaluation des systems informatiques. Dissertation. Juin (1980)
152. Nikolov, N.: Generalized nets and semantic networks. Adv. Model. Anal. AMSE Press **27**(1), 19–25 (1995)
153. Nikolova N.: Generalized net representation of backpropagation neural networks. Adv. Model. Anal. AMSE Press **1**(1), 27–34 (1998)
154. Nikolova, M., Szmidt, E., Hadjitodorov, S.: Generalized nets with decision making componets. In: Proceedings of the International Workshop on Generalized Nets, pp. 1–5, Sofia, 9 July 2000
155. Noe, J.: PRO–nets: for modelling processes and processors, Techn. Rep. 75–07–15, Department. of Computer Science University of Washington, Seattle (1975)
156. Nutt, G.: The Formulation and Application of Evaluation Nets. Dissertation. Computer Science Group, University of Washington, Seattle (1972)
157. Orozova, D., Sotirova, E.: Generalized net model of the applying data mining tools. In: Proceedings of the 10th International Workshop on Generalized Nets, pp. 22–26, Sofia, 5 Dec 2009
158. Orozova, D., Sotirova, E., Chountas, P.: Generalized net model of the knowledge discovery in medical databases. Bioautomation **13**(4), 281–288 (2009)
159. Orriols–Puig, A., Bernado–Mansilla, E.: Mining imbalanced data with learning classifier systems, In: Bull, L., Ester, B.–M., Holmes, J. (eds.) Learning Classifier Systems in Data Mining, pp. 123–145. Springer, Berlin (2008)
160. Parvathi, R., Sotirov, S., Gluhchev, G., Atanassov, K.: A generalized net model of intuitionistic fuzzy image preprocessing. Comptes Rendus de l'Academie bulgare des Sciences, Tome **64**(3), 333–338 (2011)

161. Pasi G., Atanassov, K., Melo Pinto, P., Yager, R., Atanassova, V.: Multi–person multi–criteria decision making: intuitionistic fuzzy approach and generalized net model. In: Proceedings of the 10th ISPE International Conference on Concurrent Engineering “Advanced Design, Production and Management Systems”, pp. 1073–1078, Madeira, 26–30 July 2003
162. Pasi, G., Yager, R., Atanassov, K.: Intuitionistic fuzzy graph interpretations of multi–person multi–criteria decision making: generalized net approach. In: Proceedings of Second International IEEE Conference Intelligent Systems, Vol. 2, pp. 434–439, Varna, 22–24 June 2004
163. Pechenizkiy, M., Puuronen, S., Tsymbal, A.: Does relevance matter to data mining research?, In: Lin, T.Y., Xie, Y., Wasilewska, A., Liau, C.–J. (eds.) Data Mining: Foundations and Practice, pp. 251–275. Springer, Berlin (2008)
164. Pena-Ayala, A. (ed.): Educational Data Mining. Springer, Cham (2014)
165. Pencheva, T.: Generalized nets model of crossover technique choice in genetic algorithms. Issues in Intuitionistic Fuzzy Sets and Generalized Nets **9**, 92–100 (2011)
166. Pencheva, T., Atanassov, K., Shannon, A.: Modelling of a roulette wheel selection operator in genetic algorithms using generalized nets. Int. J. Bioautomation **13**(4), 257–264 (2009)
167. Pencheva, T., Atanassov, K., Shannon, A.: Modelling of a stochastic universal sampling selection operator in genetic algorithms using generalized nets. In: Proceedings of the 10th International Workshop on Generalized Nets, pp. 1–7, Sofia, 5 Dec 2009
168. Pencheva, T., Atanassov, K., Shannon, A.: Generalized nets model of offspring reinsertion in genetic algorithms. Annual of Informatics Section of the Union of Scientists in Bulgaria **4**, 29–35 (2011)
169. Pencheva, T., Atanassov, K., Shannon, A.: Generalized net model of selection function choice in genetic algorithms. In: Recent Advances in Fuzzy Sets, Intuitionistic Fuzzy Sets, Generalized Nets and Related Topics. Vol. II: Applications. Warsaw, Systems Research Institute, Polish Academy of Sciences, pp. 193–201 (2011)
170. Pencheva, T., Atanassov, K., Shannon, A.: Generalized nets model of rank–based fitness assignment in genetic algorithms. In: New Trends in Fuzzy Sets, Intuitionistic Fuzzy Sets, Generalized Nets and Related Topics. Vol. II: Applications. Warsaw, Systems Research Institute, Polish Academy of Sciences, pp. 127–136 (2013)
171. Pencheva, T., Roeva, O., Shannon, A.: Generalized net models of crossover operators in genetic algorithms. In: Proceedings of the 9th International Workshop on Generalized Nets, vol. 2, pp. 64–70, Sofia, 4 July 2008
172. Peneva, D., Tasseva, V., Kodogiannis, V., Sotirova, E., Atanassov, K.: Generalized nets as an instrument for description of the process of expert system construction, IEEE 760–763 (2006)
173. Petkov, T., Sotirov, S.: Generalized net model of the cognitive and neural algorithm for adaptive resonance theory 1. Int. J. Bioautomation **17**(4), 207–216 (2013)
174. Petkov T., Sotirov, S.: Generalized net model of slow learning algorithm of unsupervised ART2 neural network, IWIFSGN’2013 Twelfth International Workshop on Intuitionistic Fuzzy Sets and Generalized Nets, pp. 61–70, Warsaw (2014)
175. Plaginakos, V., Tasoulis, D., Vrahatis, M.: A review of major application arreas of differential evolution, In : Chakraborty, U. (ed.) Advances in Differential Evolution , Studies in Computational Intelligence, vol. 143, pp. 197–238. Springer, Berlin (2008)
176. Petri, C.–A.: Kommunication mit Automaten. Dissertation. University of Bonn, 1962; Schriften des Inst. fur Instrument. Math., no. 2, Bonn (1962)
177. Popova, B., Atanassov, K.: Opposite generalized nets. I. Adv. Model. Anal. AMSE Press **19** (2), 15–21 (1994)
178. Popova, B., Atanassov, K.: Opposite generalized nets. II. Adv. Model. Anal. AMSE Press **19** (2), 23–28 (1994)
179. Prestwich, S.: The relation between complete and incomlpete search, In : Blum, C. et al., (eds.) Hybrid Metaheuristics, Studies in Computational Intelligence, vol. 114, pp. 63–83. Springer, Berlin (2008)

180. Radeva, V., Krawczak, M., Choy, E.: Review and bibliography on generalized nets theory and applications. Adv. Stud. Contemp. Math. **4**(2), 173–199 (2002)
181. Ramchandani, C.: Analysis of Asynchronous Concurrent Systems by Timed Petri Nets. Dissertation. MIT, Cambridge, Mass. (1973)
182. Riedemann, E. Mayer, U.: Verallgemeinerte, modifizierte Petri–Netze als Modells fur MIMD–Rechner, Abteilung Informatik, University of Dortmund, Forschungebericht Nr. 143 (1982)
183. Rokach, L.: A survey of clustering algorithms, In: Maimon, O., Rokach, L. (eds.) Data Mining and Knowledge Discovery Handbook, 2nd edn, pp. 269–298. Springer, New York (2010)
184. Roeva, O.: Bat algorithm in terms of generalized net. In: Proceedings of 15th International Workshop on Generalized Nets, pp. 1–6, Burgas (2014)
185. Roeva, O., Atanassov, K., Shannon, A.: Generalized net for selection of genetic algorithm operators. Annual of Informatics Section of Union of Scientists in Bulgaria, **1**, 117–126 (2008)
186. Roeva, O., Atanassov, K.: Generalized net model of a modified genetic algorithm. Issues in intuitionistic fuzzy sets and generalized nets **7**, 93–99 (2008)
187. Roeva, O., Atanassov, K., Shannon, A.: Generalized net for evaluation of genetic algorithm fitness function. In: Proceedings of the 8th International Workshop on Generalized Nets, pp. 48–55, Sofia, 26 June 2007
188. Roeva, O., Melo–Pinto, P.: Generalized net model of Firefly algorithm. In: Proceedings of 14th Int. Workshop on Generalized Nets, pp. 22–27, Burgas, 29 November 2013
189. Roeva O., MichalÃ-kovÃ¡, A.: Generalized net model of intuitionistic fuzzy logic control of genetic algorithm parameters. In: Proceedings of the 9th International Workshop on IFSs, BanskÃ¡ Bystrica, Notes on Intuitionistic Fuzzy Sets, **19**(2), 71–76 (2013) 8 October 2013
190. Roeva, O., Pencheva, T.: Generalized net model of a multi–population genetic algorithm, Issues in Intuitionistic Fuzzy Sets and Generalized Nets. In: Kacprzyk, J., Krawczak, M., Szmidt, E. (eds.) Wydawnictwo WSISiZ, Warszawa, vol. 8, pp. 91–101 (2010)
191. Roeva, O., Pencheva, T., Atanassov, K.: Generalized net of a genetic algorithm with intuitionistic fuzzy selection operator, New Developments in Fuzzy Sets, Intuitionistic Fuzzy Sets, Generalized Nets and Related Topics, Volume I: Foundations, IBS PAN SRI PAS, (Systems Research Institute, Polish Academy of Sciences), Warsaw, pp. 167–178 (2012)
192. Roeva, O., Pencheva, T., Atanassov, K., Shannon, A.: Generalized net model of selection operator of genetic algorithms. In: Proceedings of the IEEE International Conference on Intelligent Systems, pp. 286–289, London, UK, 7–9 July 2010
193. Roeva, O., Pencheva, T., Shannon, A., Atanassov, K.: Generalized Nets in Artificial Intelligence. Vol. 7: Generalized nets and Genetic Algorithms. "Prof. M. Drinov" Academic Publishing House, Sofia (2013)
194. Roeva, O., Shannon, A.: A Generalized net model of mutation operator of the breeder genetic algorithm. In: Proceedings of the 9th International Workshop on Generalized Nets, vol. 2, pp. 59–63, Sofia, 4 July 2008
195. Roeva, O., Shannon, A., Pencheva, T.: Description of simple genetic algorithm modifications using generalized nets. In: Proceedings of the IEEE 6th International Conference IS 2012, Vol. 2, pp. 178–183, Sofia, Bulgaria
196. Rud, O.P.: Data Mining Cookbook. John Wiley and Sons, Danvers (2001)
197. Schiffers, M., Wedde, H.: Analysing program solutions of coordination problem by GP–nets, Math. In: Winkowski, J. (ed.) Foundations of Computer Science, Lecture Notes in Computer Science, vol. 64, pp. 462–473 (1978)
198. Seifert, J.: Data Mining: an Overview, CRS Report for Congress, Order Code RL31798 (2004)
199. Shannon, A., Atanassov, K., Orozova, D., Krawczak, M., Sotirova, E., Melo-Pinto, P., Petrounias, I., Kim, T.: Generalized Nets and Information Flow Within a University. Warsaw School of Information Technology, Warsaw (2007)

200. Shannon, A., Sorsich, J., Atanassov, K.: Generalized Nets in Medicine. Academic Publishing House "Prof. M. Drinov", Sofia (1996)
201. Shannon, A., Sorsich, J., Atanassov, K., Nikolov, N., Georgiev, P.: Generalized Nets in General and Internal Medicine. "Prof. M. Drinov" Academic Publishing House, Sofia, Vol. 1 (1998); Vol. 2 (1999); Vol. 3 (2000)
202. Shapiro, S.: A stochastic Petri nets with applications to modelling occupancy timed for concurrent task systems. Networks **9**, 375–379 (1979)
203. Shmueli, G., Patel, N., Bruce, P.: Data Mining for Business Intelligence. John Wiley and Sons, Hoboken (2007)
204. Simovici, D., Djeraba, Ch.: Mathematical Tools for Data Mining, 2nd edn. Springer, London (2014)
205. Sotirov, S.: Modeling the algorithm backpropagation for learning of neural networks with generalized nets, Part 1. In: Proceedings of the 4th International Workshop on Generalized Nets, pp. 61–67, Sofia, 23 Sept 2003
206. Sotirov, S., Krawczak, M., Kodogiannis, V.: Generalized nets model of the Grossberg neural network. Part 1, Issues in Intuitionistic Fuzzy Sets and Generalized Nets, vol. 4, pp. 27–34. Warszawa (2004)
207. Sotirov, S.: Generalized net model of the accelerating backpropagation algÐ¾rithm, Jangjeon Mathematical Society, pp. 217–225 (2006)
208. Sotirov, S.: Generalized net model of the art neural networks. part3, developments in fuzzy sets, intuitionistic fuzzy sets, generalized nets and related topics. Vol. II: Applications, System Research Institute, Polish Academy of Science, pp. 257–246. Warsaw (2008)
209. Sotirov, S.: Generalized net model of the time delay neural network, Issues in intuitionistic fuzzy sets and generalized nets, vol. 9, pp. 125–131. Warszawa (2010)
210. Sotirov, S.: Modeling the backpropagation algorithm of the Elman neural network by generalized net. In: Proceedings of the 13th International Workshop on Generalized Nets, pp. 49–55, London (2012)
211. Sotirov, S.: Modelling distributed time–delay neural network by generalized net, developments in fuzzy sets, Intuitionistic Fuzzy Sets, Generalized Nets and related topics. Vol. II: Applications, System Research Institute, Polish Academy of Science, pp. 231–238, Warsaw (2010)
212. Sotirov, S., Atanassov, K.: Generalized Nets in Artificial Intelligence. Vol. 6: Generalized nets and Supervised Neural Networks. "Prof. M. Drinov" Academic Publishing House, Sofia (2012)
213. Sotirov, S., Dimitrov, A.: Neural network for defining intuitionistic fuzzy estimation in petroleum recognition. Issues in intuitionistic fuzzy sets and generalized nets **8**, 74–78 (2010)
214. Sotirov, S., Kodogiannis, V.: Generalized net model of the Grossberg neural networks. Part 2. Issues in Intuitionistic fuzzy sets and generalized nets, vol. 5, pp. 130–138. Warsaw School of Information Technology, Warsaw (2007)
215. Sotirov, S., Kodogiannis, V.: Generalized net model of the Elman neural network. In: Proceedings of the 11th International Workshop on GNs and 2nd International Workshop on GNs, IFSs, KE, London, pp. 21–26, 9–10 July 2010
216. Sotirov, S., Krawczak, M.: Modeling the algorithm backpropagation for learning of neural networks with generalized networks. Part 2. Issues in Intuitionistic Fuzzy Sets and Generalized Nets, vol. 3, pp. 65–69. Warsaw School of Information Technology, Warsaw (2006)
217. Sotirov, S., Krawczak, M.: Modeling the work of self–organizing neural networks with generalized networks. Issues in Intuitionistic Fuzzy Sets and Generalized Nets, vol. 3, pp. 57–63. Warsaw School of Information Technology, Warsaw (2006)
218. Sotirov, S., Krawczak, M.: Generalized net model of the art neural networks, Part 1, Issues in intuitionistic fuzzy sets and generalized nets, vol 7, pp. 67–74, Warszawa (2008)

219. Sotirov, S., Krawczak, M.: Generalized net model of the art neural networks. Part 2, Issues in Intuitionistic Fuzzy Sets and Generalized Nets, vol. 7, pp. 75–82. Warszawa (2008)
220. Sotirov, S., Krawczak, M.: Generalized net model of recurrent neural network. In: Proceedings of the 11th International Workshop on GNs and Second Int. Workshop on GNs, IFSs, KE, pp. 14–20 London, 9–10 July 2010
221. Sotirov, S., Krawczak, M.: Modelling layered digital dynamic network by a generalized net, Issues in intuitionistic fuzzy sets and generalized nets, Vol. 9, pp. 84–91. Warsaw (2011)
222. Sotirov, S., Krawczak, M., Atanassov, K.: Modelling of brain–state–in–a–box neural network with a generalized net. In: New Trends in Fuzzy Sets, Intuitionistic Fuzzy Sets, Generalized Nets and Related Topics, Vol. 2, pp. 153–159. Applications, SRI, Polish Academy of Sciences (2013)
223. Sotirov, S., Krawczak, M., Atanassov, K.: Generalized net model for parallel optimization of multilayer perceptron with momentum backpropagation algorithm. In: Proceedings of the 5th International IEEE Conference "Intelligent Systems", pp. 281–285, London (2010)
224. Sotirov, S., Krawczak, M., Kodogiannis, V.: Modeling the work of learning vector quantization neural networks. In: Proceedings of the 7th International Workshop on Generalized Nets, pp. 39–44, Sofia, 14–15 July 2006
225. Sotirov, S., Kukenska, V., Hristova, M., Vardeva, I., Staneva, L., Barzov, J., Dimitrov, S., Stoqnova, S.: Modeling the nonlinear autoregressive network with exogenous inputs with a generalized net, Developments in Fuzzy Sets, Intuitionistic Fuzzy Sets, Generalized Nets and related topics. Vol. II: Applications, System Research Institute, pp. 223–230 Polish Academy of Science. Warsaw (2010)
226. Sotirov, S., Orozova, D., Sotirova, E.: Generalized net model of the process of the prognosis with feedforward neural network. In: Proceedings of the XVI–th International Symposium on Electrical Apparatus and Technologies, SIELA, Vol.1, pp. 272–278 (2009)
227. Sotirov, S., Sotirova, E.: Generalized net model of the integrated system for early forest–fire detection, IWIFSGN'2013 Twelfth International Workshop on Intuitionistic Fuzzy Sets and Generalized Nets,pp. 103–114. Warsaw (2014)
228. Sotirova, E., Petkov, T., Surchev, S., Krawczak, M.: Generalized net model of clustering with self organizing map, Developments in Fuzzy Sets, Intuitionistic Fuzzy Sets, Generalized Nets and Related Topics, pp. 239–244. Foundations and Applications Warsaw, Poland (2011)
229. Sotirova, E., Atanassov, K., Chountas, P.: Generalized nets as tools for modelling of open, hybrid and closed systems: an example with an expert system, Advanced Studies in Contemporary Mathematics, October 2006, Vol. 13, No 2, 2006, 221–234
230. Spurgin, A., Petkov, G.: Advances simulator data mining for operators' performance assessment, In: Ruan, D., Chen, G., Kerre, E., Wets, G. (eds.) Intelligent Data Mining: Techniques and Applications, pp. 487–514. Springer, Berlin
231. Starke, P.: Petri-Netze. VEB Deutscher Verlag der Wissenschaften, Berlin (1980)
232. Stoeva, S., Atanassov, K.: Generalized net representation of production systems interpreters. In: Proceedings of the 15th Spring Conference of the Union of Bulg. Math., Sunny Beach, pp. 456–464 (1986)
233. Surchev, S., Sotirov, S.: Modeling the process of the color recognition with MLP using symbol visualization, IWIFSGN'2013. In: Proceedings of the 12th International Workshop on Intuitionistic Fuzzy Sets and Generalized Nets, pp. 115–124, Warsaw (2014)
234. Surchev, S., Sotirov, S., Korneta, W.: Bio–inspired artificial intelligence: A generalized net model of the regularization process in MLP. Int. J. Bioautomation **17**(3), 151–158 (2013)
235. Surchev, S, Sotirov, S.: Modelling the process of color recognition using multilayer neural network. Issues in Intuitionistic Fuzzy Sets and Generalized Nets **10**, 143–151 (2014)
236. Sumathi, S., Sivanandam, S.: Introduction to Data Mining and Applications,Berlin (2006)
237. Valk, R.: Self–modifying nets. Inst. für Informatik, Univ. Hamburg, Bericht IFI–HH–B–34/77 (1977)

238. Witten, H., Frank, E.: Data Mining: practical Machine Learning Tools and Techniques, Morgan Kaufmann (2005)
239. Xing, B., Gao, W.-J.: Innovative Computational Intelligence: a Rough Guide to 134 Clever Algorithms. Springer, Cham (2014)
240. Yao, Y., N. Zhong, Y. Zhao, A conceptual framework of data mining, In: Lin, T.Y., Xie, Y., Wasilewska, A., Liau, C.–J. (eds.) Data Mining: Foundations and Practice, pp. 501–515. Springer, Berlin (2008)
241. Zerros C., Irani, K.: Colored Petri nets: their properties and applications, Systems Engineering Lab. TR 107, University. of Michigan (1977)

# Induction of Modular Classification Rules by Information Entropy Based Rule Generation

**Han Liu and Alexander Gegov**

**Abstract** Prism has been developed as a modular classification rule generator following the separate and conquer approach since 1987 due to the replicated sub-tree problem occurring in Top-Down Induction of Decision Trees (TDIDT). A series of experiments have been done to compare the performance between Prism and TDIDT which proved that Prism may generally provide a similar level of accuracy as TDIDT but with fewer rules and fewer terms per rule. In addition, Prism is generally more tolerant to noise with consistently better accuracy than TDIDT. However, the authors have identified through some experiments that Prism may also give rule sets which tend to underfit training sets in some cases. This paper introduces a new modular classification rule generator, which follows the separate and conquer approach, in order to avoid the problems which arise with Prism. In this paper, the authors review the Prism method and its advantages compared with TDIDT as well as its disadvantages that are overcome by a new method using Information Entropy Based Rule Generation (IEBRG). The authors also set up an experimental study on the performance of the new method in classification accuracy and computational efficiency. The method is also evaluated comparatively with Prism.

## 1 Introduction

Automatic generation of classification rules has been an increasingly popular technique in commercial applications such as a decision making system. Classification rule generation approaches can be subdivided into two categories:

H. Liu (✉) · A. Gegov
School of Computing, University of Portsmouth, Buckingham Building,
Lion Terrace, Portsmouth PO1 3HE, UK
e-mail: Han.Liu@port.ac.uk

A. Gegov
e-mail: Alexander.Gegov@port.ac.uk



V. Sgurev et al. (eds.), *Innovative Issues in Intelligent Systems*,
Studies in Computational Intelligence 623, DOI 10.1007/978-3-319-27267-2_7

'divide and conquer approach' [1], which is also known TDIDT and induces rules in the intermediate form of a decision tree, and 'separate and conquer approach', also called 'covering approach', which induces 'If-Then' rules directly from training instances. Both approaches can be traced back to 1960s [2, 3] and achieve a comparable accuracy. A well-known example of classification rule generators following 'divide and conquer approach' is C4.5, a version of decision trees, introduced by Quinlan [1]. It is a widely used method but its representation has a serious potential drawback as mentioned in [4] in that a tree may contain many redundant terms (attribute-value pairs) as there are rules which cannot be easily represented by a tree structure if these rules have no common attributes. This problem is called replicated subtree problem and will be mentioned in Sect. 2 in details. The existence of this problem motivated the development of Prism method [5] with the aim to generate modular classification rules using the separate and conquer approach.

In this paper, Sect. 2 introduces the replicated subtree problem and the principle of TDIDT and Prism in generating classification rules. It also describes the advantages of Prism compared with TDIDT and its disadvantages that are going to be avoided in the new method that the authors proposed. Section 3 explains the proposed method and mentions about dealing with some common issues arising in classification tasks such as clash, conflict resolution and continuous attributes. Section 4 describes the setup of experimental study and presents the results. Section 5 evaluates the performance of this proposed method analysing its strengths and limitations. Section 6 summarises the contribution of the current work to real world applications and highlights future work aimed at overcoming the limitations identified in Sect. 5.

## 2 Related Work

As stated in Sect. 1, a decision tree may result in a replicated subtree problem. The Prism algorithm has been developed to solve this problem. This section describes the principle of the two algorithms and compare them in accuracy and computational efficiency as well as identifies some limitations of Prism in the following subsections.

### *2.1 Decision Tree Algorithm*

Decision trees have been a popular method as a means of generating classification rules and they are based on the fairly simple but powerful TDIDT algorithm [4]. The basic idea of this algorithm can be illustrated in Fig. 1.

One popular method of attribute selection for step (a) illustrated in Fig. 1 is based on average entropy of an attribute [4], which is to select the attribute that can minimize the value of entropy for the current subset being separated, thus maximize

**IF** all cases in the training set belong to the same class
**THEN** return the value of the class
**ELSE**

(a) Select the attribute A to split on*
(b) Sort the instances in the training set into non-empty subsets, one for each value of attribute A
(c) Return a tree with one branch for each subset, each branch having a descendant subtree or a class value produced by applying the algorithm recursively for each subset in turn.

*When selecting attributes at step (a) the same attribute must not be selected more than once in any branch.

**Fig. 1** TDIDT Tree generation algorithm [4]

information gain. Some examples of decision trees using entropy as attribute selection technique include ID3, C4.5 and C5.0.

Entropy was introduced by Shannon in [6], which is a measure of the uncertainty contained in a training set, due to the presence of different classifications [4]. It can be calculated in the following way [4]:

- To calculate the entropy for each attribute-value pair in the way that the entropy of a training set is denoted by E and is defined by the formula: $E = -\sum_{i=1}^{K} p_i \log_2 p_i$ summed over the classes for which $p_i \neq 0$ (p denotes the probability of class i) if there are k classes.
- To calculate the weighted average for entropy of resulting subsets.

For the conditional entropy of an attribute-value pair, $p_i$ denotes the posterior probability for class i when given the particular attribute-value pair as a condition. On the other hand, for initial entropy, the $p_i$ denotes the priori probability for class i. The information gain is calculated by subtracting the initial entropy from the average entropy of a given attribute.

## 2.2 *Replicated Subtree Problem*

In a Ph.D. project at the Open University, Cendrowska criticised the principle of generation of decision trees which can then be converted into a set of individual rules, compared with the principle of generation of classification rules directly from training instances [4]. She comments the following:

> The decision tree representation of rules has a number of disadvantages…[Most] importantly, there are rules that cannot easily be represented by trees.

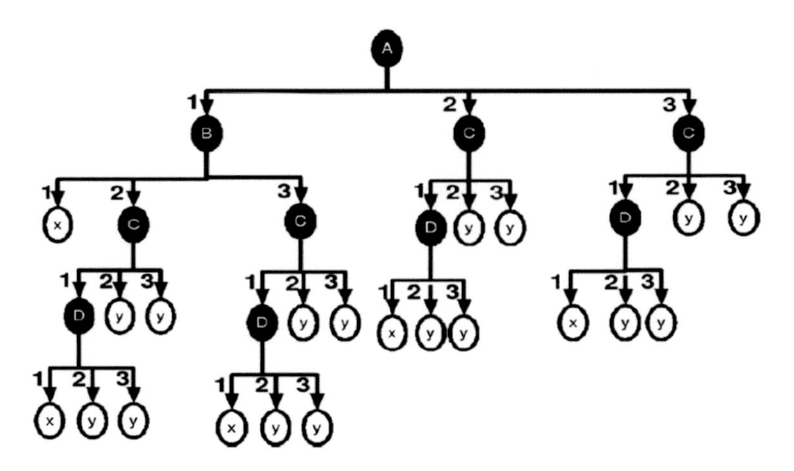


**Fig. 2** Cendrowska's replicated subtree example [7, 8]

She gave an example with the following rule set:

Rule 1: If a=1 and b=1 Then class=x
Rule 2: If c=1 and d=1 Then class=x

This kind of rule set may not fit into a tree structure. However, if they are forced to fit into a tree structure, then it is required to add other terms to at least one of the two rules, which would require at least one other rule to cover instances excluded by the addition of other terms.

For the above example, we need to have four attributes a, b, c and d, each of which has three possible values 1, 2 and 3, and is to be selected for partitioning at the root node.

The rule set would be listed as follows and it is illustrated in Fig. 2:

If a=1 and b=1 then class=x
If a=1 and b=2 and c=1 and d=1 then class=x
If a=1 and b=3 and c=1 and d=1 then class=x
If a=2 and c=1 and d=1 then class=x
If a=3 and c=1 and d=1 then class=x

It was pointed out in [9] that ID3 is difficult to manipulate for expert systems as it is required to examine the whole tree in order to extract rules about a single classification. It has been partially solved by converting a tree to a set of individual rules but there are some rules that are not easily fit into a tree structure as is the replicated subtree problem mentioned above. In a medical diagnosis system, this problem may lead to unnecessary surgery [5, 9]. The reasons identified in [9] are the following:

- The decision tree is attribute oriented
- Each iteration in the generation process chooses the attribute on which to be split aiming at minimizing the average entropy. i.e. measuring the average uncertainty. However, this doesn't necessarily mean that the uncertainty for each rule is reduced.
- An attribute might be highly relevant to one particular classification and irrelevant to the others. Sometimes only one value of an attribute is relevant.

## *2.3 Prism Algorithm*

As mentioned in Sect. 1, the development of Prism algorithm was motivated due to the existence of the replicated subtree problem. Original Prism was introduced by Cendrowska in [5] and its basic idea can be illustrated in Fig. 3.

The original Prism is based on the assumption that all attributes are discrete. When there are actually continuous attributes, one way to deal with that is discretisation of attribute values prior to generating rules as described in [4, 7, 8] such as ChiMerge [10]. In addition, Bramer's Inducer software [11] provides implementation that deals with continuous attributes as described in [4, 7, 8]. On the other hand, the original version of Prism doesn't take clash problem into account. However, the Inducer software implementations of Prism algorithm provide the strategy of dealing with a clash set as the following [4] (Fig. 4):

Another problem considered in Prism called tie-breaking is that there are two or more attribute-value pairs which have the equally highest probability. The original Prism decides to choose arbitrarily whereas Bramer improved it by choosing the one with the highest total frequency [4].

In addition, with the motivation of improving the computation efficiency, Bramer pointed out that the original Prism always deletes instances covered by rules generated so far and then resets the training set to its original state once the

For each classification (class= i) in turn and starting with the complete training set each time:

1. Calculate the probability class =I for each attribute-value pair.
2. Select the attribute-value pair with the largest probability and create a subset of the training set comprising all the instances with the selected attribute-value pair (for all classifications)
3. Repeat 1 and 2 for this subset until a subset is reached that contains only instances of class i. The induced rule is then the conjunction of all the attribute-value pairs selected.
4. Remove all instances covered by this rule from the training set.

Repeat 1-4 until all instances of class I have been removed

*For each rule, no one attribute can be selected twice during generation

**Fig. 3** Basic Prism algorithm [4]

If a clash occurs while generating the rules for class i:

1. Determine the majority class for the subset of instances in the clash set.
2. If this majority class is class I, then compute the induced rule by assigning all instances in the clash set to class i. If it is not, discard the whole rule.
3. If the induced rule is discarded, then all instances that match majority class should be deleted from the training set before the start of the next rule induction. If the rule is kept, then all instances in the clash set should be deleted.

**Fig. 4** Dealing with clashes in Prism

generation for a target class is finished. This undoubtedly increases the number of iterations resulting in high computational cost [12]. Therefore, Bramer developed two new versions of Prism called PrismTC, which chooses the majority class as the target class for each rule being generated, and PrismTCS, which chooses the minority class as the target class for each rule being generated, respectively. Both versions select a new class as the target class after a rule generated and do not reset the dataset to its original state as well as set an order in which rule is being applied to predict unseen instances. Prism TCS does not restore the training set to its original state and thus is faster than Original Prism but also provides a similar level of classification accuracy [8, 12]. However, Bramer's experiments show that PrismTC doesn't compare well against Original Prism and PrismTCS [8, 12].

## 2.4 Comparing Prism with Decision Tree Learning

Bramer described in [13] a series of experiments to compare the performance of Prism against that of decision tree learning on a number of datasets. He concluded the following:

- Prism algorithm may give classification rules at least as good as those generated from TDIDT algorithms.
- There are generally fewer rules but also fewer terms per rule generated, which is likely to aid their comprehensibility to domain experts and users. This result indicates that Prism generally performs consistently better than TDIDT in terms of accuracy.
- The main difference is that Prism generally prefers to leave a test instance unclassified rather than to give it an incorrect classification.
- The reasons why Prism is more noise tolerant than TDIDT may be due to the generation of fewer terms per rule in most cases.
- Prism generally has higher computational efficiency than TDIDT and it may be further improved by parallelisation.

As mentioned in Sect. 2.2, there is a case that only one value of an attribute is relevant to a particular classification and that ID3 (a version of decision tree) is not taken into consideration. Deng pointed out in [9] that the Prism method is attribute-value-oriented and pays much attention to the relationship between an attribute-value pair and a particular classification, thus generating fewer but more general rules than a decision tree learning method.

## 2.5 Limitations of Prism

Prism algorithm also has some disadvantages. One of them is that the Original version of Prism may generate a rule set which may result in a classification confliction in predicting unseen instances. Let us see the example below:

Rule 1: If x=1 and y=1 then class=a
Rule 2: If z=1 then class=b.

Therefore, what should the classification be for an instance with x=1, y=1 and z=1? One rule gives class a, the other one gives class b. We need to give a method of choosing only one classification to classify the unseen instance [4]. This method is known as a conflict resolution strategy. Bramer mentioned in [4] that Prism uses the 'take the first rule that fires' strategy in dealing with the conflict problem and therefore it is required to generate the most important rules first as much as possible. However, the original Prism cannot actually introduce an order to a rule according to its importance as each of those rules with a different target class is independent of the others. As mentioned above, this version of Prism would restore the training set to its original size after the completion of rule generation for class i and before the start for class i+1. This indicates the rule generation for each class may be done in parallel so the algorithm cannot directly rank the importance among rules. Thus the 'take the first rule that fires' strategy may not deal with the confliction well. The PrismTCS doesn't restore the dataset to its original state unlike original Prism and thus can introduce the order to a rule according to its importance. This problem is partially resolved but PrismTCS may potentially lead to underfitting of a rule set. PrismTCS always chooses the minority class in the current training set as the target class of the rule being generated. Since the training set is never restored to its original size as mentioned above, it can be proved that one class may always be selected as the target class until all instances of this class have been deleted from the training set because the instances of this minority class covered by the current rule generated should be removed prior to generating the next rule. This case may result in that the majority class in the training set may not be necessarily selected as the target class to generate a list of rules until the termination of the whole generation process. In this case, there is not even a single rule having the majority class as its consequence (right hand side of this rule). In some implementations, this problem has been partially solved by assigning a default class (usually the majority class) in predicting unseen instances when there is not a single rule that can cover this instance. However, this should be based on the assumption that the training set is complete. Otherwise, the rule set may still underfit the training set as the conditions of classifying instances to the other classes are probably not strong enough. On the other hand, if a clash occurs, both the original Prism and PrismTCS would prefer to discard the whole rule than to assign the majority class to the rule. As mentioned above, Prism may generally generate more general and fewer rules than decision tree learning methods. One reason is potentially due to discarding rules. In addition, the clash may happen in two main ways as follows:

(1) One of the instances has at least one incorrect record for its attribute values or its classification [4].
(2) The clash set has both (or all) instances correctly recorded but it is impossible to discriminate between (or among) them on the basis of the attributes recorded and thus it may be required to examine further values of attributes [4].

When there is noise present in datasets, Prism may be more tolerant than decision tree learning as mentioned above. However, if the reason why a clash occurs is not due to noise, then it may result in underfitting of the rule set by discarding rules as it will leave many unseen instances unclassified in prediction stage. Prism would decide to discard the rules in some cases probably because it uses the so-called 'from outcome to cause' approach. As mentioned in Sect. 2.3, each rule being generated should be pre-assigned a target class and then the conditions should be searched by adding terms (antecedents) until the adequacy conditions are met. Sometimes, it may not necessarily receive adequacy conditions even after all attributes have been examined. This indicates the current rule covers a clash set that contains instances of more than one class. If the target class is not the majority class, this indicates the search of causes is not successful so the algorithm decides to give up by discarding the incomplete rule and deleting all those instances that match the target class in order to avoid the same case to happen all over again [7, 8]. This actually not only increases the irrelevant computation cost but also results in underfitting the rule set.

# 3 Information Entropy Based Rule Generation Method

As mentioned in Sect. 2, Prism has some obvious limitations. For this reason, a new rule generation method IEBRG using the separate and conquerapproach is introduced. This method avoids the underfitting of data sets and does not lead to any redundant computational effort.

## *3.1 Essence of the Method*

This method is attribute-value-oriented like Prism but it uses the 'from cause to outcome' approach. In other words, it doesn't have a target class pre-assigned to the rule being generated. The main difference with respect to Prism is that IEBRG focuses mainly on minimising the uncertainty for each rule being generated no

1. Calculate the conditional entropy of each attribute-value pair in the current subset
2. Select the attribute-value pair with the smallest entropy to spilt on, i.e. remove all other instances that contain no attribute-value pair.
3. Repeat step 1 and 2 until the current subset contains only instances of one class (the entropy of the resulting subset is zero).
4. Remove all instances covered by this rule.

Repeat 1-4 until there are no instances remaining in the training set.

* For each rule, no one attribute can be selected more than once during generation.

**Fig. 5** IEBRG algorithm

matter what the target class is. A popular technique used to measure the uncertainty is information entropy as mentioned in Sect. 2.1. Once the antecedents added so far can reduce the entropy of the subset to 0, there is no uncertainty remaining in the subset to decide a classification assigned to this rule as the consequence. The basic idea of IEBRG with using entropy can be illustrated in Fig. 5.

## 3.2 *Justification of the Method*

As mentioned in Sect. 2.5, all versions of Prism need to have a target class pre-assigned to the rule being generated. As mentioned in Sect. 2.2, an attribute might be not relevant to some particular classifications. Sometimes only one value of an attribute is relevant. Therefore, the Prism method chooses to pay main attention to the relationship between an attribute-value pair and a particular class. However, the class to which the attribute-value pair is highly relevant is probably unknown, as can be seen from the example in Table 1 below with reference to the lens 24 dataset. This dataset shows that P (class=3|tears=1)=1 illustrated by the frequency table for attribute "tears". The best rule generated first would be "if tears=1 then class=3".

This indicates that the attribute-value "tears=1" is only relevant to class 3. However, this is actually not known before the rule generation. According to the PrismTCS strategy, the first rule being generated would select "class=1" as the target class as it is the minority class (Frequency=4). Original Prism may select class 1 as well as it is of a smaller index. As described in [4], the first rule generated

**Table 1** Lens 24 dataset example

| Class label | Tears=1 | Tears=2 |
|---|---|---|
| Class=1 | 0 | 4 |
| Class=2 | 0 | 5 |
| Class=3 | 12 | 3 |
| Total | 12 | 12 |

by Original Prism is "if astig=2 and tears=2 and age=1 then class=1". Therefore, the computational performance is slightly worse than expected and the resulting rule is more complex. Sometimes, the Prism method may even generate an incomplete rule reaching a clash as mentioned in Sect. 2.5 if the target class assigned is not a good fit to some of those attribute-value pairs in the current training set. Then the whole rule may be discarded resulting in underfitting and redundant computational effort.

In order to find a better strategy for reducing the computational cost, the authors proposed the method mentioned in Sect. 3.1. In this technique, the first iteration of the rule generation process for the "lens 24" dataset can make the resulting subset's entropy reach 0. Thus the first rule generation is compete and its rule is represented by "if tears=1 then class=3".

In comparison with the Prism family, this algorithm may reduce significantly the computational cost. In addition, in contrast to Prism, the IEBRG method deals with clashes (introduced in Sect. 3.3) by assigning the majority class in the clash set to the current rule. This may potentially reduce the underfiting of rule sets thus reducing the number of unclassified instances although it may increase the number of misclassified instances. On the other hand, the IEBRG may also have the potential to avoid clashes occurring better than Prism.

### 3.3 Dealing with Clashes

There are two principal ways of dealing with clashes mentioned in [4] as follows:

(1) Majority voting: to assign the most common classification of the instances in the clash set to the current rule.
(2) Discarding: to discard the whole rule currently being generated

In this paper, the authors choose 'majority voting' as the strategy of dealing with this problem as the objective of this paper is mainly to validate this method and find its potential in improving accuracy and computation efficiency as much as possible.

### 3.4 Dealing with Tie-Breaking and Conflict

The tie-breaking problem is solved by deciding which attribute-value pair is to be selected to split the current subset when there are two or more attribute-value pairs that equally match the selection condition. In the IEBRG method, this problem may occur when two or more attribute-value pairs have the equally smallest entropy. The strategy is the same as the one applied to Prism by taking the one with the highest total frequency as introduced by Bramer in [4].

The classification conflict problem may occur to modular classification rule generators such as Prism. Similarly, the IEBRG may also face this problem. The authors choose the 'take the first rule that fires' strategy which is already mentioned in Sect. 2.3 because this method may potentially generate the most important rules first. Let us see the example below:

Rule 1: if x=1 and y=1 then class=1;
Rule 2: if x=1 then class=2;

This seems as if there is a conflict problem but the two rules can be ordered as rule 1 is more important. In other words, the two rules can be represented in the following two ways:

Rule 2: if x=1 and y≠1 then class=2;

This indicates that after the first rule has been received, the term 'x=1' can directly reduce the entropy of the subset to 0 as all uncertainty has been removed from the current set after removing those instances covered by the first rule. Thus the first rule is more important than the second one.

# 4 Experimental Setup and Results

In this experimental study, the authors use 10 datasets retrieved from the UCI repository [14] as illustrated by Tables 2 and 3. The authors set to compare the classification accuracy and computation efficiency performed by Prism and IEBRG. For accuracy measure, the authors choose to use cross-validation [4]. With regards to efficiency, the authors choose the whole dataset as training set to build the model and then use the same dataset to do testing. The efficiency is measured in terms of the number of rules and the number of terms (antecedents) per rule during modelling stage. For the Prism algorithm, the authors also count the number of discard rules as they actually rise the computational cost. The accuracy and efficiency are illustrated in Tables 2 and 3, respectively

**Table 2** Classification accuracy: Prism versus IEBRG

| Dataset | Prism (%) | IEBRG (%) |
|---|---|---|
| Lens24 | 75 | 75 |
| Vote | 96 | 93 |
| Weather | 67 | 71 |
| Contact-lenses | 67 | 75 |
| Breast-cancer | 72 | 75 |
| Lung-cancer | 74 | 78 |
| Nurse | 43 | 61 |
| Car | 70 | 71 |
| Kr-vs-kp | 65 | 84 |
| Tic-tac-toe | 67 | 46 |

**Table 3** Computational efficiency: Prism versus IEBRG

| Dataset | Prism | | | IEBRG | |
|---|---|---|---|---|---|
| | Count (rules) | Count (terms) per rule | Count (discard rules) | Count (rules) | Count (terms) per rule |
| Lens24 | 9 | 2.25 | 1 | 4 | 1.25 |
| Vote | 19 | 5.74 | 1 | 7 | 1.71 |
| Weather | 3 | 2.0 | 1 | 2 | 1.0 |
| Contact-lenses | 4 | 2.75 | 2 | 4 | 1.4 |
| Breast-cancer | 12 | 1.33 | 0 | 7 | 1.0 |
| Lung-cancer | 3 | 1.0 | 0 | 4 | 1.0 |
| Nurse | 70 | 7.9 | 272 | 121 | 6.99 |
| Car | 5 | 4.0 | 55 | 23 | 3.96 |
| Kr-vs-kp | 77 | 6.74 | 0 | 9 | 1.0 |
| Tic-tac-toe | 164 | 8.06 | 88 | 91 | 6.65 |

# 5 Evaluation

Table 2 shows that IEBRG performs better than Prism in classification accuracy in most cases. Table 2 also shows that the classification accuracy performed by IEBRG in 'Tic-tac-toe' dataset is slightly worse than that done by Prism. The reason is not entirely clear but it may be due to the presence of noisy data as Prism is generally more tolerant to noise as mentioned in Sect. 2.4. The authors need to take further investigations about the tolerance to noisy data for IEBRG. In addition, if the noise is actually present in a training set, this may be handled by feature selection techniques and/ or pruning strategies.

In terms of computational efficiency, Table 3 shows that IEBRG performs better than Prism. It is obvious from the comparison of number of rules and number of terms per rule that IEBRG generates fewer but more general rules than Prism in most datasets. Although IEBRG generates more rules than Prism in three datasets namely 'Lung-cancer', 'Nurse' and 'Car', it is obvious from the comparison of the number of terms per rule and number of discarded rules that IEBRG gives more general rules than Prism in two of the three cases and Prism discards more rules in 'Nurse' and 'Car' datasets. In both datasets, Prism not only increases significantly the redundant computational effort but also losses accuracy compared with IEBRG due to discarding many rules. This also shows that IEBRG performs better than Prism in dealing with clashes by assigning the majority class to a rule rather than discarding rules if the reason why clashes occur is not due to the presence of noise. This is probably because the attributes recorded in the dataset are not sufficient and the uncertainty remaining in the set may become 0 if one extra attribute is added to be examined. In this case, majority voting would potentially be the dominant strategy through looking at the results shown by Tables 2 and 3. On one hand, this may avoid underfitting of rule sets. On the other hand, this can increase the probability of getting correct classifications. However, it is also dependent on application domains. For example, in domains where the decision making system operates under uncertainty, it may be required to leave instances unclassified such as safety critical systems.

## 6 Conclusion

This paper has reviewed TDIDT and Prism as well as identified some limitations of Prism. A new modular rule generation method, called IEBRG, has been proposed and validated. The experimental study has shown that IEBRG has the potential to avoid underfitting of rule sets and to generate fewer but more general rules as well as to perform relatively better in dealing with clashes. IEBRG can also generate the most important rules first so that it can effectively avoid the classification conflict problem by taking the 'take the first rule that fires' strategy. However, the experiments are all based on datasets that contain no continuous attributes. Therefore, the authors will make further experiments against continuous attributes which can be dealt with by some strategies such as ChiMerge [10] and another strategy as described in [4, 7, 8]. Furthermore, the authors will investigate the overcoming effects of noise by using some pruning methods such as J-pruning, which has been applied to Prism [15] and decision trees [16], and Jmax-pruning, which has been applied to Prism [7, 8]. Both methods have shown good performance in experimental studies and are based on J-measure, which was introduced into the classification rules literature by Smyth and Goodman [17] who strongly justified the use of J-measure as information theoretic means of quantifying the information content of a rule.

## References


1. Quinlan, J.R.: C4.5: Programs for Machine Learning. Morgan Kaufman (1993)
2. Hunt, E.B., Stone, P.J., Marin, J.: Experiments in Induction. Academic Press, New York (1966)
3. Michalski, R.S.: On the quasi-minimal solution of the general covering problem. In: Proceedings of the Fifth International Symposium on Information Processing, Bled, Yugoslavia, pp. 125–128 (1969)
4. Bramer, M.A.: Principles of Data Mining. Springer, London (2007)
5. Cendrowska, J.: PRISM: an algorithm for inducing modular rules. Int. J. Man Mach. Stud. **27**, 349–370 (1987)
6. Shannon, C.: A mathematical theory of communication. Bell Syst. Tech. J. **27**(3), 379–423 (1948)
7. Stahl, F., Bramer, M.: Induction of modular classification rules: using Jmax-pruning. In: 30th SGAI International Conference on Innovative Techniques and Applications of Artificial Intelligence, Cambridge (2011)

8. Stahl, F., Bramer, M.: Jmax-pruning: a facility for the information theoretic pruning of modular classification rules. Knowl.-Based Syst. **29**(2012), 12–19 (2012)
9. Deng, X.: A Covering-Based Algorithm for Classification: PRISM. CS831: Knowledge Discover in Databases (2012)
10. Kerber, R.: Chimerge: discretization of numeric attributes. In: AAAI'92 Proceedings of the tenth national conference on Artificial intelligence, pp. 123–128 (1992)
11. Bramer, M.A.: Inducer: a public domain workbench for data mining. Int. J. Syst. Sci. **36**, 909–919 (2005)
12. Stahl, F., Bramer, M.: Computationally efficient induction of classification rules with the PMCRI and J-PMCRI frameworks. Knowl. Based Syst. **35**(2012), 49–63 (2012)
13. Bramer, M.A.: Automatic induction of classification rules from examples using N-Prism. In: Research and Development in Intelligent Systems XVI, pp. 99–121. Springer, Berlin (2000)
14. Blake, C.L., Merz, C.J.: UCI repository of machine learning databases. Technical Report, University of California, Irvine, Department of Information and Computer Sciences (1998)
15. Bramer, M.A.: Using J-pruning to reduce overfitting of classification rules in noisy domains. In: Proceedings of 13th International Conference on Database and Expert Systems Applications— DEXA 2002, Aix-en-Provence, France, 2–6 Sept 2002
16. Bramer, M.A.: Using J-pruning to reduce overfitting in classification trees. In: Research and Development in Intelligent Systems XVIII, pp. 25–38. Springer, Berlin (2002)
17. Smyth, P., Goodman, R.M.: Rule induction using information theory. In: Piatetsky-Shapiro, G., Frawley, W.J. (eds.) Knowledge Discovery in Databases, pp. 159–176. AAAI Press (1991)

# Proposals for Knowledge Driven and Data Driven Applications in Security Systems

**Vladimir S. Jotsov**

**Abstract** The topic of the presented research is contemporary threats leading to serious problems in the nearest future. Advantages and disadvantages of typical synthetic advanced analytics methods are investigated, and obstacles are revealed to their distribution in IT, especially in security systems (SS). Security education issues also have been discussed because the high-quality learning inevitably leads to independent research. Original results for juxtaposing statistical versus logical intelligent information processing methods aiming at possible evolutionary fusions are described, and recommendations are made on how to build more effective applications of classical and/or elaborated novel methods: Kaleidoscope, Funnel, Puzzle, and Contradiction. Characteristic peculiarities, advantages, and problems with coordination and control of the investigated group of methods are demonstrated. It is shown that their combination makes possible not only data driven applications, but more complex knowledge guided control of evolutionary computing using nonstandard sets of constraints. The focus is mainly on advanced analytics applications named Puzzle methods and their interactions with other described methods. It is studied aiming at collaborative statistical and logical research based on quantitative method applications, deep processing and effective management of accumulated knowledge. It is shown that applications of intelligent technologies advance the efficiency of statistical applications by using original set of evolutionary methods for data and knowledge fusion. It is shown that all the demonstrated advantages may be successfully combined with well-known methods from big data, advanced analytics, knowledge discovery, data/web/deep data mining or other modern fields. Also, it is shown how the considered applications enhance the quality of statistical inference, reveal the reasons of its effective use, improve the human-machine interaction between the user and system and hence serve the process of gradual but a sustainable improvement of the results. The usage of ontologies is investigated with the purpose of information transfer by a sense in

V.S. Jotsov (✉)
University of Library Studies and IT (ULSIT),
P.O. Box 161, 1113 Sofia, Bulgaria
e-mail: v.jotsov@unibit.bg
URL: http://www.unibit.bg



V. Sgurev et al. (eds.), *Innovative Issues in Intelligent Systems*,
Studies in Computational Intelligence 623, DOI 10.1007/978-3-319-27267-2_8

security multiagent environments or to reduce the computational complexity of practical applications. Applications from many fields using the same set of methods have been displayed aiming to show the strength of the domain independent part of the research.

# 1 Introduction

The need to create more effective and universal tools to protect computer resources grows every year in systems for detection and prevention from intrusions (Intrusion Detection Systems IDS, Intrusion Prevention Systems IPS). For this reason, different applications of intelligent data/knowledge processing are initiated based on a combination of methods from statistical and logical information processing [1, 2]. Other elaborations with growing influence in this domain are artificial immune systems and multiagent systems [3]. The unifying factor is the longer life cycle, such system design (elaborations) require bigger teams and more time for introduction, adjustments, and realization. Due to the complex structure the prevention from direct attacks against these systems is very challenging. Still, the influence of these trends over the information security systems grows continuously due to a series of the following reasons:

- A considerable number of open source tools is elaborated for every of the quoted trends. Nowadays intruders use more and more inventive and sophisticated methods and tools to organize attacks. In this situation, it is a problem of time and to some extent of money for intruders to use multiagent applications, intelligent data processing or to use disadvantages of the code of existing data analytics/web mining/artificial immune systems. Then, the disproportion between the used tools for attacks and for the prevention of the typical user will be so large that it might have a global impact over the World Wide Web, the financial scope or the societies and governments as a whole. Therefore, it is necessary to immediately direct more resources and efforts towards the cited vanguard technologies. As a result, we are witnesses of a tendency of raised attention to these trends in the framework of the European security projects.
- Nowadays the dynamic character of intrusions shows that security of WWW offices consecutively decreases in the course of several years.

Modern applications of statistical methods are effective and convenient to use *at the expense* of information encapsulation. In other words, it is impossible to construct tools acquiring new knowledge or to solve other problems of logical nature using contemporary statistical applications. If the quoted methods have been split into two groups, quantitative and qualitative, then statistical methods belong to the first group and logical ones belong to the second group. On the other hand, even if some event A has a very high probability, let's say, P(A) = 0.9999, it is still far from being proved. For this reason, the mechanical union of statistical and logical results is of no perspective. We do not attempt to propose any isolated solutions, instead we offer a combination of original methods that is well *adjustable to the existing ones*. Our research includes a novel evolutionary metamethod for joint control of statistical and logical methods where the statistical approach is widely applied on the initial stage of the research when the information about the problem is scanty and it is possible to choose the solution region arbitrarily [4]. The accumulation of knowledge gradually makes logical applications more and more effective and more *versatile* than the probabilistic ones, as well as fuzzy estimates and similar applications. SMM (Synthetic MetaMethod) metamethod is introduced here. It controls the process of consecutive replacement of data analytics applications by better ones. The difference from the classical analytical methods lies in the fact that the design of systems which are controlled by synthetic (meta)methods is not just a science, sometimes it is an art. On of the main tasks of this chapter is the description of synthetic features: how to use algorithms (functions) in data driven approaches; how to build knowledge-driven applications, what are the possibilities for future blocks of thinking machines. If we make an analogy with the set of traditional analytical methods and fashion clothes then the synthetic method will apply the design of the display window with the fashion clothes. In the common case, during intelligent data processing, no results convergence is expected but this does not hamper practical applications of these systems. In other words, bad and good designers will arrange the display window in quite different ways. Hence, there is no guarantee that every user understands the technology enough deep, and that his access to the system will have positive results.

The cited innovations are elaborated and demonstrated here for the following reason. The problems presented above show that there is a need to introduce elements of machine creative work in intelligent security systems (ISS). It is demonstrated in the paper that this goal is accessible if the usage of possibilities for human-machine contact and a set of comparatively simple intelligent technologies are done in the right way. On the other hand, the innovations can hardly be described in a single book chapter using the traditional academic style. For this reason here we avoided where possible the technical details and formalizations, and for the sake of the contents reduction descriptions by analogy may be used together with illustration visual aids and other nonstandard approaches.

One of the important questions researched here is the following: how can security agents operate autonomously? In the first place, contemporary agents function properly because of the usage of neural networks where the agent most conscientiously copies the acts of the teacher. Well-trained agents should be enough autonomous, but we believe that there exist more effective ways to make creatively autonomous intelligent machines. The other, more simple machines detect anomalies using well-known statistical applications which count for the normal situation/traffic and other mean statistical values related to the operation of the guarded place. There is a variety of other applied approaches where the system is overloaded by heuristic information, but they are not discussed here due to their evident weakness for ISS.

Here a new way is introduced for security agents' operation. The way seems to be rather complex, and it is discussed on different levels: information preprocessing, the elements of the metamethod SMM, then communication issues between them, and in the end experimental discussions are considered. Few application areas have been shown for the discussed domain-independent methods. It is shown that same methods or procedures could be purposed to retrieve rather different results when they are controlled in a different manner. Metacontrol level is introduced aiming at better efficiency and purposed for data driven results. The usage of ontologies to model the domain is recommendable, but not obligatory. On the other hand, as it is shown in the next section, the multiagent system functions more effectively if a system of ontologies is included. But ontologies also do not contribute to a great extent in order to understand the sense of matters by the agents or with respect to transmitting the sense of matters during communications between the agents. On the way to produce an analogy to how agents think evolutionary methods are introduced to process data/knowledge. The fact is used that thinking/understanding is an evolutionary, evolving process. The problem is how to direct evolving methods with no exaggerations of heuristics, statistical information and other relative methods far from the logical processing. Our approach is rather untraditional. For more than 20 years methods have been elaborated to detect and to resolve semantic conflicts, or their simpler-to-process forms, contradictions. A method for machine self-learning is built based on them. Searching and solving conflicts/contradictions the agent constantly improves its knowledge and at the same time it may solve other, collateral problems. This addition makes the processing data driven, and we hope that in future it will become knowledge driven. Detection of semantic conflicts is based on using special models that can also be improved gradually. The agent may request an external help to solve the conflicts, but this takes place only in extraordinary situations.

At the same time, it is shown how to change/defeat the reasoning component of security agents. Different logical methods are considered that are rather analogous to means-ends analysis using constraint satisfaction, variable fitness functions, brainstorming, and cognitive graphics. The combination of new methods to a great extent mechanizes creative efforts and also it serves agents' operation to improve the abilities of security experts working with similar types of systems. Suggested innovations serve the most effective application of data analytics/data mining/Web mining/collective evolutionary components in multiagent systems. They are well combined especially for applications of evolving methods with classical neuro-fuzzy, statistical applications, genetic algorithms, etc. methods for the domain. For example such systems critically accept teacher's acts in cases of supervised learning: they may precise or argue teacher's acts and in this way they can learn more effectively and deep.

A wide application of intelligent agents is forecasted in the field of information security systems. This will lead to situations when the agent has no possibility to learn from the expert/teacher but should swiftly learn from other agents or should self-learn without any teacher. Then the role of the above considered critical learning will significantly increase. Contemporary recession times make governments pay more attention to the technology effectiveness problems accumulated for decades. It is easy to follow that the sustainable development of regions depends on using smart, intelligent technologies and know-how accumulated for years. They should be embedded in well-organized super-systems. The tendency at national levels is using more and more sophisticated tools and standards. However, the high-cost technologies do not necessarily lead to new or high quantitative results. Hence, there is a risk of ineffective use of expensive resources.

In this chapter, an advanced analytics method is considered aiming to improve the quality of the deep knowledge analysis, acquisition, and learning. More details can be found in a book [5]. Intelligent software plays more and more important role in contemporary learning systems and their realizations from industry to security and even to blended education where Puzzle methods reinforce TRIZ results [6–8]. The considered tools are domain-independent, but are especially effective in rapidly changing environments of security systems (SS), financial systems, etc. Suggested innovations serve the more effective application of advanced analytics, (deep) data/web mining and/or collective evolutionary components. The latter should be used to combine logical and statistical results in one system, as stated in [5]. Contemporary training of intelligent agents reaches such a high level that some elements of intelligent systems became practicable also in mathematical proofs, statistics, etc. Such applications are given in the last section of this chapter, before section Conclusions. Here applications to one of the best data miners are explored. The research is based on series of Puzzle methods for intelligent knowledge/data processing elaborated within ULSIT-Sofia. The classical statistical applications in data mining are a well-described area. Many of the theoretical base results from the field have been obtained since 1980-ies. Since then a few problems have been revealed like, the probability 0.9999 doesn't mean that the hypothesis has been proved, and all of them lead to new types of constraints in data mining applications. Some sources proclaim the division of advanced analytics methods into the statistical and logical ones. These accumulated problems can't be resolved using only traditional data mining/advanced analytics methods. Under these conditions, we have developed an inexpensive technology for deep knowledge elicitation and management based on a minimum of technological solutions that do not require high costs of purchasing and support [5]. Its main method named Puzzle method is described in the next two sections of the paper. Applications to SAS Enterprise Miner 12.1 are considered in Sect. 5. Instead of the division of data analytics into parts we offer a series of unifying methods strengthening the results from existing data miners.

## 2 On the Usage of Ontology Constraints

Intelligent data processing by itself is a complex process of transforming data into knowledge. If data and the knowledge accumulated in the system are not structured in advance, then in the majority of cases their processing will be ineffective or it will produce poor results. For this reason, ontologies are used [5]. The frequent tool to create and manage ontologies is Protégé OWL and its later versions. In the majority of contemporary cases, ontologies are presented as graphs.

The system could understand the sense in client's actions in different situations if and only if ontologies are introduced. Some difficult to determine phrases, e.g. suspicious actions, actions to distract attention, etc. are best expressed via ontologies (Fig. 1), e.g. ontology O is 'suspicious activities near to the ATM' and ontology $O_1$ is 'the incidence of on-line thefts of identity is increased'.

This is the way to search necessary goals not only in linear or non-linear constrained areas as in classical case but also by crossing two or more previously introduced ontologies (Fig. 1). For brevity, we shall discuss just the most popular ontologies of the vocabulary type. For example, the first ontology is 'what it sounds like the noise of mechanical devices' and the second ontology is 'how the sound grew'. The crossing of both ontologies contains the solution concerning the purpose: if the noise from a mechanical device increases and it intrudes a forbidden perimeter then ATM should activate an alarm. In the described in Fig. 1 case, the proximity between the meanings of two ontology components is estimated not just by a direct congruence but also using different assessments of the concepts contained in the nodes of the graph. For example, the coincidence of fuzzy meaning 'mechanical noise' is based not on a definite frequency of the sound but on the image of its repeatability. Fuzzy granules can be formed in the crossing.

Both ontologies are shown as graphs and they represent the sense of things of the standard is-a type 'two objects are linked by the relation R' where object2 is a feature of object1 or object1 is a predecessor of object2. In Fig. 1 with curved areas crossings between the two ontologies are denoted where the searched solution is located. If both ontologies concern sounds, multimedia or some other type of unimaginable via graphs information then their crossing is found otherwise but this does not affect or change the proposed search algorithm. Using operations with ontologies instead of classic set operations gives the advantage of faster information processing, natural usage of fuzzy boundaries that are hard for formalization

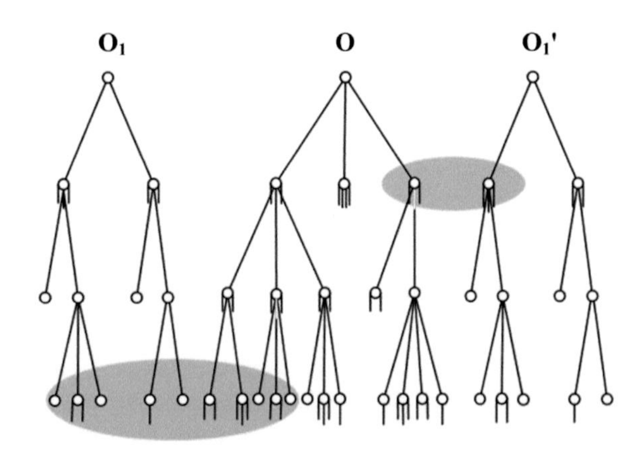


**Fig. 1** Search for causes of the intrusion at the intersection of two ontologies

otherwise, and so on. The following situation is presented in Fig. 1: $O_1$ and $O_1'$ are one and the same ontology but in different context or situation or an environment. The graph of $O_1$ is moved related to the graph of *O when the situation changes* which in turn changes also the sought solution depicted on the graph by the elliptical area. If the system explores the increasing mechanical noise, but the noise is used for diversion and during this time the intruder uses the possibility to attack the ATM or other defended equipment. The case with the sought intrusion is not in the left but in the right elliptical zone in Fig. 1. In this case, the system must include the alarm at least due to following hypothesis: detected group attack.

On Fig. 2a risk ontology is represented. As every knowledge from the Knowledge Base (KB), the ontology could be incomplete. Obviously, there exist many types of risks not included in the graph from Fig. 2a. On Fig. 2b is represented an incomplete ontology for different IT specialties. The usage of both ontologies leads to the appearance of semantic conflicts, and their resolution should gradually append the existing ontologies from Fig. 2a, b.

The fragment of ontology from Fig. 3 illustrates the types of ISS: intrusion detection systems, intrusion prevention systems, and others. The presented ontologies give an idea about the links between the objects and the relations between them. The presented graphs are widely used to search information on the Internet, e.g. if the keywords are ISS then there is no need to write 'anomaly detection' because the machine will make the necessary links between the concepts. Another widely used relation is 'has-a' meaning 'the object possesses the properties …'. Examples of these relations are the meanings of the attributes in the lower part of the figure, e.g. 'low cost'. Based on these two relations there are many applications of descriptive logics for the needs of the Semantic Web and others. Numerous other relations are introduced in projects like KAON and other initiatives that include operations with quantifiers and other logical operations.

The researched proposal to structure information is as follows: the usage of tools should be close to the so called means-ends analysis. For this reason, we widely introduce and use relations WHY, HOW, ORIGIN, IMPORTANT, and WHAT FOR. In this way, new possibilities are formed to operate with knowledge applicable by

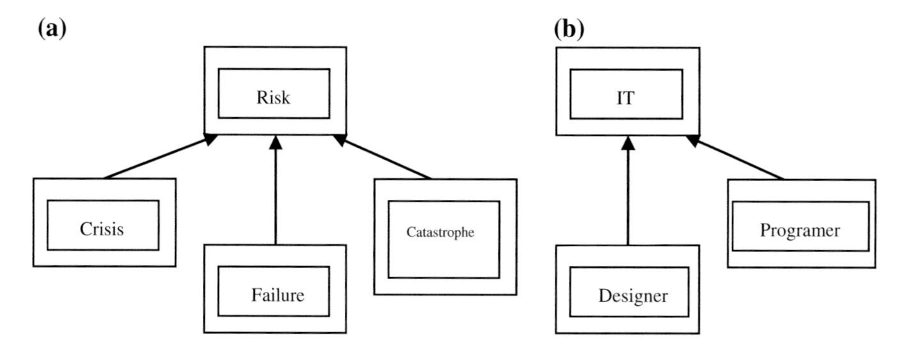


**Fig. 2** **a** Risk ontology. **b** Ontology for different IT specialties

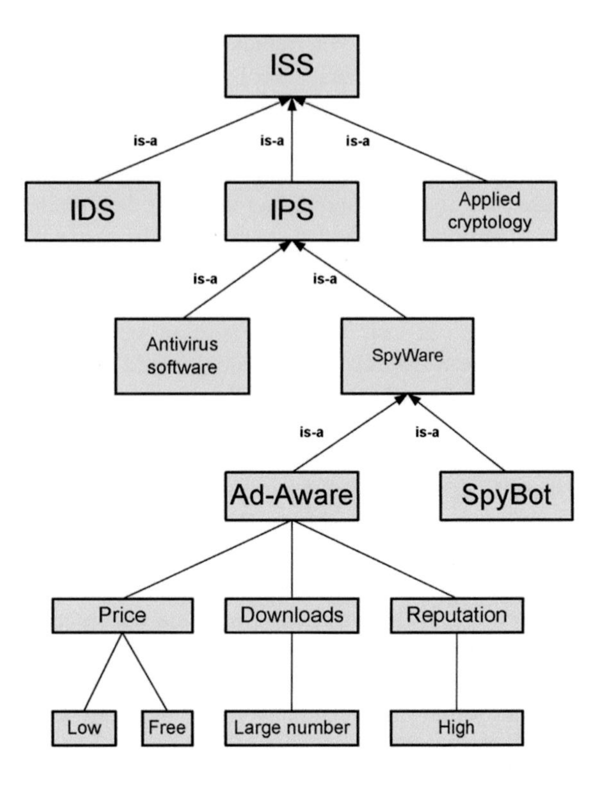


**Fig. 3** ISS ontology

traditional tools. For example, let's propose an input in the system with a large number of differential equations. Let's have a relation 'WHAT FOR' applicable to the following block: 'descriptions of trajectories during signal processing'. Using the relation WHY, we shall determine that 'this way is more effective than traditional methods because …'; via the relation ORIGIN we shall be able to compare the offered fragment of knowledge with other popular methods, tools and approaches and so to discover the roots/reasons for the problem. Applying the relation 'WHAT IS THE WAY', we shall discover texts with explanations how to use differential equations and other technical information. Using the enumerated tools, there will be a better structuring of information and prerequisites will be created for its more qualitative processing: in this way, we shall be able to operate effectively with large blocks of formalized knowledge without going in technical details.

In the Web services the usage of ontologies goes through the application of OWL or of W3C consortium standards like WSMO and others. The technology of these applications is complicated. It includes specific concepts like grounding, orchestration, and choreography. However, this is the way to apply a contemporary ontology processing—addition, deletion or update without breaking the links with other existing pieces of knowledge. In this way, the problems connected to the semantics of the researched goal/task are taken under consideration. Any form of knowledge may change its meaning depending on the context, the point of view, etc. The version of Protégé-OWL is frequently used in the creation of ontologies.

The necessity to introduce the following new relations in ontologies for intelligent applications of security is demonstrated. Let's discuss again the example with a large number of differential equations introduced in the system. It will be possible to structuralize information in a better way in the presented tools, and prerequisites will be created for its more qualitative processing. At the same time, we shall avoid too detailed formalizations, and ontologies using relations type **IMPORTANT** can contribute to using this good practice in ISS educational processes. The more relations are examined, the more precise categorization of objects will be. Substantially all other used for the moment relations just illustrate this: in what way the relation arguments are connected. Here the following relation is used '$\mathbf{R}_i$ links the arguments (the objects) from A to N in a way determined in the definition':

$$\mathbf{R}_i(A, B, C, D\ldots N)$$

The set of $\mathbf{R}_i$ is used to determine that 'A is connected with C', and so on, and the ways of connecting have been set in the definition of $\mathbf{R}_i$. It is suggested to introduce such $R_i^*$(A, B, C, D … Z), the usage of which will form automatically answers not only of the type 'B is connected with T' but **WHY** 'B is connected with T', **HOW/IN WHAT WAY**, **WHEN** the pointed link is actual, etc. Innovation notions do not deny previous ontological elaborations, they just supplement them; so the sense of matters is shown much more clearly/unambiguously. It can be changed by the context, and we can detect and track this process using considered relations. At the first sight the relation **IMPORTANT** seems to be rather subjective: the importance of any knowledge is determined by the situation, by the object or by other factors. But if an information is not graded by importance then the agents will 'be drown' in this sea of information provided the fact that the knowledge is not well structured. To resolve the problem, structuring is performed via the introduced new relation **IMPORTANT.**

## 3 Basic Methods Under the Control of SMM

### *3.1 Puzzle Method*

The basic methods of the suggested metamethod SMM are presented below. Let the constraints connected to the defined problem form a line in the space described by the Eq. (1).

$$\frac{x - x_1}{x_2 - x_1} = \frac{y - y_1}{y_2 - y_1} = \frac{z - z_1}{z_2 - z_1} \tag{1}$$

If the mentioned curves/lines have common points of intersection and if they lie in a common plain so that a closed figure (triangle, tetragon, etc.) is formed then the search space is significantly reduced and it is *searched* much easier. The practical usage of the classical technology, as well as the constraint satisfaction, is complicated by the following. The viewed plains are not only linear/nonlinear in the common case, but they also include ontology constraints or at least some fuzzy estimates. The usage of fuzzy logic significantly raises the algorithmic complexity of the problem and it can make the application ineffective. Even when the usage of constraints significantly reduces the number of the inspected solutions, for example up to 10, this does not mean that the problem is solved and that all that must be done is to explore the possibilities one by one.

**Example**. Let's admit that the goal is research of the intrusion that has led to an outflow of information from the department. By determining a system of five constraints, the problem is localized and it is reduced to the set: a combination of worms+troyans, worms+keyloggers, worms+rootkits, worms+phishing software, worms+social engineering. Also, let's admit that the solution is in the third element in the series (the combination of a worm that has assisted the followed installing a rootkit program). Here the solution seems to be rather simple and it could be detected even applying the classical rule for the excluded third: the third element seems to be the most suitable for the described situation while the rest are not so applicable. The problem with this task is that by restricting the search space, we do not always automatically reach the enumeration of the elements inside the domain. The knowledge from the subject domain is in practice always incomplete and/or incorrect. Therefore, the knowledge about the specified elements must be acquired otherwise the constraints do not lead to the discovery of the best solution. After finding the necessary constraints and narrowing the domain, the solution still depends on the available knowledge about the problem. Let's propose that there are known worms+troyans, worms+keyloggers, and worms+phishing software as instigators of violations in this case but there is no knowledge about rootkit programs. Then the research of the constraints of the restricted domain will not lead to finding a solution, just a check will be performed about the presence of troyans, keyloggers, and phishing software; due to the negative results the research will continue in wrong directions whether or not a new representative of the cited three types of threats has appeared. This may take precious time and increase the effect from the damages. The most effective outcome, in this case, is to formulate the area of the popular solutions, and also to make a consultation with experts whether worms+troyans, worms+keyloggers, and worms+phishing software are all of the possible solutions in the case and in this way to reach the unknown solution worms +rootkits. In the example, the unifying factor, the common area is 'opening side doors after an infection with a worm'. It is evident that the very formulation of the area leads to an acceleration of the solution process. The idea of this example is to show that introducing ontologies which in this case mean 'opening side doors after an infection with a worm' replaces hard to solve systems of linear, nonlinear and fuzzy constraints and makes the search process for the solution more natural and predictable.

The example will expand on the next step in the following way. Let's admit that the solution worms+rootkits is still unknown but the knowledge accumulated up to the moment, *point* to it: they are found on the disk directories with specific names, the system behaves in a characteristic way in certain modes, some functions are disabled, it is impossible to activate the Task Manager under Windows, etc. By assembling the 'puzzle' and looking for the links between its elements we should discover the necessary solution. Let's admit that $k_i \in K$ are the elements of the solution, or, same, of the *goal* **G**, and $w(k_i)$ are the weights of the elements on the condition that the weight of the goal equals to one. In different situations, a concrete measure is chosen and that is why the order of weights' processing is different. A popular example is measuring the length to the goal, how many symbols it contains, and then $w(k_i)$ equals to the length of the piece $k_i$ from the segment G. Then S from formula (2) shows the reliability of the inference 'G is true'. If $k_i \in K$ are continuous values then in (2) there must be applied integral calculus.

$$S = \frac{\sum_i k_i}{1 - \sum_i k_i} \tag{2}$$

It follows from (2) that the bigger part of the goal G is confirmed by the knowledge, the easier is the goal confirmation. The detection of the point of intersection between the available knowledge and the unknown goal uses the below-considered method of the PUZZLE that binds the unknown knowledge (the goal) to the known knowledge (the accumulated up to the moment information). More details are given in Sects. 4 and 5.

## 3.2 The Basics of the Funnel Method

Below, we discuss in brief the next proposed set of FUNNEL methods. Figure 4 presents the main elements of the method: a system of constraints in the form of a funnel around a center which is a goal (fitness) function which points to the desired

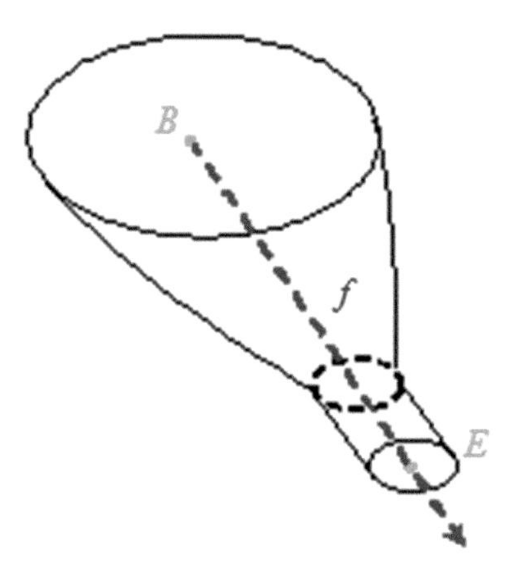


**Fig. 4** The funnel set of constraints

direction for information output or to search for new knowledge. As it is evident, the goal of this method is the gradual narrowing of the space of the feasible solutions, together with the progress of the dynamic information processes. Usually, the FUNNEL method operates properly when combined with the other methods introduced here and that is why its peculiarities are viewed in detail in the next subsections where the interactions between the methods are examined.

For example, it is convenient to concentrate on Fig. 4 shown above over one of the peak values of the diagram by fixing a funnel above it. The funnel function $f$ is analogous to the fitness function from genetic methods. At the beginning, in point B (beginning), the resolution area (search area or goal investigation area or observation area etc.) is very broad and the only restriction to resolution movements is that they should follow directions $f \pm 90°$. Gradually the restrictions changes in accordance with the current funnel relief and near the point E (end) the restrictions are very hard. Many funnel models may be applied to resolve different goals. In many of them the search trajectory depends from function $f$ just as Earth objects movements depend on gravitation: a weak but inevitable force.

## 3.3 Kaleidoscope Method Applications

The next method under SMM control is named KALEIDOSCOPE [5]. It is used mainly to visualize the results. The difference from other analogical methods [9, 10] is the usage of cognitive approaches to human-machine communication. Similarly to the classical kaleidoscope, the machine makes things that it performs better than humans: it arranges knowledge, it searches repeated elements and other regularities in the data, and it presents the results to the humans. It is the user that makes the results creative and interesting for other researchers in the domain; the machine just assists him/her analogously to the kaleidoscope which cannot estimate the beauty and the interest in the drawings produced during rotations and the arbitrary movements of the glass pieces inside of it. In both cases, the human estimates the results and the developers apply methods to attract the user's attention and to make the work with the system equipped with elements from Game Theory, less boring and consequently more effective.

KALEIDOSCOPE consists of a visualization approach, a method for information transfer-by-sense and applied approaches for maintenance of a natural-style dialog. It is shown in [5] how different machine-made visualized patterns lead to user's perceptions that cannot be represented by the machine and when this is helpful during the research or educational process. Aiming to show domain-independent approaches, the examples are introduced from Number Theory to language expressions and to nonclassical logic applications.

The Number Theory examples aren't chosen on a random basis. The first knowledgeable, significant research in this direction is since the dawn of artificial intelligence. The models from this field are very simple formulas, they include no heuristics or other handicaps. On the other hand, here we deal with very serious notions like infinity, which is usually explored through inductive research.

For example, on Figs. 5 and 6 are shown multiplication cycles for numbers divisible by 11 in a set of 8 arithmetic progressions ($\{1+30k\}_{\mathbf{k=0}}^{\infty}$, $\{7+30k\}_{\mathbf{k=0}}^{\infty}$, $\{11+30k\}_{\mathbf{k=0}}^{\infty}$, $\{13+30k\}_{\mathbf{k=0}}^{\infty}$, $\{17+30k\}_{\mathbf{k=0}}^{\infty}$, $\{19+30k\}_{\mathbf{k=0}}^{\infty}$, $\{23+30k\}_{\mathbf{k=0}}^{\infty}$ and $\{29+30k\}_{\mathbf{k=0}}^{\infty}$) left as a result of sieving out the first three prime numbers and their co-multipliers (composites) as well. The figures provoke questions **why** the images are repeated periodically across 11, or, respectively 19 rows/lines, **why** their tour sequence is in a reverse order (11–19 = 30–11) related to an imaginary center—the progression $\{15+30k\}_{\mathbf{k=0}}^{\infty}$ and many other questions the answers to which led us to interesting theorems from Number Theory, cryptology applications, and many other results. In the case of KALEIDOSCOPE the machine performs mechanical arrangement searches of repeated elements, etc. by presenting the results in a dialog with the human; it is expected that the human will see the informal elements of the picture and will draw his/hers conclusions. The correct organization of the dialog and the presentation of the results is the essence of the KALEIDOSCOPE method.

Also, questions appeared: are the other sets of arithmetic progressions containing all prime numbers p > N fruitful as the set depicted in Figs. 5 and 6? If not then **why**? Even the simplest question why the progression step equals multiplication of excluded divisors 2, 3 and 5 brought useful results. In the mathematical sense, it is very obvious, but for intelligent agent development such questions and solutions to the problem (proofs) are very important. All of the considered questions initiated logical treatment processes and error correction processes concerning the evolution of Number Theory results from questions on the visualized materials via heuristic hypotheses to a set of Number Theory lemmas and Theorems. Some the chosen results are displayed in Sect. 5.4.

It is known that the prime number function π(x) is irregular. Not surprising is the fact that a number of primes in each of depicted on Fig. 5 arithmetic progressions tend to be also irregular (approx. π(x)/8). Anyway, there exist no simple proof to this simple to find hypothesis. The main question appeared: if π(x)/8 is the middle value for calculation of prime numbers from *any* of the depicted progressions from Fig. 5, then what is the maximum possible deviation from this value π(x)/8? In the end, it appeared that the deviation function is π(y), and it is same irregular function as π(x) where y equals a square root of x. Then for every enough large **x**, π(y) is a negligible value compared to π(x), hence the number of prime numbers in every quoted progression is practically the same. Hence, they are replaceable. Hence, if some features from one arithmetic progression have been explored, for the sake of simplicity it can be shifted up by another progression, and the error level is π(y).

Previous experience, the accumulated knowledge is the best indicator whether the research and discovery process goes wrong or is in the correct direction. Really, many peculiarities could be discovered exploring Fig. 5. The exploration process isn't an urgent one, and there is a huge gap between the visual discovery and the desired proof. Just to mention, the machine should extrapolate the finite area from Fig. 5 into infinite sets of numbers, and cope with other difficulties. The successful

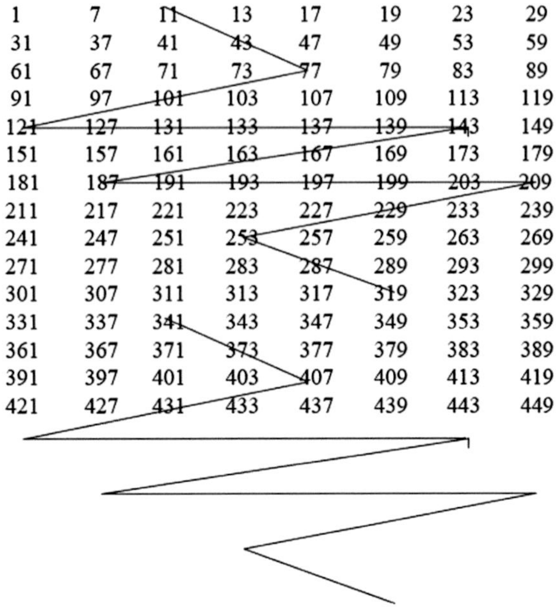

**Fig. 5** Multiplication cycle for divisor 11

| 1 | 7 | 11 | 13 | 17 | 19 | 23 | 29 |
|---|---|---|---|---|---|---|---|
| 31 | 37 | 41 | 43 | 47 | 49 | 53 | 59 |
| 61 | 67 | 71 | 73 | 77 | 79 | 83 | 89 |
| 91 | 97 | 101 | 103 | 107 | 109 | 113 | 119 |
| 121 | 127 | 131 | 133 | 137 | 139 | 143 | 149 |
| 151 | 157 | 161 | 163 | 167 | 169 | 173 | 179 |
| 181 | 187 | 191 | 193 | 197 | 199 | 203 | 209 |
| 211 | 217 | 221 | 223 | 227 | 229 | 233 | 239 |
| 241 | 247 | 251 | 253 | 257 | 259 | 263 | 269 |
| 271 | 277 | 281 | 283 | 287 | 289 | 293 | 299 |
| 301 | 307 | 311 | 313 | 317 | 319 | 323 | 329 |
| 331 | 337 | 341 | 343 | 347 | 349 | 353 | 359 |
| 361 | 367 | 371 | 373 | 377 | 379 | 383 | 389 |
| 391 | 397 | 401 | 403 | 407 | 409 | 413 | 419 |
| 421 | 427 | 431 | 433 | 437 | 439 | 443 | 449 |

strategy is to build hypothetical theorems like the one from the previous section, and then find ways to prove them one by one. This approach is data-driven, it resembles the means-ends analysis, but the knowledge-driven part had been performed by the human. How could we make conclusions that the research direction is right one, and the time and resource are not wasted (cf. the next section on Puzzle methods: same questions are very important there)? Typically, after the accumulation of new results, proofs appeared of well-known coefficients or formulas, or of other results obtained statistically. Those results had been kept w/o any proof, and during the considered research the new proofs revealed their semantics, origin, their background. At the same time, the quoted ‘avalanche’ of new results was the best indicator: the direction is right. Just to mention, many of the considered in Sect. 5 practical Number Theory results helped us to answer the question **why** probability methods are so effective in different Number Theory branches. The quoted research examples clearly show the vision for the effective evolution of a *bundle* of contemporary statistical and logical methods where statistics are activated in the beginning, in knowledge-poor environments. Well-known statistical results, coefficients, hypotheses are used. Later more versatile logical applications reveal deep foundations of the quoted processes, and in the end statistical and/or logical theories may be applied at the highest levels of the research. Many famous elementary Number Theory hypotheses are solvable using the quoted prime number distribution formula. All of the considered theoretical results are provable using intelligent tools.

**Fig. 6** The cycles for composite numbers divisible by 19

| 1 | 7 | 11 | 13 | 17 | 19 | 23 | 29 |
|---|---|---|---|---|---|---|---|
| 31 | 37 | 41 | 43 | 47 | 49 | 53 | 59 |
| 61 | 67 | 71 | 73 | 77 | 79 | 83 | 89 |
| 91 | 97 | 101 | 103 | 107 | 109 | 113 | 119 |
| 121 | 127 | 131 | 133 | 137 | 139 | 143 | 149 |
| 151 | 157 | 161 | 163 | 167 | 169 | 173 | 179 |
| 181 | 187 | 191 | 193 | 197 | 199 | 203 | 209 |
| 211 | 217 | 221 | 223 | 227 | 229 | 233 | 239 |
| 241 | 247 | 251 | 253 | 257 | 259 | 263 | 269 |
| 271 | 277 | 281 | 283 | 287 | 289 | 293 | 299 |
| 301 | 307 | 311 | 313 | 317 | 319 | 323 | 329 |
| 331 | 337 | 341 | 343 | 347 | 349 | 353 | 359 |
| 361 | 367 | 371 | 373 | 377 | 379 | 383 | 389 |
| 391 | 397 | 401 | 403 | 407 | 409 | 413 | 419 |
| 421 | 427 | 431 | 433 | 437 | 439 | 443 | 449 |
| 451 | 457 | 461 | 463 | 467 | 469 | 473 | 479 |
| 481 | 487 | 491 | 493 | 497 | 499 | 503 | 509 |
| 511 | 517 | 521 | 523 | 527 | 529 | 533 | 539 |

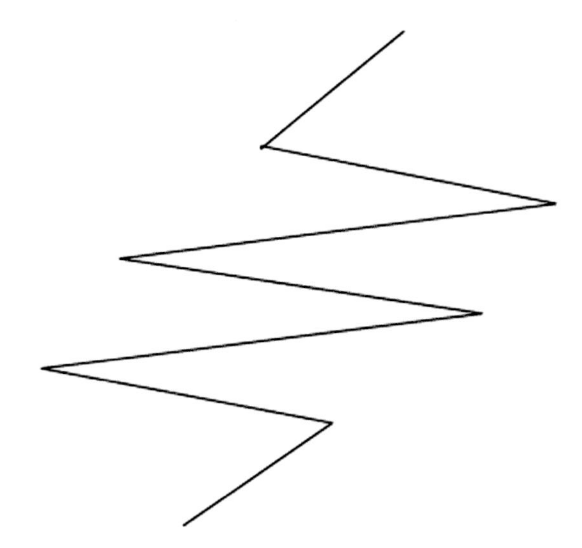

As a result of same explorations, new cryptography-oriented formulas for ciphers had been invented. The considered visualization methods lead to rather simple formulas, as a whole they didn't lead to a discovery of new algebraic objects, but they provoked an elaboration of a large set of formulas and, in the end, theorems from elementary Number Theory. This set is constantly growing and leads to unexplored, but presumably underestimated fields. It is obvious that above shown results and environment (model) will strengthen the security level of ISS applications.

For above mentioned logical, and, in the end, algebraic proofs we widely used conflict resolution methods. The next one from the methods discussed here is the inference from conflict/contradictory information. This method is presented with more details in [5]. On the first stage, based on the check for preliminary input knowledge as models of conflicts, the system determines whether there is a conflict. If the sides of the conflict are more than two, then the problem is reduced to the resolution of a set of 2-sided cases. Depending on the situation, the methods for the solution are automatic or interactive.

It is the conflict resolution that allows to eliminate the incompleteness from situations similar to the one depicted on Fig. 5/6, and also to correct insignificant elements and/or to present the situation more clearly.

In this section, basic methods of the synthetic metamethod SMM are presented. The next section is dedicated to the coordination of their operation under the control of SMM.

## 3.4 Conflict Identification

Any lack of collaboration in a group of agents or the any intrusion could be defined as an information conflict with the existing models. Many methods had been elaborated where a model is given and every non-matching it knowledge is assumed as contradictory or as a form of semantic conflict. Let's say, in an anomaly intrusion detection system, if the traffic has been increased, it is defined as a conflict to the existing statistical data and an intrusion alert has been issued. The below considered approach is to discover and trace different logical connections to reveal and resolve conflict information. The constant inconsistency resolution process gradually improves the system database (DB) and/or KB, and leads to better intrusion detection and/or prevention. Models for conflicts are introduced and used, and they represent different forms of ontologies.

Let the strong (classical) negation be denoted by '$\neg$' and the weak (conditional, paraconsistent negation be '$\sim$'. In the case of an evident conflict (inconsistency) between the knowledge and its ultimate form, the contradiction (semantic conflict, etc.), the conflict situation is determined by the direct comparison of the two statements (the *conflicting sides*) that differ one from another by just a definite number of symbols '$\neg$' or '$\sim$'. For example A and $\neg$A; B and not B (using $\neg$ equivalent to 'not'), etc.

In the case of implicit (or hidden) negation between two statements, A and B can be recognized only by an analysis of type (3) models.

$$\{\mathrm{U}\}[\eta : \mathrm{A}, \mathrm{B}] \tag{3}$$

where $\eta$ is a type of negation used in the model, U is a statement with a validity including the validities of the concepts A and B (*the sides*), and it is possible that more than two conflicting sides may be present. It is accepted below that the contents in the formula (3) in brackets U is called *a unifying feature*. In this way, it is possible to formalize not only the features that separate the conflicting sides but also the unifying concepts logically joining the sides. For example, the intelligent detection may be either automated or of a human-machine type but the conflict cannot be recognized directly and without the investigation of the following model.

$$\{\text{detectors}\}[\neg\text{: automatic, interactive}].$$

The formula (3) formalizes a model of the conflict the sides of which implicitly negate each another. In the majority of the situations, the sides participate in the conflict only under definite conditions: $\chi_1, \chi_2, \ldots \chi_z$.

$$\{\mathrm{U}\}[\eta\text{: } \mathrm{A}_1, \mathrm{A}_2, \ldots \mathrm{A}_\mathrm{p}]\,\langle\tilde{\chi}_1 * \tilde{\chi}_2 * \ldots * \tilde{\chi}_z\rangle. \qquad (4)$$

where in addition to what is explained in (3), $\tilde{\chi}$ is a literal of $\chi$, i.e. $\tilde{\chi} \equiv \chi$ or $\tilde{\chi} \equiv \eta\chi$, * is the logical operation of conjunction, disjunction or implication. Only classical logical cases had been explored till now.

The present research allows a transition from models of conflicts to ontologies in order to develop new methods for revealing and resolving semantic conflicts, and also to expand the basis for cooperation with the Semantic Web community and with other research groups. This is the way to consider the suggested ontology-type models from (3) or (4) as one of the forms of static ontologies.

The following factors $\chi$ have been investigated:

T time factor: non-simultaneous events do not bear a conflict

M place factor: events that have taken place not at the same place, do not bear a conflict. In this case, the concept of place may be expanded up to a coincidence or to differences in possible worlds

N a disproportion of concepts emits a conflict. For example, if one of its parts is a small object and the investigated object is very large (huge), then and only then it is the conflict case

O identical object. If the parts of the contradiction are referred to different objects, then there is no conflict

P the feature should be the same. If the conflicting parts are referred to different features, then there is no conflict

S a simplicity factor. If the logic of user actions is executed in a sophisticated manner, then the conflict is detected

W a mode factor. For example, if the algorithms are applied in different (wrong) modes, then the conflict is detected

MO conflict to the model. The conflict appears if and only if (*iff*) at least one of the measured parameters does not correspond to the meaning from the model. For example, the traffic is bigger than the maximal value from the model.

**Example**. We must isolate errors that are done due to lack of attention from tendentious faults. In this case following model (5) is introduced:

$$\{\text{user : faults}\}[\sim: \text{accidental, tendentious }]\langle \text{T}, \neg\text{M}, \text{O}; \neg\text{S}\rangle \quad (5)$$

It is possible that the same person does sometimes accidental errors and in other cases tendentious faults; these failures must not be simultaneous on different places and must not be done by the same person. On the other hand, if there are multiple errors (e.g. more than three) in short intervals of time (e.g. 10 min), for example, during authentications or in various subprograms of the security software, then we have a clear case of a violation, and not a series of accidental errors. In this way, it is possible to apply comparisons, juxtapositions, and other logical operations to form security policies thereof.

## *3.5 Main Ways for Semantic Conflict Resolution*

Recently we shifted conflict models with ontologies that give us the possibility to apply new identification/resolution methods. For pity, the common game theoretic form of conflict detection and resolution is usually heuristic-driven and too complex. We concentrate on the ultimate conflict resolution forms using contradictions. For the sake of brevity, the resolution groups of methods are described schematically.

The conflict recognition is followed by its resolution. The schemes of different groups of resolution methods have been presented in Figs. 7, 8, 9, 10 and 11.

In situations from Fig. 7, one of the conflicting sides does not belong to the considered research space. Hence, the conflict may be not be immediately resolved, only a conflict warning is to be issued and processed in the future (nothing urgent). Let's say, if we are looking for an intrusion attack, and side 2 matches printing problems, then the system could avoid (postpone) the resolution of this problem. This conflict is not necessary to be resolved automatically, experts may resolve it

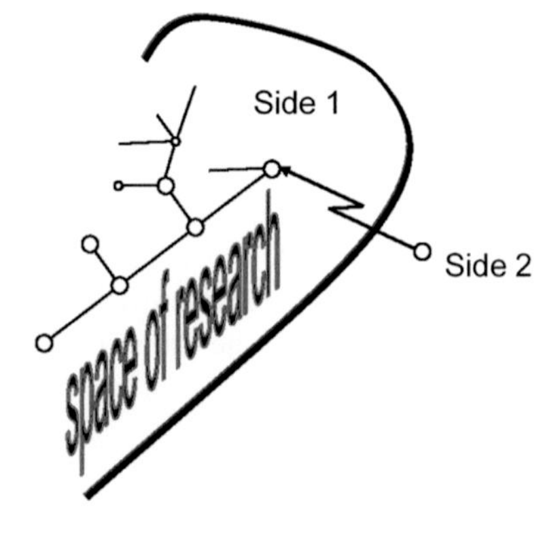


**Fig. 7** Avoidable (postponed) conflicts when *Side 2* is outside of the research space

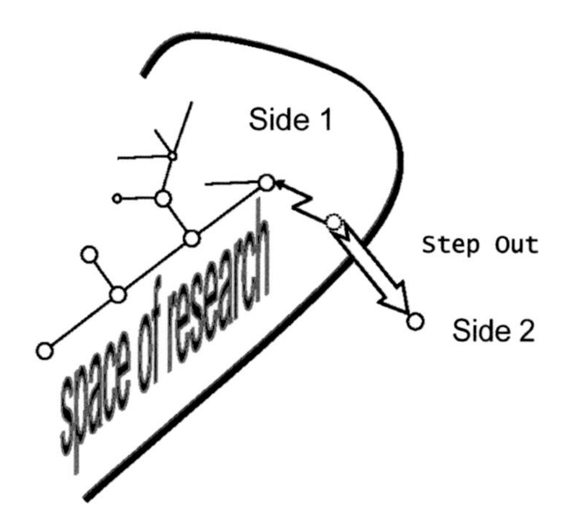


**Fig. 8** Conflict resolution by stepping out of the research space (postponed or resolved conflicts)

later using the saved information. In Fig. 8, a situation is depicted where the conflict is resolvable by stepping out from the conflict area. This type of resolution is frequently used in multiagent systems where conflicting sides step back to the pre-conflict positions and one or both try to avoid the conflict area. In this case a warning on the conflict situation also should be issued. The 'step out' process depend of the explored goals.

The situation from Fig. 9 is automatically resolvable without issuing a warning message. Both sides have different priorities, let's say side 1 is introduced by a security expert, and side 2 is introduced by a non-specialist or is a machine hypothesis with lowest possible priority. In this case, side 2 has been removed immediately. A situation is depicted on Fig. 10 where both sides have been derived by an inference machine by using deduction. In this case, the origin for the conflict could be traced, and the process is using different human-machine interaction methods. Every conflict is depicted as a lightning. The more are the revealed conflicts, the more easy is the main conflict resolved because every conflict is a form of restriction (constraint).

At last, the conflict may be caused by incorrect or incomplete conflict models (3) or (4), in this case they have been improved after same steps as shown in Figs. 6, 7, 8, 9 and 10. The resolution leads to knowledge improvement as shown in Fig. 11, and this evolutionary improvement gradually builds a machine self-learning. Of course in many situations from Figs. 6, 7, 8, 9, 10 and 11 the machine prompts the experts just in the moment or later, but this form of interaction is not so boring,

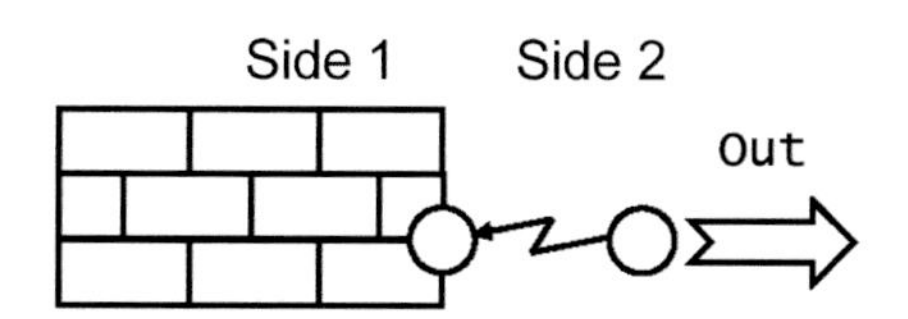


**Fig. 9** Automatically resolvable conflicts

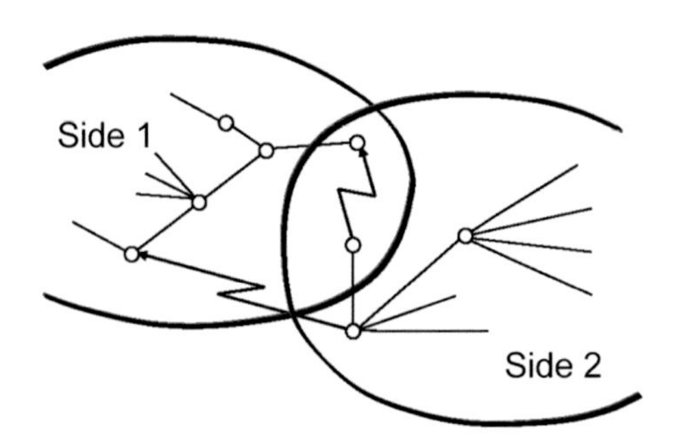


**Fig. 10** Conflicts resolvable using human-machine interactions

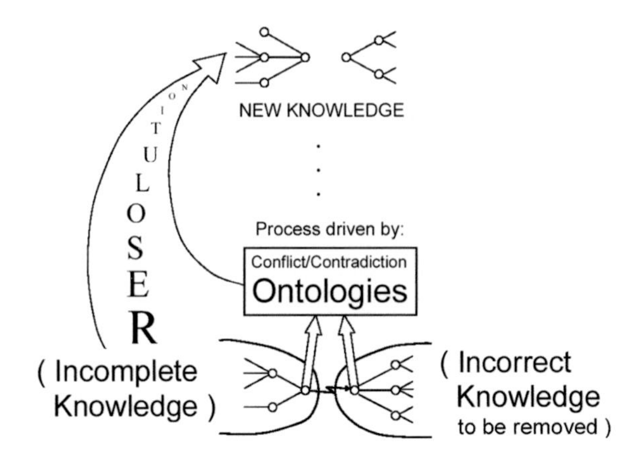


**Fig. 11** Learning via resolving conflicts or with contradictions

because it appears in time and it is a consequence of a logical reasoning based on knowledge (3) or (4). Our research shows that automatic resolution processes may constantly stay resident using free computer resources OR, in other situations, they may be activated by user's command. In the first case the knowledge and data bases will be constantly improved by continuous elimination of incorrect information or by improving the existing knowledge as a result of revealing and resolving contradictions. As a result, our conflict resolution methods have been upgraded to a machine self-learning method, i.e. learning without a teacher which is very effective in the case of ISS. This type of machine learning is novel and original in both theoretical and applied aspects.

## *3.6 Self-organization of Evolving Systems Via Resolution of Conflicts or Contradictions*

The common drawback of many modern information security tools is that they cannot process data in an evolutionary mode without the direct human (user/expert) interference. The considered evolution is without known fitness function but with an adjustable, variable, 'data-driven' one. On the other hand, subjects that cannot autonomously process data and knowledge evolutionarily, are out of perspectives to learn *how to think*.

The presented evolutionary processing of data or knowledge is based on two main procedures. Objects including agents and their surroundings should be modeled. For this purpose, the usage of the introduced ontology relations like HOW, WHY, IMPORTANT, and other earlier discussed becomes greater and more important. The richer the arsenal of the ontology means used, the more effective the system operation is. The second information process registers changes of the incoming information with respect to the models. This is the place to use various methods to compare and to calculate distances via the usage of the different metrics; to reduce the application complexity of the agents. It is recommended to apply more elementary methods as juxtapositions, generalizations, etc. Let us assume, that, for example, the change of time for attribute $A_1$ of ontology $O_i$ is denoted $\Delta A_1(\Delta t)$:

$$\Delta A_1(\Delta t) = \|A_1(t + \Delta t) - A_1(t)\| \quad (6)$$

If $\Delta A_1(\Delta t) \leq T$ where T is an a priori given threshold value then it is accepted that the change is insignificant. In this case, there is no need to expect reactions from the agent. On the other hand when $\Delta A_1(\Delta t)$ is compatible with T then it is necessary to inspect a greater time slice $\Delta t$ to explore the dynamics of the change, asking whether there is a gradual accumulation of cardinal changes. Attributes may change without using any temporal factor from (6). For example, in the case $\Delta A_1 = \| A'_1 - A_1 \|$, the formula may be deciphered as follows. The critical mass of attacking computers from botnets will block the provider, but there is no data **when** this will happen. Hence, there is no need to include the temporal factor in the goal description.

Another example is a world-wide chart of cyber criminality. Based on it, the agents can draw useful information about the regions with lowest and highest criminality and they can search the reasons for it. With the same success the domain may be shifted: if an example from Number Theory is considered, then areas with most density of composite or prime numbers are established and the reason for such distribution is analyzed. The presented evolutionary information processing is based on the detection, analysis and isolation of conflicts/inconsistencies/contradictions thus passing through them, aiming improved models. In spite of the theoretical nature of the application (cryptography) or the heuristic problems (information input about the reasons for cybercriminality in various regions world-wide), the attention is concentrated over extremities, e.g. the regions with highest or lowest criminality. Analysis of the data with extreme values most often lead to conflicts which in knowledge-poor environments must be resolved in an evolutionary way. Evolving methods from previous sections give new ways to resolve conflicts/contradictions which are not discussed above. Their application is successful not only in separate agents but also in multiagent systems (MAS) that can be a type of collective evolutionary systems. With that, as compared to popular swarm and other strategies, the coordination applied in the present research is not controlled heuristically but via constraint satisfaction for the constraints imposed by conflicts or contradictions. For more details see descriptions of the semantics of Puzzle applications, as well as of the introduced evolving methods.

Let's assume that the model M* will be used where attributes of one or more objects are from the following set $A = \{A_1,\ldots, A_N\}$. The very model M* can be presented as one of the parameters of the individual from the population, so it must be replaced by M** which is chosen due to its similarity estimate with respect to M* or according to other properties in the accepted classical methods of genetic algorithms (GAs). It is possible to change the examined set in M**. With that, after one or more iterations the set A will contain just elements that are *invariant* to model changes. The greater the distance between M* and the current model, the more important the role of the invariant attributes will be. On the other hand, there is a difference from classical GAs in the way that there is a possibility of the evolution, mutations and crossovers to be independent of probabilistic estimates or from different forms of heuristics, they can depend on the constraints instead.

In the interpretation proposed above, machine-type thinking is a reaction to a change (of time, environment, etc.) and the greater the change is, the greater attention will be paid to explore the reasons leading to its appearance. According to this logic, training is not only a memorizing teacher's activities as it is in the classical case but also uses resolutions of inconsistencies, conflicts, contradictions via evolutionary information processing; in other words, the strategy 'doubt in everything' is used in the proposed agents. The transition from M* to M** inspected above is a single case of this strategy. In this way, elements of unsupervised training are added and even of self-organization in the case of a (un)supervised training. Various practical applications of the strategy are investigated because the trained agent often cannot change the training scenario or it cannot ask the supervisor counterexamples. As a result of the proposed evolutionary way of training, acquired knowledge is constantly refined or corrected. Here a far analogy with antiviral software can be observed: the agent could gradually improve its KB by using such technology and constantly checking every piece of it.

### 3.6.1 Self-learning Via Knowledge Comparisons

In our opinion, it is almost impossible to construct necessary data driven tools by using only classical logical operators. Below is represented a rather alternative way to the resolution of this problem.

The proposed self-learning procedure is composed of three basic steps. During the first stage, the conflict is detected by using models from (3) or (4). In a second stage, the contradiction is resolved by using one of the above schemes. As a result of actions taken after the second stage set of knowledge K is converted to K′, by K a subset of incorrect knowledge $W \subseteq K$ can be eliminated aiming to correct a subset of knowledge with incomplete description of the subject area $I \subseteq K$, to add a subset of new knowledge about the problem $N \subseteq K$ and replenish a subset of knowledge to be specified $U \subseteq K$. The last of the aforementioned subsets include deferred issues discussed in the version described in Fig. 7, knowledge of possible non-compliance with expertise (knowledge in question) and other knowledge subjects to further research found on the basis of heuristic information.

Using ontologies of metaknowledge pieces or more complex forms of control strategies, the removal of knowledge and completing KB is a non-trivial task. In the Semantic Web, in particular for the purpose WSMO concepts orchestration and choreography of ontologies are introduced. Removing at least one of the relationships between the knowledge might lead to inaccuracy in one or more subsets of information in K. So after that second stage, the third stage is introduced: checking the relations which include the removed or modified knowledge. New knowledge of the subsets W, N, I, U is also checked for consistency by the procedure described in the section above. Upon successful completion of the process a new set of knowledge K′ is formed, which is of better quality than K and this criterion is the result of unsupervised machine self-learning driven by a priori defined models of conflict and managing strategies with or without the use of metaknowledge.

Relationships in connection with the sides to the conflict or other knowledge from K are not always imaginable through the implications, standard type links is-a/has-a and many other known connections. Figure 12 presents two sets of objects associated with both sides of the explored conflict. Some of the objects are associated with one side by implication $\Rightarrow$, $\subset$, $\supset$, is-a or other relation. The majority of the objects do not have this kind of connections, but enter into the area around each side marked by dotted area. For example, let the following relation between two concepts or two objects is confirmed by statistical means. Attacks are often accompanied by port scanning, but the fact that there is a scan does not mean that there will be an attack. In this case, the interpretation is ‘object has any bearing on the research side of the contradiction, in other words, the nature of the relationship is not fully understood. It is therefore necessary to place an objective ‘elucidation of causes and the nature of the relationship’. In most cases, after examination of the purpose the relation is reduced to a known relationships mentioned above, or for instance is defeated (canceled) and if the exception to the rule is true, then some of the relations in Fig. 12 can eliminate or change its direction. If the goal (objective) ‘to clarify the causes and nature of the relationship’ is unsolved for now, the knowledge connected by this relation forms a set of heuristic confirmations or denials of knowledge of the figure and can be used for conflict resolution, and so is the classic case of knowledge conflict.

Figure 10 represent decision trees in which by definition all logical links are in the form of classical implications. On Fig. 12 one of the relations in the area around

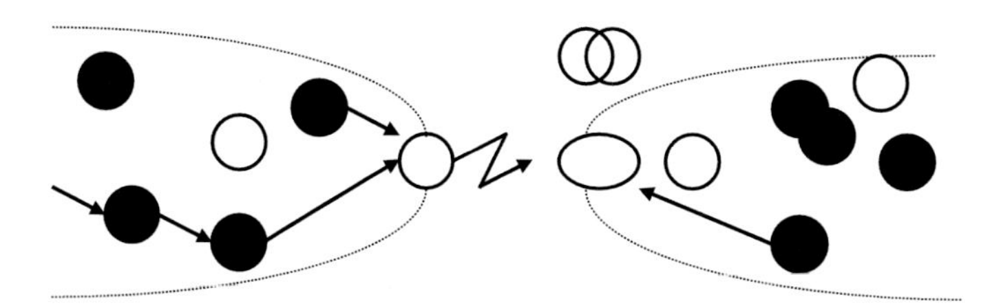

**Fig. 12** Sets of concepts relating to conflicting sides

the left side of the figure could be is-a, and the other is the implication, classical case. It is possible here to use classical and non-classical means in one scheme. In this case, the discovery of the causes behind the conflict resolution is stumped when using classical formal logical approaches and in this case it is necessary to use the classical applications. Here original methods and applications are offered for information cancellation, etc. It is known that each KB over a thousand knowledge units inevitably contains at least one semantic conflict. Conflicts are detected and resolved in accordance with the above considered ways. Each of the methods from each resolution way is described by specific algorithms, but a result of their use is not always algorithmically predictable. Thus the solutions have intrinsic properties of intelligent data processing (knowledge acquisition/discovery, data analytics, data mining): results depend on the data or knowledge. In this way data driven/knowledge driven applications have been constructed. *Passing through the contradictions* improves the quality of the accumulated information in the criteria mentioned above. Accordingly, the non-essential is mostly automatically removed, which leads to increasing the quality of information. Thus one of the effects of a famous principle that was formulated during the XII century is realized. The principle has been examined since the dawn of computing, and has not been fully realized till today. Occam's razor is the following: choose a simple, elegant, rejecting the unnecessary and complicated. The method of finding contradictory information allows removing the deficiencies in crisis situations and correcting the non-essential and/or providing a clearer situation as was done in the transition from Figs. 13 and 14.

- Figure 13 is a summary graph showing the trends in intelligent security systems (ISS). Presented are the average values for the parameters in the current ISS price ($), time (years) and quality (%).

The last of quoted parameters operates the fuzzy concept of "quality", hence drawing the figure are used not only data but also knowledge. Figure 14 presents the same information by excluding various non-important data/knowledge and small deviations from the trends in order to clarify the strategic picture (core

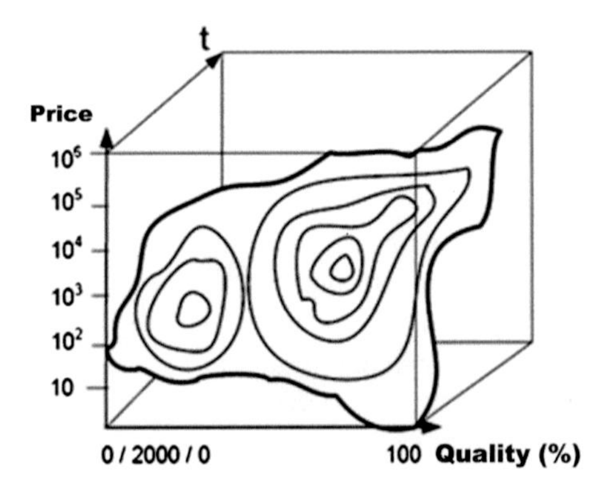


**Fig. 13** Comparison charts on different ISS

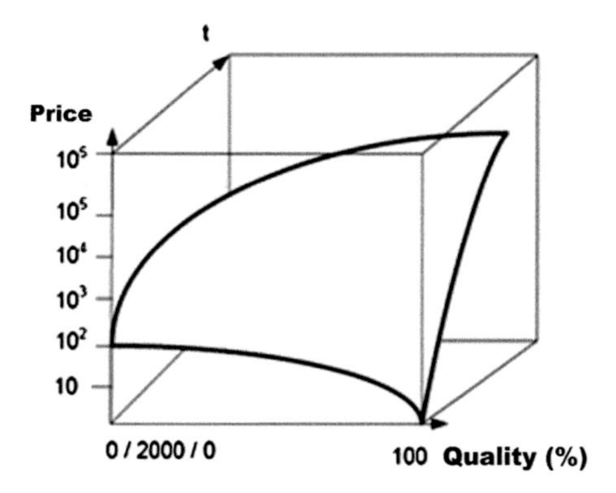


**Fig. 14** Same trends after the elimination of non-essential elements

knowledge) from the field. This type of processing can be discussed as another Occam Razor application.

The statistics of Fig. 14 show that over time poor designed or difficult adaptable systems gradually drop out of the general trend and the crisis accelerates this process. The price of a quality system is constantly growing, but even the most expensive of them today provide fewer safety guarantees compared to the past. Therefore, significant changes are needed in the security software to address the situation.

In this chapter are proposed new methods for extracting information from KB or through man-machine dialogue. Under the overall control of the logical meta-method, intelligent analysis of information SMM can be applied successfully to the known statistical methods. Conflict resolution does not always clarify all issues related to it. Depending on the way to detect and solve the conflict the answer to "what did the conflict lead to" or "why did it appear" or "how to repair knowledge leading to it" are found. But questions such as "Are there other conflicts and contradictions", "whether the discrepancy is caused by an expert error or expert's working style", "how long is the quality of knowledge poor", " how long will agents clash in a given situation" "why do agents enter contradictions" and many other similar questions still can't be used.

Placing such questions as targets, agents have more opportunities to improve the KB and to advance without a teacher, although in certain circumstances an expert assistance may be required. In this case our, ontological development will help to introduce new types of relations type like the described ORIGIN, HOW, WHY, and more. After the study of the set of above-mentioned targets, the relations can be automatically included in the system of existing ontologies. Thus, the connection agent-ontology closes and provides new opportunities for deep processing of knowledge of the agents. Conflicts or contradictions between knowledge pieces help to create restrictions in the formation of ontologies, and this can help specify the individual attributes of the ontology as well.

Parallel operation with distributed artificial intelligence, for example by using Petri nets, or fuzzy intuitionistic logics, brings significant advantages in the realization of the described methods. Figure 15 presents results of self-assessment of the agent's actions based on its used knowledge. The truth values are used from

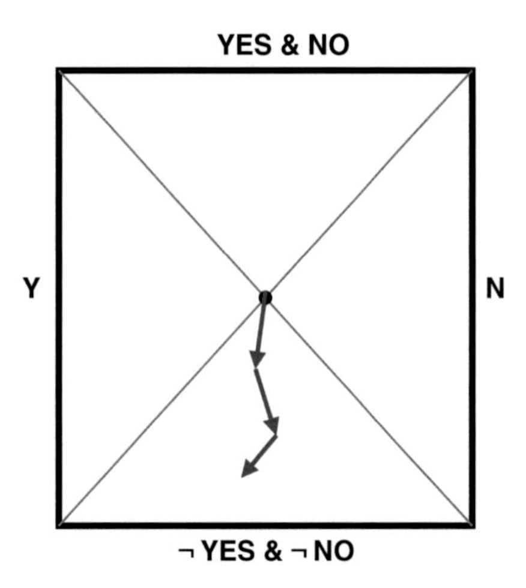


**Fig. 15** Parallel processing by comparing knowledge

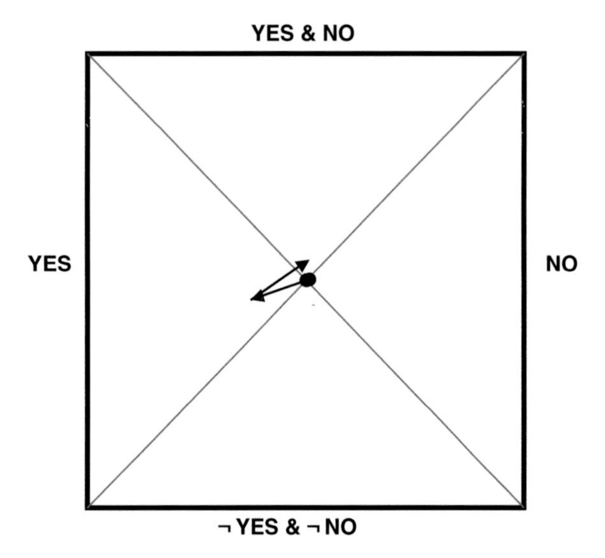


**Fig. 16** Evaluation of information in agent 2

four-valued logics. The figure shows that the intermediate calculations do not lead to the desired final result, answer "Yes" or " No" to the objective, or at least one of the four possible answers. Let another agent 2 involved in the conflict have assessment knowledge of Fig. 16. Let the results of Figs. 15 and 16 be compared after a preset period of time to resolve the conflict, for example after 5 min.

The comparison shows that agent 1 has a lot more progress in resolving the conflict in comparison to agent 2. Additional information relating to a comparative analysis of the parallel action of agents gives more opportunities to solve the problem in comparison to the analysis of the agents one by one. On the other hand, if we imagine same types of results for both agents, value Y reached on both Figs. 15 and 16, still we can prefer agent 1 if it reached the value Y in two ways (stored two different proofs of the same).

The results are mostly in the form of fuzzy estimates. They allow reducing considerably the algorithmic complexity of the problem of resolving conflicts or contradictions in multiagent systems.

### 3.6.2 Self-learning by Comparing the Selected Knowledge and KB

The procedure for the self-learning is composed of the three stages, as in the previous subsection. Differently, the algorithmic complexity of the solutions is significantly reduced. During the first stage, the contradiction is detected by using models from (3) to (4), where one of the parties is the checked knowledge Z, and the other side can be any other knowledge of KB. Second stage: the conflict is resolved by using one of the above schemes of the previous subsection, where one of the parties is fixed; it is the chosen for inspection knowledge Z. As a result of actions taken after the second stage set of knowledge K is converted to K′, where a subset of incorrect knowledge $W \subseteq K$ can be eliminated by K aiming to correct a subset of knowledge with incomplete description of the subject area $I \subseteq K$, to add a subset of new knowledge about the problem $N \subseteq K$ and replenish subset of knowledge to be specified $U \subseteq K$. The last of the aforementioned subsets includes deferred issues discussed in the version described in Fig. 7 knowledge of possible non-compliance of expertise (knowledge into question). After that second stage, the third stage is activated, check the relations, which includes the removed or modified knowledge and new knowledge of the subsets W, N, I, U are checked for consistency by the above-described procedure(s). Upon successful completion of the process is formed a new set of knowledge K′.

Self-organization in Evolving Systems Using Conflict Resolution

Processes for solving contradictions between two software agents are mostly evolutionary. In general, the goal (conflict resolution) is presented in the form of fitness function(s) and applying genetic algorithms (Gas) the necessary solution is found. Meanwhile, the examined above methods to resolve conflicts bring significant opportunities for improvement of existing methods of evolution calculations.

Evolution in agents occurs analogously to the evolution of living organisms, where significant change has taken place by chance, but only if necessary. Contrary to the normal conditions of existence (work) is a vivid example of the need to change, i.e. the engine of evolutionary change. When the fitness function is a task of conflict resolution, then the above methods have the following advantages compared to other methods of evolutionary computing. For more information on the terminology of evolving hardware, genetic algorithms, etc., book [5] is recommended.

1. The fitness function 'how to resolve the conflict' can't be set a priori because no one can predict all the imperfections of knowledge/data that can lead to formation or change of that function. Thus, in the process of operation of the proposed use automatically appears and varies (changes) the multitude of fitness functions.

2. Mutations, crossover and other most significant factors influencing the evolutionary process change as follows. Probabilities of their occurrence, for example, 2 % of mutations and 8 % for crossover can significantly change and most importantly, are not programed to appear randomly because it leads to too set or at least predictable, not data driven character of the evolution. In their place will be used ontology restrictions or other limitations posed by the emergence and/or conflict solution. Restrictions construct areas with a high probability of mutations respectively crossovers. Accordingly, in crossover individuals avoid mutations or enter mutagenic areas and these principles are only changed in exceptional situations, such as the appearance of insoluble conflicts.
3. Fitness function (or a set of fitness functions) can be altered or change direction during the solution of the problem (variable fitness function). The change takes place by the action of additional information. For example, if an agent has in its list of objectives "to find Agent X to jointly protect the computer network" and after the last attack is clear that agent X contributes to the attack, this agent X should not be looking for collaboration but seeking to avoid it. The case-based scenarios should be applied here.
4. When an individual or a model is successfully applied to a specific fitness function, then the individual model or fitness function can be replaced in order to create conflict. Similarly, collaborative behavior can be overridden to a competitive one. Artificial changes in order to create and resolve new conflicts by the above funds are used for checking the solutions and adaptation of successful solutions in other areas. One of the effects of evolution going through conflicts, crises and contradictions is clearing of non-essential details and unnecessary complications, in other words in this way can be realized a new application or achieved the effects of the quoted.
5. Occam's Razor.

### *3.7 The Basics Under SMM Coordination of Other Advanced Analytics Methods*

SMM is a synthetic method. Results from synthetic methods are more influenced by the coordination and the control among the components than by the components themselves. Usually, the components of the synthetic methods are various algorithms from traditional analytic methods but we also investigated communications between several synthetic methods under the control of one synthetic metamethod. The effect from an unqualified control in the domain of information security is very significant and it leads to a complete crash.

It is recommended that at any time during the operations with SMM a direct control of the user's processes should be possible. For example, to detect the causes that led to a breakthrough in the system, the user inputs a {PUZZLE, FUNNEL, KALEIDOSCOPE, CONTRADICTION} and in this way he/she engages all resources to detect the cause. The cited sequence of commands is applied by default. In this case, every method operates independently, computer resources are allocated equally and it is the user intervention that may change the course of the investigation. It is provided that the results can be collected in two ways. The first way is the *blackboard* option. Here results are easily transferred between different methods and also to the user via a central node that is called the *blackboard*. In the second option methods from multiagent systems are used and results are transferred via the so-called pheromones. This is information left by the agent which is kept intact for some time and which may be read only by agents that are close enough in space. The second option is especially good for transfer of operative information. Multiagent applications are discussed at the end of this section.

Another version for coordination is based on the use of statistical data and the personalization of information. For example, let's assume that statistics show that user X works most effectively with the PUZZLE method and also that he/she is interested least of all of the data visualization. In such case for him/her it is just the PUZZLE method that must be executed with the top priority, and the KALEIDOSCOPE must be activated only on special demands. On the other hand, as based on experts' estimates there must be a common system of priorities tied to the presented methods where the priorities are linked to the number of successive runs of the respective method. In this case, the methods with higher priorities will occupy more computational resources.

The role of tests for conflict resolution(s) and the KALEIDOSCOPE method play a special role in the system. KALEIDOSCOPE is a visualization tool for results by constructing special modes for the human-machine contact. Therefore, it must be activated after terminating the execution of any other method under the control of SMM. If the results do not require special technologies for their representation, then the method may be disabled manually.

In the example presented below the coordination set {FUNNEL-KALEIDOSCOPE} is used in a way shown on Figs. 17 and 18. Here the KALEIDOSCOPE effectively executes the user interface; in such a way real-time solutions are obtained.

**Fig. 17** Funnel overflow case

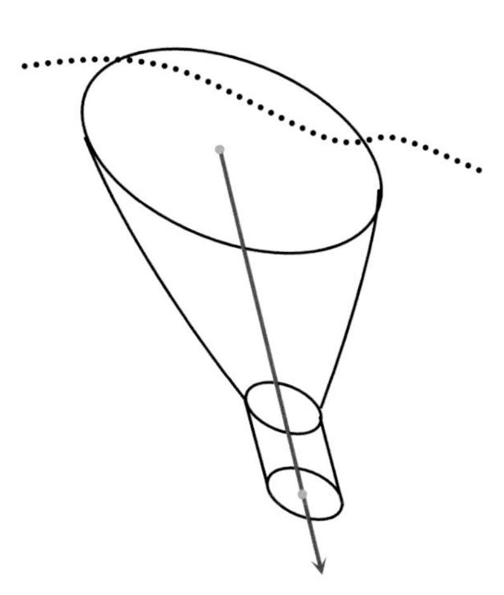

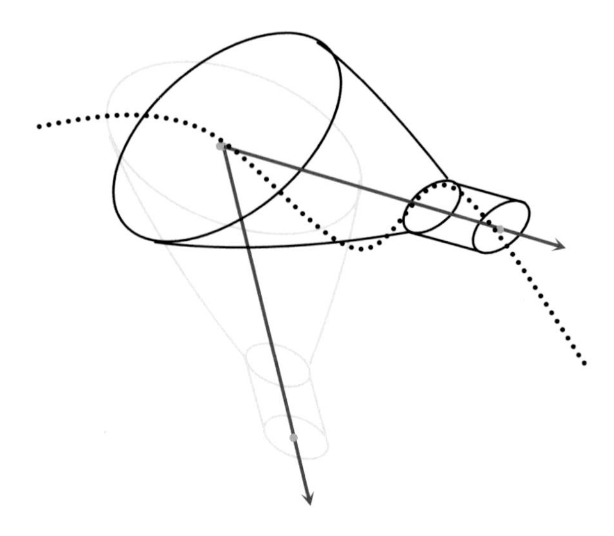

**Fig. 18** An adjusted FUNNEL case

Let's assume that the goal function/fitness function which are equivalent in the terms of evolutionary computation be marked by a central (lighter/red) line on Fig. 17. Let's assume that the solution process of the defined problem is presented by a dotted trajectory on the figure. The intelligent data processing from the example is a type of approaches that are data driven by nature. One of the properties of these approaches is the one that the solution may deviate from its way to the goal when lateral unexpected solutions are sufficiently interesting. A case with a deviation from the predicted trajectory is presented in the following way. Let the function G be denoted by the central (red) line on Fig. 17. The set of data accumulated in the system and knowledge M is filled via applying various logical inference rules, machine learning and/or other ways for knowledge acquisition. Each inference and any other change in the set M is traced via the gradient of information where the coordinate system is tied to the goal direction G. The sum of gradients depends on the depicted on the figure via the dotted line. In the case when the change of M is inconsistent, for example the gradient calculus often change their direction because there are not enough significant intermediate results, then the dotted trajectory will change its direction when it reaches the funnel edge or, in other words, new knowledge will be searched only in the direction delimited by the position of the solution in the funnel which is presented for the point—an intermediate solution Y in (7):

$$A \leq grad(Y) \leq B. \quad (7)$$

The influence of G over the process of changes in M may be compared to the one of the gravity. Gravity is by no means one of the strongest force on Earth, but we overcome it by walking, etc. On the other hand elevating (a motion along the anti gradient) requires a lot of energy. If the energy is not exhausted then the first cosmic speed is reached i.e. the earth attraction is completely overcome. Therefore, limitations A and B from (7) are useless if in the flow, the dotted trajectory in Fig. 17, there is much accumulated moment of inertia. There is an equivalent of the inertia in the case linked to the dotted trajectory in Fig. 18. The following relations are introduced in relation to the problem defined. Let $s_i$ be an intermediate solution that is interpreted as a point from the trajectory and then by analogy with (7) the following constraints (8) are introduced:

$$A \leq \sum_{j=1}^{i-1} grad(s_j) \leq B \tag{8}$$

This is based on the following explanations. Gradients were the necessary interval between A and B for steps from the first up to $i-1$-th. Then, the bigger $i-1$ is, the bigger the inertial property ad hoc the solution $S_i$ will be. When $i$ is greater than an a priori threshold value then the constraints of the FUNNEL change their effect.

If the sum of gradients is bigger than an a priori set threshold value $T_1$, then the cited limitations may be overcome (9):

$$\sum_{j=1}^{i} grad(s_j) \geq T_1 \tag{9}$$

The next version includes a new constant value $T_2$ and a solution $S^*$ which is different from the defined goal G but which is marked by experts as sufficiently interesting. It can also be marked by the machine because for example, it is the missing link to solving other earlier defined problems. $S^*$ does not participate in the series of $S_i$ but if the distance between them is calculated via an additional p-adic metrics, then it is possible to avoid the limitations given by the FUNNEL:

$$\|s^* - s_i\| \leq T_2 \tag{10}$$

In the three cases, cited above the 'inertia stream' pierces the FUNNEL under conditions given in the formulas as it is shown in Fig. 17. Otherwise, all consecutive solutions will be inside the FUNNEL as it is presented in the description of the method above. In the case of a break, the FUNNEL stops running. Therefore, if the FUNNEL method does not adapt to the new conditions of the information flow as it is shown on Fig. 18, then the method ceases to perform its main function and starts to gradually narrow the limitations to direct the set of dynamically changing information to the required place. During the tuning from Fig. 18, this problem does not exist. On the other hand, by solving the problem from Figs. 17 and 18 we had to construct more than one system of fitness functions where the first (classical) function is directed downwards and the second one operated temporarily only between the beginning and the end of the depicted FUNNEL. In this way, conflicts are avoided between classical evolving methods and the requirements of contemporary intelligent data processing.

As a whole, apart from the method control mode, SMM main strategy is the evolutionary one 'survival of the fittest'. The human influence as in the presented default variants is a scarce case, in other conditions every schematically described method from Puzzle, Funnel, Inconsistency, and Kaleidoscope group together with logical and statistical applications produces a large set of comparable results. Their priority can be estimated using different classical means. After that, some applications are reinforced, and other should be defeated, and not deleted. Delete operation is seldom used at the synthetic level. At the agent level, same defeat operation is frequently used in any data driven application. The basics of the proposed defeasible reasoning in MAS are considered in the next subsection.

## 3.8 Defeasible Reasoning/Inference in MAS

The defeasible inference is an important part of SMM metacontrol strategy. The algorithms for the change of agent's knowledge should be also changed, especially in the case of an ontology-represented knowledge. Agents, especially in information security systems, often have to rewrite or append different forms of knowledge, let's say belief-desire-intention (BDI). These changes shouldn't be carried out in a 'disorganized manner', *as is*, since soon in a different situation the old knowledge or a piece of it may be valuable again. The more effective way is to add new knowledge by defeating the old one in a new situation. It is shown below that almost every relation in a rule using classical implications may be defeated. Eleven main defeasible schemes have been introduced, and they can't describe all the proposed defeasible inference and features of defeasible reasoners. Simply they are the most frequent ones.

Let the unity of classes V be comprised by the subsets $S_1$, $S_2$, … where $S \in V$. Every subset of type S includes elements $x_{s:1}$, $x_{s:2}$, …, that form a new model. The original set S is related to one of the classes $S_i \in V$. The final result from the analysis of S is idenitified with one of the classes $S_i$ from U. The output is an answer of the type $V_s$ = (T; F;?) with three truth values: "true", "false" and "uncertainty" (?, don't know). In the case with an answer $V_s$ = ? or $V_s$ = F, the set S may be identified with more than a single known class $S_{i1}$, $S_{i2}$, … ($i1 \neq i2$ …). The answer $V_s$ = T is obtained if and only if the examined class S coincides with $S_i$.

Amongst the classes $S_i$, there exists an interdependence of the type "ancestor—successor", (*is-a*, e.g. $S_i$—an ancestor of $S_{i1}$). Thus, it is possible to form simple types of semantic nets—with one type of a relation. It is necessary to note that the elements $x_{si1:1}$, $x_{si1:2}$, … produce the differences between the class $S_{i1}$ and the other successors of the common ancestor $S_i$. All differences that appear in the comparison process of $S_{i1}$ with other classes that are not direct successors of $S_i$ are determined after the application of the inheritance (heredity) mechanism.

The conclusion (response) $V_s = T$ is formed when the corresponding conjunction terms for all ancestors $S_i$ and also for $S_{i1}$, are of the following form: $A'_1 \Lambda A'_2 \Lambda \ldots A'_n$, where $A_k$ is $x_k$ or $\neg x_k$; $A'_k$ may coincide with $A_k$ or it may include (using a disjunction) $A_k$ and analogical terms for other variables.

Let's assume that the rules of a Horn type describe some domain:

$$B \leftarrow \underset{i \in I}{\Lambda} A_i \tag{11}$$

During the usage of a truth/binary or even Boolean-like logic in the quoted rules, if at least a single variable $A_i$ is not 'true', then the truth of B is indefinite, i.e. B may mean "true" or "false". In the case when the corresponding exclusion from the conjunction (11) of the rule is based on the inclusion of a term with any $A_k$ ($k \in I$) then the inference procedure changes. In the case when the exclusion E (C, $A_k$) and C is true and $A_k$ is false, then the right side of rule B may be true (as an exception). The extended inference models with exclusions are introduced and generalized in the following form.

$$\frac{B \leftarrow \Lambda_{i=1}^{z} A_i, C, E(C, A_k), \neg A_k \leftarrow C}{B \leftarrow A_1 \wedge A_2 \wedge \ldots A_{k-1} \wedge \neg A_k \wedge \ldots A_z}, \tag{12}$$

$$\frac{C, B \leftarrow \Lambda_{i=1}^{z} A_i, E(C, A_k)}{B \leftarrow A_1 \wedge \ldots A_{k-1} \wedge A_{k+1} \wedge \ldots A_z}, \tag{13}$$

$$\frac{C, B \leftarrow \Lambda_{i=1}^{z} A_i, E(C, A_k)}{B \leftarrow A_1 \wedge \ldots A_{k-1} \wedge (A_k \vee C) \wedge A_{k+1} \ldots A_z}, \tag{14}$$

It is clear from formulas (12)–(14) that the exclusions are a kind of special, nonimplicative rule inclusions with their effective fields. The interpretation of the formula (12) is based on the following: if an exclusion E(C, $A_k$) exists that is related to one of the rules with a conclusion B and $A_k$ is its effect, then the conjunct $A_k$ must be replaced by $\neg A_k$. In the case when C is not 'true' the corresponding replacement is impossible. The application of the Modus Ponens rule means that the relation between B and $\neg A_k$ leads to a formal logical contradiction.

Therefore, the formation of exclusions of the type E(C, $A_k$) may lead to a contradictory result that is provoked by an incompleteness in the description of the object field. In the case when C is true then the exclusion E(C, $A_k$) includes this meaning in the conjunct $A_k$ to defeat the meaning of the last conclusions. The result is that $A_k$ is replaced by C because the test of its meaning does not influence the output. In the case when C is true then the corresponding conjunct $A_k$ is directly replaced by C.

Rules of type (11) are united in systems:

$$\begin{cases} B_1 \leftarrow \bigwedge_{i \in I} A_{1i} \\ B_2 \leftarrow \bigwedge_{j \in I} A_{2j} \\ \ldots \end{cases} \tag{15}$$

In the general case, the causal-effective relation may be applied using non-classical operations of successions that are denoted '<-' in the paper and $B_i$ may be presented as combinations of sophisticated logical relations (see formula (16)).

The usage of exclusions (12) up to (14) may be also applied in systems (15) or (16); in the general case it reflects the interrelations between different parts of the causal-effective relations influenced by a new information (an exclusion that is attached to one or another group of relations). The new information may influence the mutual relation between the elements of the rule (11) or of systems (15); (16).

$$\begin{cases} B_1 < - \bigwedge_{i \in I} A_{1i} \\ B_2 < - \bigwedge_{j \in I} A_{2j} \\ \ldots \end{cases} \tag{16}$$

In this case, the relations of a causal-effective type are defeated or they are strengthened due to an additional information that is contained in the exclusions. The rest of the paper does not include versions (15) and (16) because in the majority of our practical applications it is sufficient to confine ourselves to rules (11), in this way the algorithmic complexity of the used combination of methods is significantly lowered. The presented exclusions by their nature are an enlarged version of defeasible inferences that are widely used in the intelligent systems. The presented work is different from the classical inference with exclusions because it is possible not only to exclude the exclusion $A_k$ that is contained in and tailored to the rule, but also because we may include a new formula in the rule, e.g. $\neg A_k$ in formula (12) or an interrelation between $A_k$ and C in (14). The research also includes versions of formulas using a non-classical negation $\sim$, versions with exclusions of implications influenced by exclusions, etc.:

$$\frac{B \leftarrow \Lambda_{i=1}^{z} A_i C, E(C, A_k), \sim A_k \leftarrow C}{B \leftarrow A_1 \wedge A_2 \wedge \ldots A_{k-1} \wedge \sim A_k \wedge A_{k1} \wedge A_{k+1} \wedge \ldots A_z}, \tag{17}$$

$$\frac{C, B \leftarrow \Lambda_{i=1}^{z} A_i, E(C, A_k)}{A_1 \wedge \ldots A_{k-1} \wedge A_{k+1} \wedge \ldots A_z}, \tag{18}$$

$$\frac{C, B \leftarrow \Lambda_{i=1}^{z} A_i, E(C, A_k)}{A_1 \wedge \ldots A_{k-1} \wedge A_k \wedge A_{k+1} \wedge \ldots A_z}, \tag{19}$$

where $A_{k1}$ is an additional condition for the transition from $\sim A_k$ to $\neg A_k$. The investigation includes schemes with multi-argument exclusions $E(C, A_k, A_l, \ldots A_s)$ that lead to the simultaneous change of several parts of the rule. The introduced method leads to three basic results: the truth of parts of the rule is altered when influenced by the exclusion (if the conditions for activation of the exclusion are enabled), formulas are included in or excluded out of the rule or the rule itself is defeated as it is shown in (17) or (19). The results from the research led to a large number of inference versions with exclusions; a part of them is included in our bibliography list.

As it was already shown we introduced a generalized concept of defeating that is based on the following facts. Object scope modeling is a dynamic process. In the act of scope-field completion by the system the old relations between separate parts of the knowledge and/or between different knowledge may be eliminated, changed or their effect may be redirected. This is accomplished by the influence of the new knowledge that complete or correct the primary existing knowledge or its interrelations. The processes are formalized in the following way.

We did a research of the situations that appear after the addition of the new knowledge to the existing knowledge basis, and we grouped them in 11 basic groups. Let P be the part of the new knowledge that influences one or more formulas (e.g. see (11) up to (14)).

I. P 'nullifies' $A_k$: it defeats its relation to the conclusion B. As a result of the defeat, $A_k$ has a meaning of 0 and no matter whether it is true or false, the true of the conclusion does not change.

$$\frac{B \leftarrow \Lambda_{i=1}^{z} A_i, P}{B \leftarrow A_1 \wedge A_2 \wedge \ldots A_{k-1} \wedge A_{k+1} \wedge \ldots A_z, \neg(B \leftarrow \Lambda_{i=1}^{z} A_i)}, \tag{20}$$

where in difference with defeasible inference schemes the first rule format existing before the appearance of P becomes false.

II. This is an extreme version of the situation from group I when all the atoms in the antecedent are defeated. Now rule (11) turns into a fact: $B \leftarrow$.

$$\frac{B \leftarrow \Lambda_{i=1}^{z} A_i, P}{B, \neg(B \leftarrow \Lambda_{i=1}^{z} A_i)}, \tag{21}$$

III. P changes the true of $A_k$ from true to false or v.v.

$$\frac{B \leftarrow \Lambda_{i=1}^{z} A_i, P}{B \leftarrow A_1 \wedge A_2 \wedge \ldots A_{k-1} \wedge \neg A_k, \neg(B \leftarrow \Lambda_{i=1}^{z} A_i)}, \tag{22}$$

IV. P defeats the existing meaning of $A_k$ and increases it to 1. The meaning of the other parts of the antecedent of (11) duly drops down to 0. Independently of the way (conjunctively or disjunctively), they are related to $A_k$ and in this situation they are defeated by the antecedent of the rule (11).

$$\frac{B \leftarrow \Lambda_{i=1}^{z} A_i, P}{B \leftarrow A_k, \neg(B \leftarrow \Lambda_{i=1}^{z} A_i)}, \tag{23}$$

V. P redirects the relation between the rule and the other knowledge in the domain.
The causal-effective relations are not exhausted by the classical implication and the next example will show that even by formal means it is possible to present different causal-effective relations. Let us have the following two rules:

$$R_1 : B \leftarrow A; \quad R_2 : N \leftarrow M. \tag{24}$$

Let both rules initially be related to the object X. Let also after the appearance of the new set of conclusions P $R_1$ be related to Y and $R_2$ to the former object X. In this case, the first rule is preserved but its effect is redirected to another object.
For example, it is known that by nature a disease is provoked either by a virus or by a bacteria. However, let us have a case when a patient manifests simultaneous symptoms of an illness both from a virus and from a bacteria. The sequent investigation (P) shows that the symptoms of a virus-provoked disease are related to the patient's throat and that the bacterial symptoms are related to the patient's lungs. The redirecting of the conclusion that contradicts the rule from the example and the discovery of the second disease solve the problem from this example. It is possible to redirect whole rules by analogy to the presented example.

VI. P breaks or amplifies the relation between the rule and the other knowledge in the domain.
The difference with the previous situation V now is either the elimination of the existing relations or the addition of new relations between the existing rules. The very rules are preserved with that.
For example, every chess-player must have a good physical condition so that he/she can present himself/herself well in the tournaments. If however the 'examined' chess-player is a computer program—this is the effect from the new information P—then the already said does not at all concern this program.

VII. P influences the conclusion from one or from a group of rules: from $R_1$: B ← A into $R'_1$: $B^*$ ← A. In this way, the old conclusion P is defeated or it is replaced by the new one $B^*$.

$$\frac{B \leftarrow \Lambda_{i=1}^{z} A_i, P}{B^* \leftarrow \Lambda_{i=1}^{z} A_i, \neg(B \leftarrow \Lambda_{i=1}^{z} A_i)}, \tag{25}$$

VIII. The appearance of P changes the antecedent of the examined rule (11). It imports a new atom on the place of $A_k$, before or after the chosen one $A_k$. In the last two cases, the new atom is conjunctively or disjunctively related to $A_k$, e.g.

$$\frac{B \leftarrow \Lambda_{i=1}^{z} A_i, N(P,J)}{B \leftarrow A_1 \wedge A_2 \wedge \ldots A_{k-1} \wedge J \wedge \ldots A_z, \neg(B \leftarrow \Lambda_{i=1}^{z} A_i)}, \tag{26}$$

This situation can be named by specifying the antecedent as a result from the new information P.

IX. $R_1$ is replaced by $R_2$ and influenced by P:

$$R_1 : B \leftarrow A; \quad R_2 : N \leftarrow Q. \tag{27}$$

The difference from the previous situation here is in the complete replacement of the rule provoked by P in accordance with the a priori defined concepts.

$$\frac{B \leftarrow \Lambda_{i=1}^{z} A_i, P}{N \leftarrow Q, \neg(B \leftarrow \Lambda_{i=1}^{z} A_i)}, \tag{28}$$

X. We have a situation from I to IX, but the obtained consequences may not be used in the antecedents of the other rules. The reasons for similar constraints are different, e.g. limiting an insecure information along long chains of rules, etc.

XI. The atoms of the investigated rule (11) remain the same, but some of the logical operations are changed affected by P, e.g.

$$\frac{B \leftarrow A_1 \wedge A_2 \wedge \ldots A_{k-1} \wedge \neg A_k \wedge \ldots A_z, N(P,J)}{B \leftarrow A_1 \wedge A_2 \wedge \ldots A_{k-1} \wedge \sim A_k \wedge \ldots A_z, \neg(B \leftarrow \Lambda_{i=1}^{z} A_i)}, \tag{29}$$

A characteristic example of a similar situation is the transformation of the strong classical negation '$\neg$' into a weak paraconsistent negation '$\sim$'.

Let's discuss the following illustrative example. In principle, it is not possible that the same person is a teacher and a student at the same time. Let's denote that 'John is a teacher' by the variable Q. Then it will not be an error if we denote that 'John is a student' by $\neg$Q.

This is valid in the prevailing number of situations, but it is inapplicable in condition (P) that John is a student in one subject in one school, but he is a teacher in another subject in other e.g. sports school or at the same school. After the advent of the new information P, it is not possible to say that 'John is a student' is $\neg$Q; now it is correct to use the weak negation and ~ Q will lead to a contradiction only in the cases when definite conditions hold—in the example the conditions are the subject for teaching and also the location for teaching.

The described situations from I to XI present a research for the influence of the new information P over different parts and relations between existing conclusions. In the majority of the discussed situations, contemporary mechanisms for defeasible inference may be used. The difference is just in the fact that P totally changes the situation existing a priori. However, if P replaces the literal in the first argument in the exception $E(C, A_k)$, then the exclusion does not change the action progress for the existing up to the advent of P things and it adds a new scheme to them that is activated if and only if P is false. The present chapter does not contain formal descriptions of all the possible variants of the situations from I to XI because the number of their combinations in all the possible applications is too large.

In the end, the whole rule may be replaced by totally different rule(s) using one or more exclusions type $E(C, A_k, A_l, \ldots A_s)$. The total change isn't the recommended one because the proposed MAS environment isn't revolutionary but evolutionary one. Total changes should be carefully checked if acceptable.

### 3.9 Specifics of Knowledge Management in the Proposed Agent-Ontological Applications

As shown in [5], almost every relation in a rule may be defeated. It is introduced a new method and algorithms for implementation of the ontologies with the aim of the quality improvement of the electronic education on ISS. For this purpose combination of the following methods operating with ontologies is used.

On the first place were searched more effective ways for the creation of ontologies with the following aim: minimization of the role of the man for creation and verification of ontologies. For the purpose were distinguished mechanically pure functions which are executed excellent by the machine from creative, human functions and in the next stages these two parts are combined.

On the basis of the represented material are created and developed ontologies connected with the objective field. The problems and hypothesis connected with courses of lectures are represented graphically also and part of the used anthologies include weakly formalized knowledge, caused by the cognitive agent. The technology of the human-machine contact in the presented research includes preparation machine stage with results which are visualized. In the presented method visualization often replaces the formal descriptions and from here comes the title of "kaleidoscope method".

From the one hand during the decision of the settled aims and manual or automatic formation of the new aims, the ontologies are clarified, changing, some of them are erased, other are supplemented. In the connection with the fixed aims are researched the possibilities for implementation of the standard orchestration and choreography by Web Service Modeling Ontology (WSMO).

The use of ontologies and other resources for modeling and visual presentation of the information for education on ISS is rather slow and expensive process. However, it is imposed because of the lack of well-prepared staff in the area: some people often are coming with the attitudes to remember and copy but it is necessary to learn to think independently, to undertake original activities, to understand the best resources, to understand the best ways and practices in the area (which are more often multidisplinary) and to be ready for lifelong learning on the subject of their study on the basis of their independent development. This way is very necessary to use mentioned in the chapter resources.

# 4 Main Features of the Proposed Puzzle Methods

## 4.1 New Trends in Constraint Satisfaction

The considered ways (sets of methods) have been designed for applications in different research systems. It is shown that if the same method is applied in a different way, it will give new results. On the other hand, teaching SS stands out from other research/university programs because it operates in rapidly changing models, requirements, concepts, tools, instruments and policies where the problem for qualitative education is particularly difficult. Deep research leads to deep teaching which is demandable in ISS courses. Teaching highlights in the brightest way the advantages of the proposed new methods and applications in ISS.

One of the most instructive facts about the audience for training is that current neural-based machines in the form of software agents or holons repeat the actions of the teacher on a much higher level, even related to an excellent student. Therefore, it is obvious to say that rote learning, simple memorization of lectures makes no sense: sooner or later machines will displace humans from similar low-tech activities; instead creative learning should be deeply rationalized through comparisons with other research accessible through the Internet, by experimenting with numerous courses and with open source products which become more pervasive in our practice together with contemporary Web-based applications. The above example illustrates how to achieve simultaneously two goals: explain essential principles of artificial neural networks (ANNs) functioning and modern principles of quality training through effective and deep understanding. Similarly, using the neural networks examples explain the best practices for knowledge management, for more tests for lagging students, etc. The application of the below-described methods makes the training/tutoring process easier, understandable and transparent. Hence, sustainable education results are obtained. Advanced data analytic methods are not easy to explain. However, their correct applications don't make the education complicated, because they are purposed to find easier and effective ways of presentation.

On the other hand, monitoring the quality of educational processes most often uses different statistical methods. We have used the same methods in an advanced analytics module together with logic applications described in this chapter. The unification of logical and statistical parts comes using evolutionary ways. Those cardinally different results can't be used just in one simple cluster of data or knowledge.

During the evolutionary process of joining statistical and logical data/knowledge, the accumulation of knowledge makes logical applications more and more effective and more universal than the probabilistic ones/fuzzy estimates/analogical applications. In the common case during data analytics-style processing, there is no convergence of the results but this does not impede practical applications of these systems. In other words, poorly trained staff will retrieve poorest results using same data mining systems. In general, the use of synthetic methods is one of the most important and demanding elements of the efficiency of intelligent applications. But the winning combination of the proposed methods in the educational field depends on the lecturer's experience, which makes the performance much simpler than just in the automatic case. At this level, the operation of the advanced analytics module is limited to implementation-voluntary advice to the lecturer. Suggested innovations serve the most effective application of education practice altogether with data analytics applications, collective evolutionary components, etc. More information can be obtained from the book [5].

The research on traditional, syntactically produced puzzle methods reveals that their algorithmic complexity is rather high and this doesn't allow automatic processing of large enough sudokus, crosswords or puzzles. In the case, an emphasis should be placed on the fact that the studied machine-based procedure takes the words from a predefined set using a random principle where only the length of the word is of importance. It was mentioned above that when the statistical applications are used alone they are not so effective. On the contrary to the quoted syntactical realizations, a semantic variant of Puzzle method is considered. It is based on the logical analysis of the existing data or knowledge interconnections. Three types of relations have been used aiming at narrowing the set of possible solutions to the problem.

Briefly, every Puzzle method aims to discover new or hidden knowledge by connecting the unknown, the sought solutions with previous experience accumulated in knowledge bases. Let the constraints of the considered problem form a curve in space as depicted in Fig. 19. The main goal of the Puzzle method is to reduce the multidimensional search space for the solution. For this purpose, we used several limitations. In [5] the study is focused on the case of using ontologies instead of a set of nonlinear constraints, etc.

Furthermore, in certain cases the research of process dynamics aimed at reducing the field allows us to derive new knowledge in the form of rules as discussed in the next subsection. This inference process and the usage of different constraints gives way to significant simplification of formal and evidence material in research/lectures, attracts the attention to details and simultaneously increases the activity of learners. Showing the process of connecting known to unknown improves the understanding and retention of the presented material.

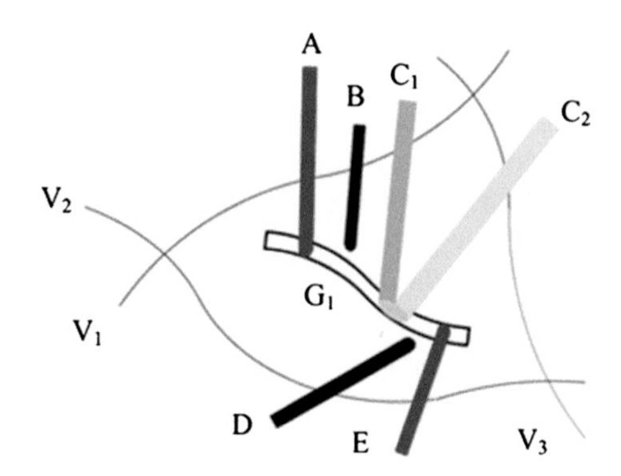


**Fig. 19** Binding, crossword and classical set of constraints

For example, if a bachelor who has graduated one of the London Universities lives in London and he/she does not want to work anywhere else, then the line (curve) restricts the search space and in this way a lot of unnecessary [re]search is avoided. It is also possible to inspect a case when the constraint is defined in the form of a surface but, as a result, a more general solution is obtained where a special interest is provoked by the boundary case of the crossing two or more surfaces. When the common case is inspected in details then in the majority of cases the problem is reduced to exploring the lines instead of curves with complicated forms obtained as a result of crossing surfaces. Therefore, below the usage in systems of constraints is investigated by using lines of first or higher orders.

The $V_i$ constraints of a linear/nonlinear form are the classic case borrowed from constraint satisfaction methods. The B&D binding constraints, dot-type variants are also possible, doesn't intercept the unknown goal but are located close to it, they reveal a 'neighboring' area around the target. The A&C&E constraints are named the crossword constraints because their interception with $G_1$ gives a part of the searched goal where Ci form larger interconnected areas in $G_1$. The usage of the considered set of Ci-s is much more effective than of the classical set of $V_i$-s. The pointing and ontological constraints are not depicted because of similar images.

Through *binding* constraints, it is convenient to implement the many non-classical causal relations of the type "A is linked to B but the connection between them is not implicative (provable)". The same could be presented through heuristics, which is not recommendable. It must be pointed out that through this type of constraints the location of the searched solutions is fixed in a way that is best combined with fuzzy methods.

Crossword constraints offer new ways of assessment (outlook) for the searched unknown solutions on the basis of the accumulated so far knowledge.

The next group of newly designed constraints used in Puzzle methods is named **pointing constraints**. It can be classified as an exotic type of binding constraints, but constructs a new constraint type because of addition of a new feature: when at least one element of pointing constraints is reached, it shows the direction to the solution, to goal $G_1$. In comparison, the binding constraints show that the goal is somewhere near, let's say, with higher probability, but doesn't show the exact direction. Let's say, the mineral pyrite signs that the gold *may be* near. Another examples: if an infrared or thermal camera shows a human-like image inside the building, we know **where** to find the person [pointing constrains], but if we are told that somebody lives in this building, we don't know exactly where is he. Also we don't know whether the quoted person is really located inside the building. The latter analogy concerns just the binding constraints.

On the other hand, many well-known statistical methods could effectively work with pointing/binding constraints and can improve the effectiveness of crossword constraint applications by calculating the probability value and the preferrable direction of the goal location.

The following example shows how the search process can be reduced using ontologies. Let's admit that the search space is presented on Fig. 20 where statistical data about ISS are generalized about different regions depending on their price and quality. Let's the goal is: necessary to select an acceptable ISS to our project. Let the right vertical, blue-colored (dark-grey), subset of feasible solutions is chosen: 'excluding ISS designed outside Europe'. Then the space of feasible solutions is to the left of the separating surface which is depicted on the figure in the rightmost corner.

In Fig. 20 another horizontal surface, depicted in the high corner in green (light gray color), is shown delimiting the search space of the solutions. In our case, it means 'systems with unknown principles of operation'. It is accepted that in the databases there is no clear distinction related to the presented criteria so the search of the feasible solutions is nonlinear, of high dimensionality, and practically it cannot be solved using traditional methods. Nevertheless, by applying ontologies the problem is solvable via the proposed Puzzle method. There are two red dots (two brighter gray color dots) depicted in the left corner on the same Fig. 16. Each of them also represents a kind of constraint but of another type which we name a

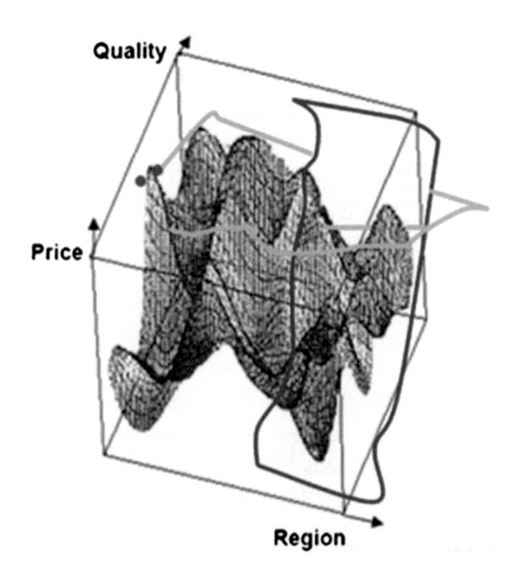


**Fig. 20** SS application example

binding constraint and it is introduced and researched by us. Its semantics is the following: it is not a solution, but it resides close to the searched solution. We can say, that the two (or more) binding dots fix up the solution surface.

## 4.2 Different Types of Binding Constraints

The following section discusses the introduction of following three types of binding constraints.

The knowledge used in the Puzzle method can be presented as parts of information (atoms), linked by different relations. Usually, these relations have been obtained by logical processing of information like structuring, extracting meaning from information blocks or other processing. In this terminology, one rule can be presented in the following way:

The conjunctions of antecedents are $A_1$, $A_2$ .... $A_z$.
The conclusion/consequent is marked as *B*.
Let all $z$ number of conjunctions are proved to be true/confirmed, then B is true, whereby the goal/problem of checking whether B is true has been solved (see Fig. 21).

When significance has not been pointed out, as in the case shown in Fig. 21, then the significance of the conjunctions are considered equal (1/z). The above bows show that each conjunctive has its individual *significance relation, mainly invisible and often informal, in proving the conclusion*. It is a real number from the interval [0, 1]. But in the common case, some conjunctivals are of great significance while others are less important, depending on the situation. Part of these invisible links between the atoms of the type conjunctive-conclusion or other parts of knowledge can be **torn** in different conditions, for example when there is additional information and during a process called *defeat*, which can change the truthfulness of each atom, reduce its significance in proving the conclusion/goal to zero or change the whole rule completely, for example, by replacing it with another one.

The considered process of defeat is started by special forms of presenting knowledge called *exceptions to rules*, knowledge of the type *E*(*C*, *Ap*), (cf. Sect. 3.8 and Fig. 20) where a prerequisite for the defeat is the argument C of the exception, which must be true in order to start this process of defeat. Therefore, both arguments of the exception enter a causal relation, which is not implicative. It is a kind

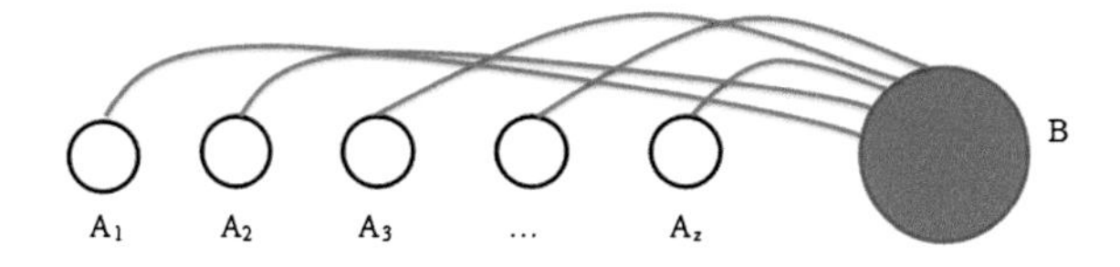


**Fig. 21** Informal relations inside a rule

of non-classic causal relation. In life, we use so many similar relations and most of them are difficult to formalize because they are not included in the classical mathematical and formal logic (Fig. 22).

The accumulation of such knowledge atoms, the compound with different classical and non-classical relations allows us to use new opportunities for reaching the set of goals.

For example, let's have goal *X* proved via classical or non-classical means, as shown in Fig. 23 and let *X* and *Y* have an unidentified causal relation, for example, *X* defeats *Y*; or for example, let *X*, and *Y* be statistically linked variables. In this case, the fact that there is a large volume of information linked to *X*, as shown in the figure, leads to imposing informal constraints over the choice of the condition of *Y*, regardless of the fact that in a classical logical (formal) sense, *X* and *Y* are not linked. Informally, proving *X* non-monotonically leads to the solution of the goal *Y*. In other analogous situations, solving *X* does not lead to proving *Y*. However, they are linked through the following non-classical relation: proving *X* shows that we are *close* to the solution of *Y*, the more rules and facts prove the truthfulness of *X*, the higher the confidence/sureness/certainty or belief in the hypothesis that *Y* is true. The described process will be called *binding*, as interpreted in Fig. 23.

Very often using methods like fuzzy logic helps the binding process, for example, in situations where indefinite notions are used or notions that change their meaning depending on the situation and the context.

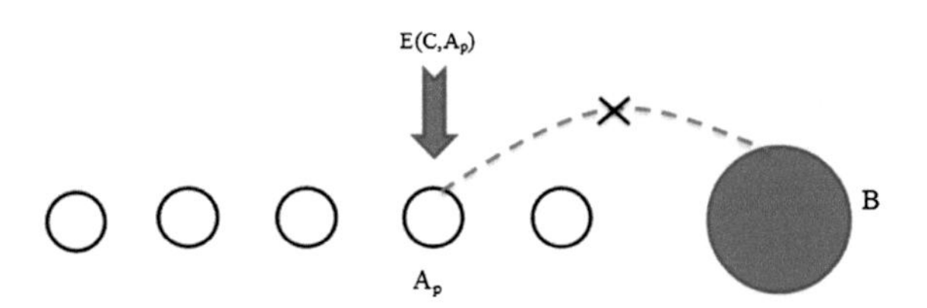


**Fig. 22** The defeat of implicative connections

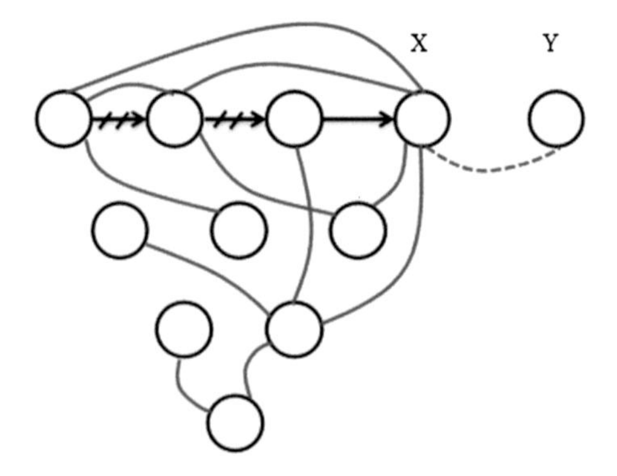


**Fig. 23** The binding process scheme

Let's see the following rule:

If N is a good specialist, but N cannot work with people, then N is not elected a manager.

In the described example, there are a lot of notions that require clarification: good specialist, ability to work with people, a leader, etc. For example, it is important to clarify that this is a leader of what, in what conditions and so on. It must be considered the fact that being able to work with people is quite a subjective judgment and it can be greatly reduced due to a number of subjective reasons. In situations where there is doubt about the judgment of the ability to work with people, it is better to use binding relations, including the non-referable part of the rule. In this case, what is left is the relation that if someone is a good specialist, they can be elected a manager. Here the link isn't implicative. Obviously being a good specialist is not enough to be elected a manager. Additional knowledge should be drawn here, increasing the confidence that if someone is suitable for a manager on the problem and in the particular situation, for example the necessary facts can be drawn from the CV, etc. In the example of using the binding, in fact, one or more conjunctivals are defeated from the antecedent of the rule.

Defeating the link of the second conjunction with the conclusion actually removes the implicative link on the whole rule, and is replaced by a binding.

## *4.3 On Crossword Constraints*

This section discusses the introduction of one type of constraint called *crossword*.

Let's assume that KB contains deeply structured knowledge, for example, ontology on the problem. In this case, if other knowledge is discovered, and is related to the ontology, but badly structured, for example, written in another language, or written via pictograms, incomprehensible to us, or information lacking in text or noised or encoded, i.e. in situations when the meaning of just a fragment of the information is comprehensible, and out of which only part of the information can be drawn, and if this part of the information is new, in a sense that it complements the ontology, then the mentioned new knowledge is added to the ontology regardless of the missing parts of the knowledge. In this case, invisible rule/clause relations are being used, similar to the one in Fig. 21 and linking the non-structured knowledge with the ontology, as well as with the atoms of the non-structured knowledge through the used constraints of the crossword. Thus, new knowledge extraction is carried out. In this case most often searching in the constrained area of options is avoided, and the solution is obtained straightforward and non-alternatively. For example we assume as a known fact that a banker has been killed in the center of Tirana. The event relates to the ontology 'murder' with relations to a banker, center, and Tirana. Here as another column we can add to the ontology *when* the murder took place, etc. Then, if someone watches a TV show on the Albanian television and does not understand the local language, they can nevertheless determine that the news is about killing a banker for example by the written time of the murder and so on, but when in this context of unfamiliar words appears the known word MOTOR…, our ontology of the murder is complemented with a new relation: a motorcycle has been used during the murder. In this case, the context of the event described in an unfamiliar language characterizes the unknown, and only the word motor and some other specific information units are understood/known or link the unknown to the known. Regardless of the lack of information, by using the crossword constraints here we will obtain an extension of the existing ontology solution: we have obtained an improved with new knowledge ontology.

By introducing new constraints or, sometimes, ontologies our goal is to show that it is possible to use causal links different from implications and that they help us search the goals in a more effective manner.

In the tutoring/ISS education case, using the crossword constraints style Fig. 19 also helps us reveal the dynamics of the problem-solving process where the resolution process is sometimes more important than the proof itself. The usage of the classic cases, Vi constraints one by one doesn't always form the necessary closed area. Revealing Vi and other constraints one by one helps represent the resolution process in its dynamics and hence make a deep inference to the problem. Even in the worst case, when the accumulated knowledge is incomplete for the problem resolution, the dynamics of what is represented analogically to Fig. 19 will show what should be the resolution. The binding/crossword constraints are the form of non-implicative rule-based relations. Their application effects are discussed in the next sections.

## *4.4 On the Usage of Pointing-Type Constraints*

The pointing constraints are best illustrated by different gradients. The principle is depicted in Fig. 24. The pointing nature of the phenomenon is that the best way to find optimal solutions (min/max value) from the depicted point x, result of the intersection of the line and the isosurface line is via (anti-)gradient. Nabla symbol is used to depict the gradient direction which is perpendicular to the depicted line.

Every gradient is a restricted case of the proposed pointing constraints. The universal, pointing cases show not only the (local/global) optimal direction, but also informal notion what is the most preferable (where is it), what should be executed first using the precedence operators, or, for example, where is the simplest solution. Beautiful, effective, perspective, and other informal notions are no problems for pointing constraint operations. In this sense, some forms of pointing constraints are similar to ontological constraints.

A set of [mutually] connected pointing constraints may form an analog to an algorithm describing the resolution to the goal.

One of the actual forms for pointing constraints is based on the logical analysis of the situation. There exist many situations, when there is only one solution to the problem, or their number is fixed and well-known (no incomplete information). Then just the a priori given pointing constraints help to reach the goals concerning

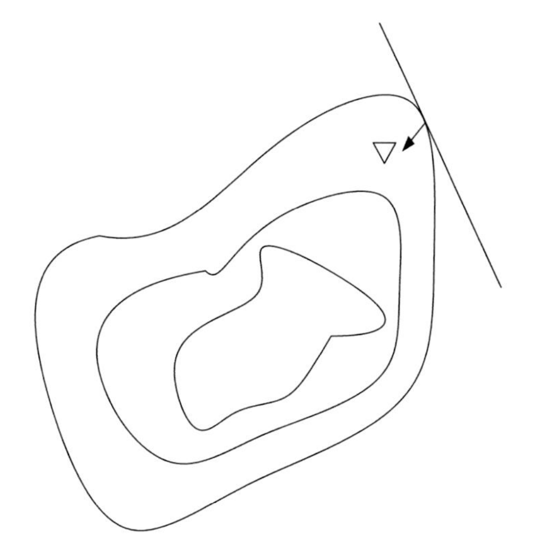

**Fig. 24** Interpretations to the specifics of gradient methods

them. For example, let somebody is asking: where are the fire exits. As an answer, an emergency plan is shown with 3 fire exits from the floor. Let as a result two ways are fixed up in the direction to the nearest fire exits from the considered room. Both ways are just one of the cases of the discussed pointing constraints.

## *4.5 Visualization Elements Using Puzzles*

Essentially the Puzzle methods include a form of visualization, which leads to a creative work in various forms of education/training. We would like to remind that the high-quality learning inevitably leads to independent research. In the present study, it is shown that in some cases the graph itself from Fig. 19 may be sufficient for decision-making by the user or the expert.

Different versions of the modeled knowledge contained in Fig. 20 provide a base to describe things that are difficult to express in words. Not surprisingly, a Chinese aphorism states 'A picture replaces a thousand words.' It should be added: '… and is understood and memorized better.' Below are given different linguistic equivalents of combinations of interpretations, similar to that of Fig. 19 and obtained by the Puzzle method:

$$\text{Things are moving in a way that}\ldots \quad (30)$$

$$\ldots\text{neither}\ldots\text{nor}\ldots\text{but it is too close to}\ldots \quad (31)$$

$$\text{Looked at this from another angle}\ldots\text{other findings}\ldots \quad (32)$$

An object can be represented as an intersection of two dynamically changing areas . . . (33)

In the future, following two concepts should be used jointly . . . (34)

etc.

where (30) illustrates the application of hard to formulate sentences, and pointing constraints may be applied, (31) is close to the vague suggestions through Puzzle methods, (32) shows the dynamic capabilities of the Puzzle method applications for example when a change in the meaning of the used knowledge depending on context, (33) and (34) depict the technical capabilities of the Puzzle applications as a unification, intersection of data/knowledge, ontologies, etc.

These natural-language fragments can be presented in plain text for learners, but further it is shown that by the Puzzle method this is executed much more naturally and better. For example, Fig. 25 shows another example of Puzzle method application revealing certain processes and relationships between objects. Concentrating the attention in the closed area of Fig. 25 and narrowing the set of analyzed elements supports the revealing of new implicative relation M-N referred to in Fig. 26. Informal causal relations are elaborated that are associated with the implication 'from M it follows that N'.

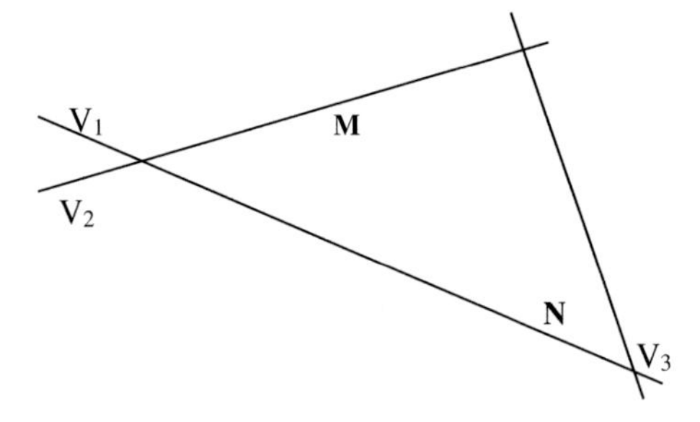


**Fig. 25** M and N are unrelated but belong to same confined area

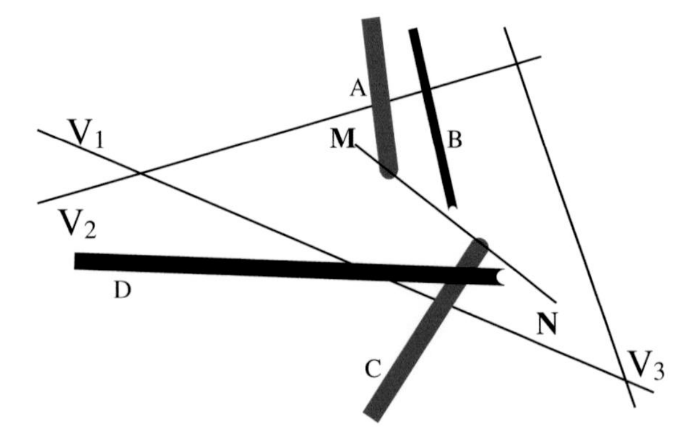


**Fig. 26** Detection of the rule connection between M and N

In Fig. 26 the system of three different constraint sets helps detect a causal relationship between M and N, with no evident correlation between them in Fig. 25. The connection occurs in imposing additional constraints in Fig. 26. The process is dynamic in nature and it is almost impossible to be properly explained only with words. Here the set of the pictures has a role of the replacement of the intuition process.

The process of finding new rules using Puzzle methods gives the trained people the necessary deep knowledge. This form of knowledge can be sometimes represented in the form of ontologies.

The image [it could be a media] from Figs. 25 and 26 is convenient and efficient to visualize, transform and use different dynamic processes through sense-based explanations. Examples and figures can lead to the conclusion that ontologies are introduced, used and dynamically changed applying the Puzzle methods. Hence, they can be improved using Puzzle methods.

Unlike the traditional oral presentation, the material visualized by the Puzzle method gives more directions for [future] own research and for detecting and correcting gaps in knowledge. For example, 'it is neither … nor …' and many other natural-language interpretations incorporate a lot of fuzziness that can sometimes be misunderstood. Careful study of graphs similar to Fig. 26 excludes presentation incompleteness and furthermore—a misunderstanding of the material. On the other hand, the whole learning process is concentrated in one place, it is not necessary to look for other Web sources to clarify the question e.g. 'why', 'what', 'how' to get results, etc. The quoted relations also have been used for pointing the lecture main points. Intelligent training tools not only accelerate but also intensify the training. Hence it is high quality.

A wide range of tools for modeling and presentation of material is used in this chapter where the modeling must be used not for ISS software agents but for people. Respectively it is easier to introduce audio or multimedia presentations, for example, ontologies for transmitting the meaning values. Sometimes to present the meaning of things it is enough to show one or several key points of the subject altogether with corresponding explanations in visual, graph, tactile or sound form. Sometimes one key picture with concise explanations is better than an entire movie. Of most importance are the descriptions to a picture that reveal the essentials, important details, and the sense of the presentation. The picture explains many details clear to people, but intelligent agents will learn almost nothing from this picture, especially if the agents are designed only for ISS purposes. Hence, many agent-based methods still are much more effective in people tutoring systems. This situation will change.

Information presentation by meaning is a key element of modern higher and university education. The proposed Puzzle methods altogether with the other tested methods greatly expand opportunities for qualitative research and teaching in the field of SS, one of dynamic interdisciplinary specialties with increased complexity and also of various types of strict and high requirements for the learners.

### 4.6 Discussion on Experiments Aiming at Creative Collaboration

All around the world there have long been developed many theories for creative improvement of instruction and training some of which stand out like brainstorming methods and TRIZ, for example, what can be found in [6]. Essentially, they activate the technical capabilities of the learners through accumulating the experience from previous technical solutions. The above-presented Puzzle methods do not deny either of the quoted research. On the contrary, we dare claim that the suggested Puzzle methods will improve both the brainstorming and TRIZ due to the following reasons:

1. TRIZ prompts the strategic technological ways to winning solutions. Puzzle methods gradually, consistently show the particular solutions for relating the unknown to the known, whereby they specify the common schemes of TRIZ. We should not forget that the nature of Puzzle methods is evolutionary, i.e. this is one of the possible ways for building not only a group human thinking but a multiagent machine thinking as well. Maybe in the future TRIZ will enter multiagent systems through a combination with Puzzle thus expanding its activity area.
2. Both Puzzle and TRIZ build effective schemes for detection and semantic conflict resolution. In practice, Puzzle methods are particularly effective in combination with other methods (the test on inconsistency; of the Kaleidoscope; of the Funnel, etc.) as shown in [5]. We claim that our methods for conflict resolution are much more specific, formal and technological than their TRIZ analogs. Here is one more reason for future cooperation.
3. It should not, therefore, be forgotten that Puzzle methods equally effectively function in intelligent systems for training people while TRIZ and other cited methods here are not applicable for machine learning.
4. We have not discovered non-technological applications of TRIZ while Puzzle methods act practically in the same way in separate fields. Thus, they combine the technological advantages of TRIZ with the wider applicability of brainstorming methods.
5. Both Puzzle and TRIZ are evolutionary in nature, which is a good basis for their future combination.
6. Well-known methods for correlation analysis or regression analysis help Puzzle methods in the search of logically linked sets of knowledge presented in the first sections here as an analog to the searched word in a multi-dimensional context. These methods help the search in Puzzle of binding/ontology/crossword constraints (limitations), strongly narrowing the search area. Here is yet another example of the power of the suggested solutions widening the application area of the modern statistical methods to drastically different logical areas.

The graphs analogous to the presented information in Fig. 19 help the visual perception of the situation in a key moment and at the same time avoiding the difficult to digest formal descriptions. After the discovery of new knowledge, the picture is complemented and a transition is made to a better or final decision, for example, the situation in Figs. 25 and 26. Gradually a picture appears of the sought solution or at least a picture of what has been achieved so far. This consequence is precious as it illustrates the creative process of finding a solution, and not just its solution. Today, machines memorize and copy much better than us people methods and solutions, which does not make them smarter. Remembering is no longer an advantage, but the creative thinking of the situation is increasingly an issue with our students and even with some researchers. That is why Puzzle methods will more often be used for security educational purposes.

It is shown in Sect. 2 that not always Puzzle gives instant solutions. In this case different logical relations may be used together with the set of crossword or other constraints and solutions. Thousands of such relations and their applications in ontologies had been investigated. The most useful are the following three: IMPORTANT, WHY, and HOW. By using them, researchers make deep structuration of all accumulated knowledge concerning the presentation. Hence, the restriction of the area of possible solutions, and a new solution appears or it becomes obvious how to obtain it. Let‘s represent the puzzle situation again like in Sect. 2 in the form of a set K of known letters from some unknown word (this is the known information on the problem) and a set of unknown letters U. Correspondingly, the elements from both sets are $k_i \in K$ and $u_j \in U$. If the set K is a subset of the set corresponding to the relation IMPORTANT, and no important part belongs to the unknown set, then the elements from K can be omitted. In this way, the introduction of logical relations may lead to the resolution of the problem without the need of new findings in the unknown set K.

The other way is to defeat one or more ki ϵ K aiming to ease the resolution process. Game theoretic approaches also accelerate the Puzzle resolution process.

The data analytics methods are not easy-to-explain. As shown in this chapter, the goal isn't only the explanation of novelties but application in security education process. We applied it in both pupil and student groups at undergraduate and graduate levels. The results were rather high: at least 50 % less poor or average results at test exams, and much more excellent results. Neither pupils nor students knew data analytics, it is taught at M.S. student level. Hence same methods that are effective in multiagent environment stimulate creative thinking results in human groups.

When methods for artificial cause of semantic conflicts are used in education, the results are very clear and indicative—these are the winning strategies of the tutor. In this situation Puzzle methods allow to follow the situation completely: we have different types of constraints, and in an unwanted situation the limitations become this or that and the weakest areas of the modeling solution are cut out... thus, Puzzle allows the improvement of the description of creative processes with materials difficult to digest in relation to the human (or machine) thinking.

It is obvious that using Puzzle methods provides new solutions in the sphere of Evolutionary Computing. On the other hand, even in the case of applications of the classic genetic algorithms (GAs), Puzzle significantly accelerates the latter and diminishes the heuristic elements in them. GAs themselves can be used to sift the versions of space layout of the three types of limitations presented in Fig. 19. Here new areas of perspective applications can be foreseen.

# 5 Realization Examples

## *5.1 SAS Applications*

Experiments have been done with SAS Enterprise Miner™—powerful software for data mining which has significant applications in science as well as in the sphere of business [11, 12]. The software offers the opportunity to make a predictive model through different methods: neuron networks, regression analysis, cluster analysis, etc., so that the prediction is improved. For this purpose, in this paper we are going to analyze the results from intelligent methods–the artificial neural networks because:

- They are intelligent means of presenting knowledge/they can be self-taught;
- They are flexible (they offer a huge variety of variables)

For the purpose of the experiment, the following nodes of SAS Enterprise Miner™ have been used.

The scope of the considered experiment comprises how through SAS Enterprise Miner™ with default settings the Puzzle method the settings of the artificial neural network can be optimized so that a random model always gives better results compared to the neural network default settings of SAS Enterprise Miner™.

Here the binding is based on operating in similar situations, where the variables are related to one object and use analogous knowledge. In the latter example of training at different speeds, other bindings are found; for example, when the training time is limitless, the winner is the one that learns faster regardless of the initial meanings of accumulated knowledge.

The conducted experiments show the advantages of ANN, trained with the Puzzle method, realized in SAS Enterprise Miner 12.1 over the default settings ANN. The results are superior not only to perceptron applications, but for RBF, autoencoders, and SOMs. Using constraints and bindings, the sensitivity of the network rises dramatically. This is one of the proofs of the successful realization of Puzzle methods.

## *5.2 Chosen Java Applications*

The experimental system related to the presented project includes an adaptation of several popular semantic reasoners via adding Java code to change the analysis and resolving contradictions and also input of different combinations of elements by the Puzzle method.

The presented system source codes are written in Java. All the ongoing research is done in Java because the multiagent FIPA-based software has been used. An application example has been shown below.

A fragment of Java class Estimator is shown in Fig. 27. An Estimator receives messages from agents capable to execute a chosen goal. Then it is calculating their competence. The first agent or set of agents with enough high competence will get the job. The competence of agents is a constantly changing value depending on job type and the number of sweepstakes collected by an agent, say, a set of realized goals, etc.

Conflict situations occur in the module from Fig. 27 when the agent can't execute its goal in an acceptable time or in other situations. Let's say, the agent's way to the goal is interrupted by a route for many other agents. The standard way frequently

```
//private MessageTemplate mt;
//the template to receive replies

private class Estimator extends CyclicBehaviour{
private int thereshold = 60;
private char sector = 'A';
   public void action() {
      MessageTemplate mt = MessageTemplate.MatchPerformative(
ACLMessage.ACCEPT_PROPOSAL);
       ACLMessage msg = myAgent.receive(mt);
     if(msg != null) {
       //PROPOSAL Message received and waiting to be processed
       String title = msg.getContent();
       ACLMessage reply= msg.createReply();
       Integer competence = (Integer) goalSet.remove(goal);
      if(competence >= threshold) {
             reply.setPerformative(ACLMessage.INFORM);
             System.out.println(goal+"in sector "+sector+ " hold by agent
"+msg.getSender().getName();
      }
      else{
             reply.setPerformative(ACLMessage.FAILURE);
             reply.setContent("not-available");
      }
      myAgent.send(reply);
    }
      else {
             block();
      }
   }
}
```

**Fig. 27** Java Estimator codes fragment

quoted in multiagent systems is 'step out' for all conflicting agents. Then another route should be found, otherwise same conflicts will repetitively occur. A growing avalanche of conflicts is also a possible variant in large MAS applications. Meanwhile, some goals may be lost. On the other hand, one or more of conflict resolution methods considered above resolve the problem using well known class Agent methods halt(); suspend(); activate(); getState(); move(Location where); and sometimes clone(Location where, java.lang.String newName). Several libraries for class Agent had been used. All of them comply with FIPA standards.

The pseudocode table of agent activities is represented below in Table 1.

The learning procedures have been improved in IDS and IPS using the following algorithm. Using the considered pseudocode is aiming at more effective learning process, preventing attacks via wrong learning in IDS leading to the corruption of the whole system. The contribution is an upgrade to the classical backpropagation algorithm. Experiments with the semantic reasoners Pellet and FACT++ show that their upgrade procedures are easy and don't demand large code transformations. Description logics (DL) are naturally combined with the considered conflict resolution models. Multiagent applications using JADE and the Puzzle schemes and the other resolution methods are under construction using Java-JADE libraries/classes/methods.

The described self-learning applications improve the tools for security administration, most types of IDS/IPS with the usage of ANNs, vulnerability detection software, crisis management system, and other agent-based software. The conflict situations are mostly avoided for the sake of efficiency, but sometimes artificial conflict situations have been created, let's say model change to enhance or reinforce the security agent learning abilities.

Application examples for other methods are much more complex and can't be described here in details. They show that the considered methods aren't so scholastic as it may seem at the first glance. Let's say, Puzzle method depicted in Fig. 19 or in Fig. 25 could be effectively used to find intersections, changing sense

**Table 1** Agent activities using JADE

| **Function Critical Learning(A)** | A = teacher's example or counterexample |
|---|---|
| **Begin {check}** | |
| **B:=0** | B = value to return |
| **C:=0** | C = activities to resolve the conflict |
| **If A=Condition Then Call identify(A,B,D)** | *if Conditions exist then check if a conflict between A and another knowledge exists* |
| **Else Call backpropagation(A)** | D = other sides of the conflict excluding A |
| **If B:=0 Then Call backpropagation(A)** | *backpropagation = the classical back propagation algorithm* |
| **Else Resolve(A,C,D)** | Resolve = known activities to resolve the revealed type of conflict |
| **Return execute(A,C)** | |
| **End {check}** | |

variations and other important issues between different agent ontologies. The Puzzle effectively works with the other methods considered in [5], and in first place with the considered above method for identification and resolution of semantic conflicts.

Many of the described procedures rely on the usage of different models/ ontologies in addition to the domain knowledge. The latter are metaknowledge forms. In knowledge-poor environments the human-machine interactions have a great role, and the metaknowledge helps make the dialog more effective and less boring to the human. The dialog forms are divided into 5 categories from 1='most informative' to 5='silent' system. Knowledge and metaknowledge fusion is always documented: where the knowledge comes from, when done, who made it, etc. This is one of the main data driven principles: any knowledge is useful and if the system is well organized, it will help us resolve some difficult situations in future.

We rely on non-symmetric reply to intruders with a motto *surprise and win*, on the usage of unknown methods code in combination with well-known methods, and on the high speed of automatic reply in some simple cases e.g. to halt the network connection when the attack is detected. If any part of the program is infected or changed aiming at reverse engineering or other goals, then the system may automatically erase itself and in some evident cracking cases a harmful reply will follow. The above-presented models of users and environment are welcome in the case. Therefore, different Puzzle method realizations are not named intrusion detection/prevention systems but intelligent security systems because they include some limited automatic reply to illegal activities.

Experimental activities in our faculty are developed by other authors—Ph.D. students and assistant professors; for this reason its detailed description goes beyond the present representation.

## 5.3 Main Principles for Building Knowledge Driven Human-Machine Interfaces

It should be left old fashioned comprehensions that the emotional teaching of the naturally-scientific disciplines is an obstacle but not a contribution. Not only the teaching but the research, especially human-machine contact may and sometimes should be emotional. The term "emotional teaching" long ago doesn't mean that it is necessary "to read lecture" enthusiastically, with a lot of emotions. Comfortable conditions should be created for work with the system or of the students that will cause the necessary emotions.

The two are the basic winning approaches for the contemporary education/research on ISS. We issued some projects to study the effective work with students, but same or similar principles can be applied in machine learning environments.

On the first place it is necessary to attract the attention of the learners to the teaching material, for example through introduction of the competition element (who first will resolve the problem) through system of bonuses and at the same time introduction of the additional work for the slow working students (themes for additional learning, additional tasks and others). To provoke a desired new research, it is a good idea to show before the potential researchers a creation process for something analogical.

The indicated element for the system of security education should be combined in one clear and in advance explained system where the excellent students are examined for a shorter time than the others. One of the aims is to avoid the monotony in a teaching of the contemporary dynamically developing disciplines where the lack of attention can cause misunderstanding or wrong understanding of the teaching material. Meanwhile, the usage of Puzzle and other methods diminish the monotonicity of shown results. On the second place, the won attention of the students has to be maintained. Without the introduction of the particular emotional elements in the lecture is difficult to attract the attention and it is practically impossible to maintain it for a long time.

In the direction emotion-aware systems the implementation of the resources for the measuring of the emotions is widely discussed, same as the personalization of the data with the aim—elaboration of the individual scenario for education, the modeling of every person working in the system.

These resources are not implemented in the research because they require more financial resources (that is way are implemented mainly in the expensive researches in ambient intelligence) and are not effective: the reactions and the external image of the person very often lie for his emotional status, the models permit to individualize education, but their elaboration is slow and expensive process. That is the way proposed not to measure the positive emotions but to cause them in the suitable conditions. To create the conditions on the first place it is necessary to exclude factors which become obstacles as:

(a) Difficult and huge material and lack of time to teach it;
(b) Complicated for understanding material with many formal descriptions or with many included fragments from the programs. The formal or program descriptions require time which can "eat" the bigger part of the lectures.
(c) It is not easy to lecture dynamically changing material with the interdisciplinary purpose to people who used to study on the traditional way of methods, algorithms, data from the area and sources for speedy learning, with requirement—everything should be written in the Bulgarian language. At the same time we run into the collective attitude—all to be done on the fast and of a poor quality way, for example, the resolves of the homework tasks to be looked in "manuals" or in other Internet sources.
(d) It is impossible to establish emotional contact without a clear description of the things, without interest and mutual confidence. Here is used the following thesis:

More convincing, better structured, clearer is emotional (cause positive emotions for the students).

However, how to describe for example intelligent technologies like computational discovery or data mining, a serious explanation which is accepted as a very difficult even by the experts in the area? In order to solve the represented problems without implementation of expensive technical resources is started with "ice breaking" through explanations that more complicated matter leads to highly appreciated knowledge, through surprises, awards, change of the environment, etc.

It is good, for example, to use the ideas of the management of our university, ULSIT, when there are things for verbal explanations to use comfortable armchairs for lessons outside. The main aim of these measures is some of the best and most active students to help to the others. It is encouraged and contra questions and comments from the side of the students if they have a direct connection with the teaching topics. It is possible for example to be used a system of bonuses from +0.25 to +1 to the examination marks of the next controls works for individually presented theme, ad hoc resolved task, etc.

The practice shows that even the smaller bonus later as consequence stimulates not very well organized students to give more time for the subject because they are convinced that they are better with something than the others.

The organizational measures for better teaching have to be multiple: control to the presences, interim control of the knowledge, grouping of the students in dependence of their results and diverse approach to the different groups, etc. Sometimes are important and the most neglected details as a teaching of the easier material in the weeks with more tests and other engagements with other disciplines, but how many of us did it? Meanwhile, the resolution of this task also concerns constraint satisfaction depicted in Fig. 19.

In general, the requirements to the lecturers/researchers of ISS are contradictory. In manual applications to the problem with the teaching of the proposed methods for conflict resolution, the following results were received.

It is suggested to the students or software agents to use and other new relation IMPORTANT with the aim of the structuring of the material according to different degrees of importance.

When are explained, for example, implementations of lidars/radars for a protection of the airports, relations WHY and IMPORTANT have to be directed to the following text: Doppler radar is significantly more expensive than the ordinary radar but permits not only to show the localization of the monitored objects but and their speed. One of the basic its function today is to discover air micro disturbances, something very dangerous to landing aircrafts, and the movement of the winds in the storms.

Of course, the detailed technical explanations of the radars are important also but they didn't give new information for the meaning of the learning objects and for their use and that is they are not connected with the relation IMPORTANT. If an agent without understanding the meaning of the things submit the texts to the man in connection with the relations WHY, ORIGIN, HOW or, for example, IMPORTANT, he will fulfill the task for correct informing and no need of the additional actions. In the scenario with two software agents, this technology will not work without additional operations.

The suggested way is from the category blended teaching, when IT widely used in the process of teaching but didn't replace the lecturer. It is necessary for the easier learning of the subject the most important things which concern the strategies, new and basic concepts, paradigms, changes in the learning area to be learned at the spot but the training tools to help for learning of the technical details with the examination of the knowledge during the exercises. This way gives an opportunity for an easier perception of the subject and to give more examples and presentations.

When the question concerns emotions, everybody reminds emoticons. Their use on the screen or on the board is accepted unserious may be as bad taste, as playing with the students. Meanwhile, my opinion is that their entering in the ISS is forthcoming and their role and statute will be changed. The emoticons more frequently are used in web forums because of their compact presentation of the information. Because of the same reasons it is suggested they be used by the agents. The symbol "!", for example, is used in very wide meaning including as "attention", "important", "danger", "check-in", "see why it is like that", "see how", etc. dependent of the situation and the context. Trough emoticons the human can on the fastest and correct way to be oriented what the agents want from him and the student will be oriented better what the teacher wants from him.

The main conclusion is as follow: The emotional contact inciting the learner agent to make creative expressions, but it is not obligatory to be based on the resources for measuring of the emotions. The emotional elements in the teaching together with the other measures of organizational character diminish the difficulties of the subject and have to leave "doors for independent research" This process should be reinforced. The use of innovative elements in the teaching of ISS is necessary. It concerns the teaching of IT as a whole. It is presented many approaches to that. Last but not least, the represented above human-machine analogs on the brainstorming method can't be successfully realized without indicated in the chapter methods. Consequently, there is a direct connection between the right teaching on ISS today and creation of the technologies of the future intelligent technologies.

## 5.4 Chosen Theoretical Applications

The software for the research includes more than 20 programs. They are written in Visual Basic and more than 200 MB corresponding Excel data communicate with them. The assistant and defensive software consist of more than 20 programs in C++.

The following denotations are introduced. $\{m + nk\}_{k=0}^{\infty}$ is an arithmetic progression (progression for short). In it m is the first member, and n is the step. $\boldsymbol{\pi(x)}$ is the total number of the primes which are elements of the set $\mathbf{P}$ ($p_i \in P$, $p_i \le x$). $\boldsymbol{\pi_{n,m}(x)}$ is the number of primes $\le$ x which are contained in the progression. $\mathbf{S_5}$ is an union of

8 progressions $\{y + nk\}_{k=0}^{\infty}$, $y \in Y$, $Y = \{1,7,11,13,17,19,23,29\}$. Every of these progressions is represented as a column in Fig. 5 or Fig. 6 if the elements of $S_5$ are shown vertically. The number environment is shown on the figures without the prime numbers because the composite numbers are of the most importance. Blanc places are introduced instead of primes. Every of the elements in $S_5$ is computed in the following way. The first number from the corresponding column, see line 1, is added to the number from the same line and the leftmost column. For example, $s_{14,2}$ = 7+390 is located in line 14 and column 2. Composite numbers in $S_5$ are represented as products of prime numbers. The primes are the result of the decomposition of the composites. In Fig. 5/6 the primes are *omitted* while the particular cases $y \in Y$ are given in brackets. For example during the investigation of the operations addition and multiplication in $S_5$ the following parameters attract the attention: primes (with just a single divisor), composites with at least 2 divisors, 8 columns which are parallel to the vertical axis $\vec{y}$ and 15 lines which are parallel to $\vec{x}$ These 4 parameters can have other designations, which will have similar meanings. The names are not significant. The parameters are established by mere observations e.g. directly on the figures.

($T_1$). The primes are determined but not shown from all the numbers in the fragment, see Fig. 5 or Fig. 6. The *very omission* introduces some new information. The figures below demonstrate the following versions of transformations in $S_5$.
($T_2$). All the composites are presented as products of prime divisors.
($T_3$). All the composites with the divisor of 11 on Fig. 5 and divisible by 19 on Fig. 6 are successively connected with straight lines.
($T_5$). Besides the graphical interpretations in Fig. 5 and 6 must be added similar pictures for the "neighbors below" 41 = 11 + 30 and 49 = 19 + 30. The result has the same succession of beat for the columns with periods 30 times 41 and 30 times 49. The illustrations resemble the Fig. 5, respectively Fig. 6 but they are more elongated due to the greater period of the 'repeating multiplication cycles'.

Using the proposed explanations steps, we can explore the well-known formulas and introduce new ones. It is known that for enough large x formula (35) is executed.

$$\pi(x) > \frac{x}{\ln(x)}. \tag{35}$$

where the result can be monotonously generalized to the whole interval [0, ∞]. But the case when x = 8 violates the formula (35). This is because of a larger deviation % for small number, 8 in the case. The introduction of additional constraint $x \geq 11$ leads to the correct result:

$$\pi(x) \underset{x \geq 11}{>} \frac{x}{\ln(x)}. \tag{36}$$

The introduced method generated new results even during the first investigations in 1986. The following strategy was formulated later. The target is to find dependencies in the arrangements of different sets of numbers, e.g. which are multiples of 11 or 19. For example, the multiplication cycle 19 is depicted on Fig. 6. One can say that at the start we have *zero information*. We introduce descriptions of well-known hypotheses, e.g. the twin primes hypothesis, Goldbach conjecture etc. in the same model. Finally, we obtained original mathematical dependencies and formulas. In practice, this approach starts with a research of the twin primes hypothesis with a difference of 2: these are couples of prime numbers 5 and 7, 11 and 13 etc. The hypothesis is based on the suggestion that there exist an infinite number of such similar pairs. The hypothesis formalization *must not be mistaken* with the goal function. It is just a model inside the considered sets of progressions. The research of the multiplication operations with prime numbers in different numerical models, e.g. in $S_5$ leads to the conclusion that the principle properties of different composite numerical unions are also prime number functions (37), and (38). This result at a first glance is very remote from the twin primes hypothesis. This result relates to the proof of Theorem 1 which was not a target in the research. Nevertheless, it may assist in the process of solving for many different **goals.** The famous Dirichlet theorem is a corollary from this Theorem 1.

$$C_{K,6,1}(x) = \sum_{p=7}^{p_{z7}} C_{K-1,6,1}\left(\frac{x}{p}\right) + \sum_{p=5}^{P_{z5}} C_{K-1,6,5}\left(\frac{x}{p}\right). \tag{37}$$

$$C_{K,6,5}(x) = \sum_{p=5}^{p_{z5}} C_{K-1,6,1}\left(\frac{x}{p}\right) + \sum_{p=7}^{P_{z7}} C_{K-1,6,5}\left(\frac{x}{p}\right). \tag{38}$$

where $p_{za} \in \{a+6k\}_{k=0}^{\infty}$, $C_{k,6,a}(x)$ are all the composites $\leq$ x from $\{a+6k\}_{k=0}^{\infty}$ which contain k prime divisors.

**Theorem 1** *Let the interval of integer numbers [0, x] is considered. Let two progressions are considered* $\{m_1+nk\}_{k=0}^{\infty}$ *and* $\{m_2+nk\}_{k=0}^{\infty}$ *and the relevant numbers are mutually prime:* $(m_1, n) = 1$, $(m_2, n) = 1$. *Denote* $\Delta\pi_{n,mi}(x)$, $i = 1.2$. *The denotation introduces the difference (delta) in the number of the primes* $\leq x$ *included in both progressions. This difference may not be greater than the number of the primes in the range* $[0, \sqrt{x}]$, *which is signed as follows:* $\Delta\pi_{n,mi}(x) \leq \pi(\sqrt{x})$.

The Theorem 1 proof isn't given for the sake of simplicity. The way to produce the proof is more important in the case because it produces many more theorems. Theorem 1 is the basic tool for the derivation of the twin primes formula:

$$P_x(p, p+2) \sim 1.320323632\frac{(\pi(x))^2}{x}. \tag{39}$$

where $P_x(p, p+2)$ is the number of twin prime couples $\leq$ x, $\sim$ means "asymptotically equal". The solutions below are related to the well-known Hardy-Littlewood hypothesis, the formalization of which is introduced in (41). The formalization check of it revealed a series of inconsistencies, so the hypothesis was transformed in (40). Finally, a new Hypothesis 1 is constructed which is stronger than the Hardy-Littlewood in the sense, that the Hardy-Littlewood hypothesis is one of the possible cases of *Hypothesis* 1.

$$P_x(p, p+d_1, \ldots p+d_{z-1}) \geq K_Z \frac{(\pi(x))^z}{x^{z-1}}. \tag{40}$$

where $P_x$ is the number of z-tuples ≤ x. They have different admissible differences between, and $K_z$ are the corresponding coefficients.

$$P_x(p, p+d_1, \ldots p+d_{z-1}) \sim K_Z \frac{x}{(\ln x)^z}. \tag{41}$$

**Hypothesis 1** Denote **z** the arithmetic progressions $\{a_1 + b_1 k\}_{k=0}^{\infty} \ldots \{a_z + b_z k\}_{k=0}^{\infty}$ with a *non-coinciding* step of progressions. Let all the corresponding (a, b) = 1. If **z**-tuples of positive integers ($c_i$, $d_i$, …$z_i$) are compared; all of them are positive integer numbers; $c_i = a_2 - a_1 + (b_2 - b_1)(i - 1 + w_1)$ … $z_i - c_i = a_z - a_1 + (b_z - b_1)(i - 1 + w_z)$, and **w** are positive integers, then there exist infinitely many such **z**-tuples ($c_i$, $d_i$, … $z_i$) in which all the numbers are primes $c_i \in P$, $d_i \in P$, … $z_i \in P$.

Hypothesis 1 is formulated as a result of the application of Theorem 1 to the formula (40). The initial version of the Theorem 1 comprised more than 30 pages with several errors and logical bottlenecks. The proof had been improved just using the Puzzle method. The obtained by now results confirm the effect in cases with *infinite* sets of integers and they reveal possibilities for solving problems with *higher complexity*.

In the fragment of security-oriented Number Theory research using Puzzle and other quoted methods including original mathematical variant for evolutionary computing, it is interesting to mention that the coefficient 1.320323632 used in (39) is well-known in Number Theory, it is considered heuristic by nature, but in this application it had been logically inferred in the form of an answer to question: what if the multiplication cycles are researched not only from $\mathbf{S_5}$ but also in larger sets of arithmetic progressions containing only primes >13 and their composites, and so on. Just this coefficient appeared as a result from *lim* transition formula.

Maybe the most demonstrative example from this theoretical by nature Number Theory but practical from the Puzzle-Conflict Resolution point of view is the following fact. Hardy-Littlewood hypothesis had been formulated as a research goal using $\mathbf{S_5}$ but the research process doesn't stop after the proof of the corresponding formula. Instead, a new conjecture had been constructed containing Hardy-Littlewood hypothesis as a particular case. Same way, the formula (39) helped the resolution of the Goldbach hypothesis but later it also helped for the proof of Hardy-Littlewood hypothesis and for many more cryptography-oriented applications. In this way, a data driven application is demonstrated. This is only a fragment of considered Number Theory applications. Many more Number Theory and logically oriented results had been obtained in the period 1986–2005, and one of them was Puzzle methods.

## 6 Conclusions

The main advantages of the proposed research are the following. A set of original ways applicable to a wide range of advanced analytics methods is introduced. The Funnel method is used instead of traditional fitness or goal functions. The main Conflict Resolution method application is hidden in the possibility for advanced unsupervised machine learning. The Puzzle method leads to formal or informal reasoning using a set of classical or new forms of different constraints. The Kaleidoscope method is purposed for visualization activities and for establishment of creative human-machine interactions. The role of the above ways to construct methods for the security bots and agents is discussed. Same methods in different combinations are effectively used to enhance security administrator possibilities or in contemporary e-learning systems in the field of Information/National Security. Applications outside the field of information security have been made since a long time, but their explanation goes out of the field of the considered research. All of the quoted ways for intelligent data processing are used under a control of semantic, data driven methods. Actually the non-algorithmic approaches had been constructed using puzzle methods, conflict resolution methods, and other considered combinations of methods. A large part of this chapter is devoted to show possibilities to construct data driven approaches using traditional algorithm/functional methods.

Analysis is represented for technologies used for machine learning in intelligent agents, and for performing information by sense, and for understanding the semantics of the retrieved information. Common disadvantages for different existing groups of contemporary applications have been revealed. To overcome the shortcomings, methods and applications are considered concerning the logical parts of knowledge discovery and data mining.

The main conclusion is that to overcome the contemporary advanced analytics, data mining, and ANN application shortcomings, methods and applications are considered concerning the advanced analytics based on logical processing. The role of the above methods for different security, financial, educational or other purposes is discussed. The proposed set of methods and applications are domain independent.

Analysis is represented for technologies used for discovery of meaning, exploring information by meaning, and for understanding the semantics of the accumulated data/knowledge information. Common advantages and disadvantages for different groups of contemporary applications have been revealed.

The same methods in different combinations are effectively used to enhance security staff possibilities or in contemporary e-learning systems in the field of Homeland Security.

It is shown how, by applying the principles of advanced analytics technology in data miners, there is a substantial rise in the quality of obtained results. Applications in SAS Enterprise Miner 12.1 leading to significant improvements have been shortly discussed.

And last but not the less important conclusion is from the application part. Same methods are shown very effective to students and machine learning using intelligent agents. In future agents will learn analogically to people, so they can teach people, too. This is the way to deep tutoring, deeper research and communications at contemporary level and deep learning in the world of fast growing knowledge.

## References


1. Jotsov, V.: Information Security Systems, 156 p. Za bukvite-o pismeneh, Sofia (2006)
2. Thuraisingham, B.: Data Mining Technologies, Techniques, Instruments and Trends. CRC Press, New York (1999)
3. Dasgupta, D.: Artificial Immune Systems. Fizmatlit, Moscow (2006)
4. Jotsov, V.: Evolutionary parallels. In: Proceedings of the First International IEEE Symposium on 'Intelligent Systems', Vol. I., Varna, Bulgaria, 10–12 Sept 2002, pp. 194–201 (2002)
5. Jotsov, V.: Intelligent Information Security Systems, 277 p. Za bukvite-O Pismeneh, Sofia (2010)
6. The TRIZ Journal. Part of the Real Innovation Network. http://www.triz-journal.com/archives/what_is_triz/ (to date)
7. Jotsov, V.: Advanced analytics methods and intelligent applications in education. In: Proceedings of the 7th IEEE International Conference on Intelligent Data Acquisition and Advanced Computing Systems: Technology and Applications IDAACS'2013, vol. I, Berlin, Germany, pp. 197–202 (2013)
8. Denchev, S., Pargov, D., Jotsov, V. (eds.): Crisis Management. Avtookazion, Sofia (2013)
9. Bonneau, G.-P., Ertl, T., Nielson, G. (eds.): Scientific Visualization: The Visual Extraction of Knowledge from Data. Springer, Berlin (2006)
10. Wang, M., Peng, J., Cheng, B., Zhou, H., Liu, J.: Knowledge visualization for self-regulated learning. Educat. Technol. Soc. **14**(3), 28–42 (2011)
11. SAS® Enterprise Miner™ http://www.sas.com/en_us/software/analytics/enterprise-miner.html (to date)
12. Estimating sensitivity, specificity, positive and negative predictive values, and other statistics. http://support.sas.com/kb/24/170.html (to date)

# On Some Modal Type Intuitionistic Fuzzy Operators

**Krassimir T. Atanassov and Janusz Kacprzyk**

**Abstract** A review of two groups of basic modal type operators, defined over the intuitionistic fuzzy sets, is given. Two new modal operators are introduced for the first time, and some of their properties are discussed. Some open problems are formulated.

**Keywords** Intuitionistic fuzzy operator · Intuitionistic fuzzy set · Modal logic

**AMS Classification** 03E72

## 1 Introduction

Intuitionistic Fuzzy Sets (IFSs) were introduced in [1] as extensions of the well-known fuzzy sets of Zadeh [14].

Over them, all operations and relations, existing for ordinary fuzzy sets, have been defined, new operations and relations have been introduced, but what is most important is that over them some kinds of modal operators have also been formulated and their properties have been studied.

Here, a short review of some of these modal operators is given, and a new type of operators is discussed.

K.T. Atanassov (✉)
Department of Bioinformatics and Mathematical Modelling,
Institute of Biophysics and Biomedical Engineering,
Bulgarian Academy of Sciences, 105 Acad. G. Bonchev Str, 1113 Sofia, Bulgaria
e-mail: krat@bas.bg

J. Kacprzyk
Systems Research Institute – Polish Academy of Sciences,
ul. Newelska 6, 01-447 Warsaw, Poland
e-mail: kacprzyk@ibspan.waw.pl



V. Sgurev et al. (eds.), *Innovative Issues in Intelligent Systems*,
Studies in Computational Intelligence 623, DOI 10.1007/978-3-319-27267-2_9

The basic definitions and properties of the intuitionistic fuzzy sets are given in [2, 3].

## 2 Intuitionistic Fuzzy Operators of a Standard and an Extended Modal Type

Let us have a fixed universe $E$ and its subset $A$. The set

$$A^* = \{\langle x, \mu_A(x), \nu_A(x)\rangle | x \in E\},$$

where

$$0 \leq \mu_A(x) + \nu_A(x) \leq 1$$

is called the intutionistic fuzzy set (IFS, for short) and functions $\mu_A{:}E \to [0, 1]$ and $\nu_A{:}E \to [0, 1]$ represent the *degree of membership (validity,* etc.*)* and *non-membership (non-validity,* etc.*)*. Now, we can also define a function $\pi_A{:}E \to [0, 1]$ by means of

$$\pi(x) = 1 - \mu(x) - \nu(x)$$

and it corresponds to a *degree of indeterminacy (uncertainty,* etc.*)*.

Below, we write $A$ instead of $A^*$.

One of the geometrical interpretations of the IFSs is shown on Fig. 1. In it, to each element $x \in E$ the point $x$ with the coordinates $\langle \mu_A(x), \nu_A(x)\rangle$ is related.

An interesting research direction, that has been recently initiated, boils down to the introduction of some novel definitions of the intuitionistic fuzzy (IF-) interpretations of the standard modal operators. For more information on modalities,

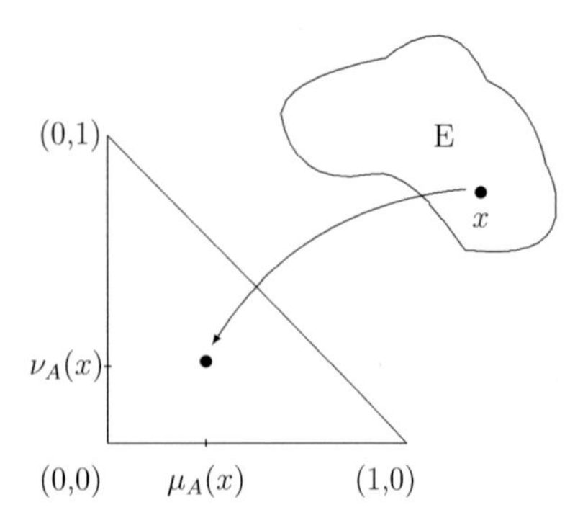


**Fig. 1** Intuitionistic fuzzy interpretation triangle

modal operators, etc., cf. [6–13], to cite just a few literature items that are representative for this vast research area. They are:

$$\Box A = \{\langle x, \mu_A(x), 1 - \mu_A(x)\rangle | x \in E\},$$
$$\Diamond A = \{\langle x, 1 - \nu_A(x), \nu_A(x)\rangle | x \in E\}.$$

for every IFS $A$.

Obviously, if $A$ is an ordinary fuzzy set, then

$$\Box A = A = \Diamond A,$$

while, for a proper IFS $A$

$$\Box A \subset A \subset \Diamond A,$$

where the inclusion is strong and for two IFSs $A$ and $B$,

$$A \subset B \quad \textit{iff} \quad (\forall x \in E)(\mu_A(x) \leq \mu_B(x) \& \nu_A(x) \geq \nu_B(x));$$
$$A = B \quad \textit{iff} \quad (\forall x \in E)(\mu_A(x) = \mu_B(x) \& \nu_A(x) = \nu_B(x));$$

These observations were further mentioned in [1] and they showed the fact that the IFSs are an essential extension of the ordinary fuzzy sets. Both operators have the geometrical interpretation of above type, shown in Figs. 2 and 3 (cf. [2, 3]), where here and below, $Ox$ is the mapping of point $x$ with respect to operator $O$.

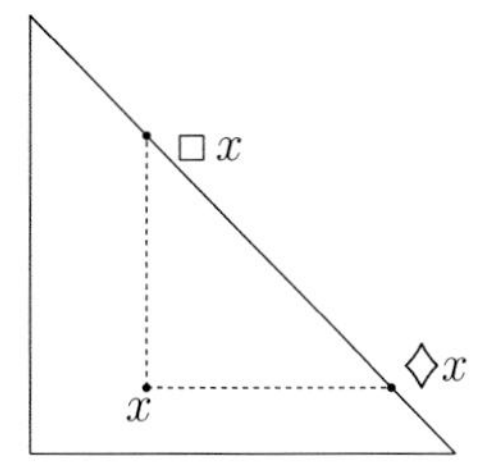


**Fig. 2** Geometrical interpretations of modal operators $\Box$ and $\Diamond$

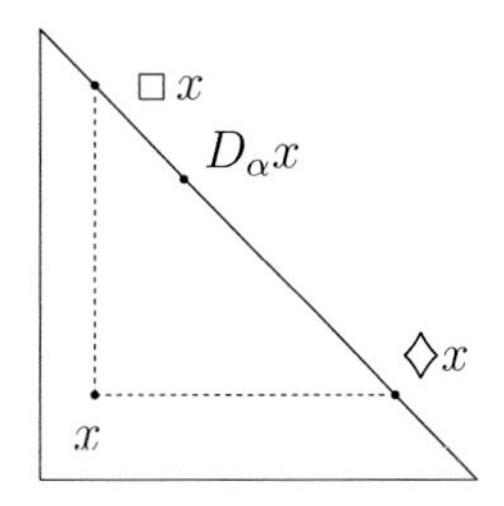


**Fig. 3** Geometrical interpretation of extended modal operator $D_\alpha$

Then, the two model operators has become subjects of further extensions. The first of these is

$$D_\alpha(A) = \{\langle x, \mu_A(x) + \alpha.\pi_A(x), \nu_A(x) + (1-\alpha).\pi_A(x)\rangle | x \in E\},$$

where $\alpha \in [0, 1]$ is a fixed real number (see Fig. 3). Obviously, $D_0(A) = \Box A$ and $D_1(A) = \Diamond A$. Therefore, $D_\alpha$ is an extension of the standard modal operators. It is interesting whether a modal operator can be defined, that is, as an analogue of $D_\alpha$.

Operator $D_\alpha$ was extended to operator $F_{\alpha,\beta}$, where $\alpha, \beta \in [0, 1]$ and $\alpha + \beta \leq 1$. It has the form (see, Fig. 4)

$$F_{\alpha,\beta}(A) = \{\langle x, \mu_A(x) + \alpha.\pi_A(x), \nu_A(x) + \beta.\pi_A(x)\rangle | x \in E\}.$$

Now, $F_{\alpha,1-\alpha} = D_\alpha$.

By analogy with the latter operator, the following five other operators (see, e.g. [2, 3]) were introduced:

$$\begin{aligned}
G_{\alpha,\beta}(A) &= \{\langle x, \alpha.\mu_A(x), \beta.\nu_A(x)\rangle | x \in E\},\\
H_{\alpha,\beta}(A) &= \{\langle x, \alpha.\mu_A(x), \nu_A(x) + \beta.\pi_A(x)\rangle | x \in E\},\\
H^*_{\alpha,\beta}(A) &= \{\langle x, \alpha.\mu_A(x), \nu_A(x) + \beta.(1 - \alpha.\mu_A(x) - \nu_A(x))\rangle | x \in E\},\\
J_{\alpha,\beta}(A) &= \{\langle x, \mu_A(x) + \alpha.\pi_A(x), \beta.\nu_A(x)\rangle | x \in E\},\\
J^*_{\alpha,\beta}(A) &= \{\langle x, \mu_A(x) + \alpha.(1 - \mu_A(x) - \beta.\nu_A(x)), \beta.\nu_A(x)\rangle | x \in E\}.
\end{aligned}$$

where $\alpha, \beta \in [0, 1]$ are fixed numbers.

The geometrical interpretations of the seven operators are given in Figs. 5, 6, 7, 8, 9.

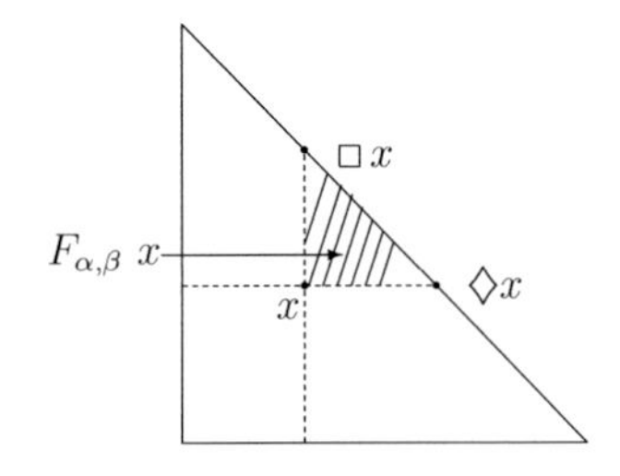


**Fig. 4** Geometrical interpretation of extended modal operator $F_{\alpha,\beta}$

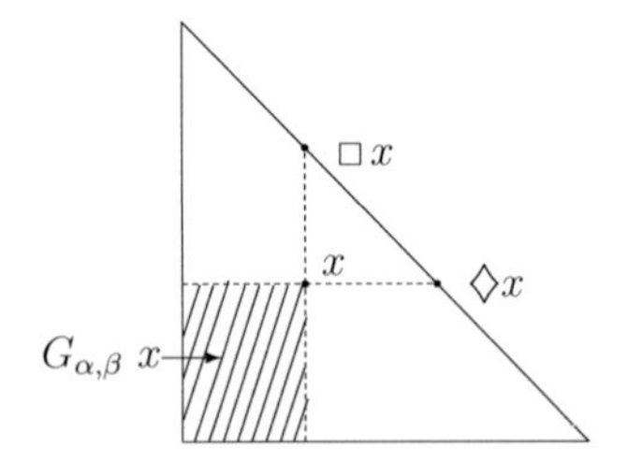


**Fig. 5** Geometrical interpretation of extended modal operator $G_{\alpha,\beta}$

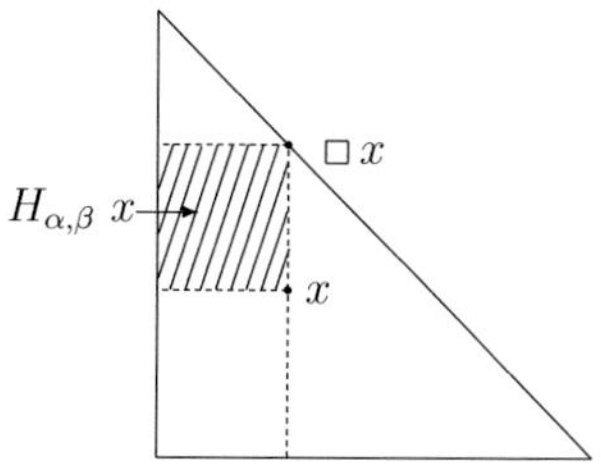

**Fig. 6** Geometrical interpretation of extended modal operator $H_{\alpha,\beta}$

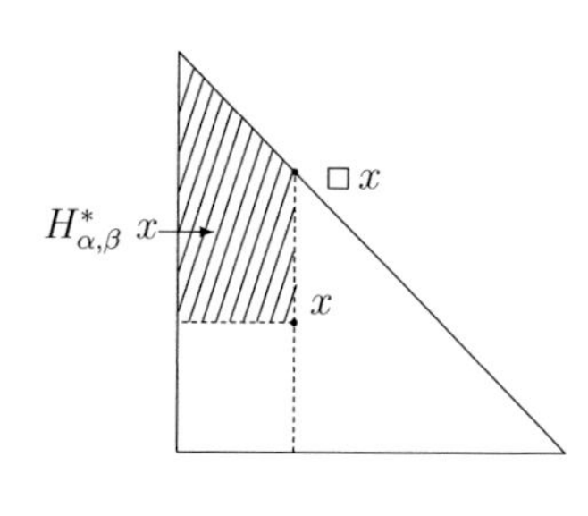


**Fig. 7** Geometrical interpretation of extended modal operator $H^{*\alpha,\beta}$

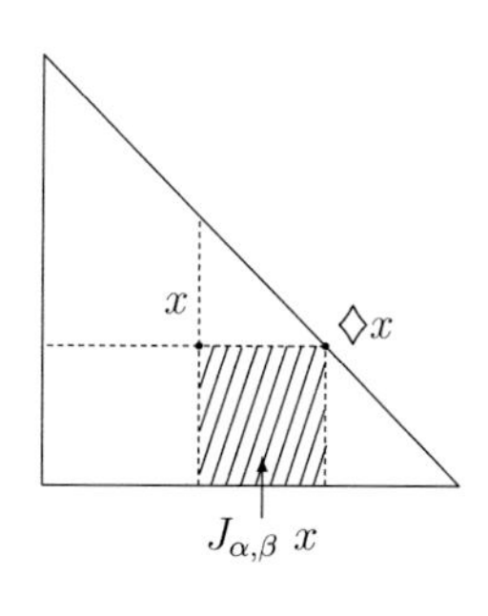


**Fig. 8** Geometrical interpretation of extended modal operator $J_{\alpha,\beta}$

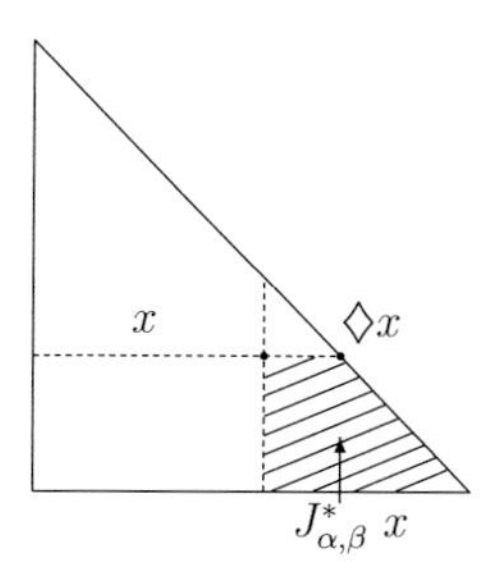


**Fig. 9** Geometrical interpretation of extended modal operator $J^*_{\alpha,\beta}$

The operators of a modal type were extended in some directions as described in [2, 3]. Here we only mention one of them:

$$\begin{aligned} X_{a,b,c,d,e,f}(A) = \{ \langle x, & a.\mu_A(x) + b.(1 - \mu_A(x) - c.\nu_A(x)), \\ & d.\nu_A(x) + e.(1 - f.\mu_A(x) - \nu_A(x)) \rangle | x \in E \} \end{aligned}$$

where $a, b, c, d, e, f \in [0, 1]$ and

$$b + e \leq 1,$$
$$a + e - e.f \leq 1,$$
$$b + d - b.c \leq 1.$$

Following [4], we must note that in [2, 3] the first inequality was omitted, but it is important to be added, because for

$$U^* = \{\langle x, 0, 0 \rangle | x \in E\}$$

we obtain

$$X_{0,1,0,0,1,0}(U^*) = \{\langle x, 1, 1 \rangle | x \in E\},$$

that is impossible. On the other hand, this condition is valid in all cases when operator $X_{a,b,c,d,e,f}$ represents some of the modal type of operators.

## 3 Intuitionistic Fuzzy Operators of Set-Theoretic Modal Type

Here, a new interpretation of the four operators of set-theoretical modal type, that are different than the ones discussed above, are given. They were introduced in [5] many years ago but since then no further research effort in that direction had occured.

Therefore, for a fixed IFS $A$ and for a fixed element $x \in E$, let us define the following operators:

$$\Delta(x) = \{\langle y, \mu_A(y), \nu_A(y) \rangle | y \in E \,\&\, \mu_A(y) = \mu_A(x) \,\&\, \nu_A(x) \leq \nu_A(y) \leq 1 - \mu_A(x)\},$$
$$\triangleright(x) = \{\langle y, \mu_A(y), \nu_A(y) \rangle | y \in E \,\&\, \mu_A(x) \leq \mu_A(x) \leq 1 - \nu_A(x) \,\&\, \nu_A(y) = \nu_A(x)\},$$
$$\nabla(x) = \{\langle y, \mu_A(y), \nu_A(y) \rangle | y \in E \,\&\, \mu_A(y) = \mu_A(x) \,\&\, 0 \leq \nu_A(y) \leq \nu_A(x)\},$$
$$\triangleleft(x) = \{\langle y, \mu_A(y), \nu_A(y) \rangle | y \in E \,\&\, 0 \leq \mu_A(y) \leq \mu_A(x) \,\&\, \nu_A(y) = \nu_A(x)\}.$$

The geometrical interpretations of the new operators are shown in Fig. 10
In [5], some of the basic properties of the four new operators are discussed.

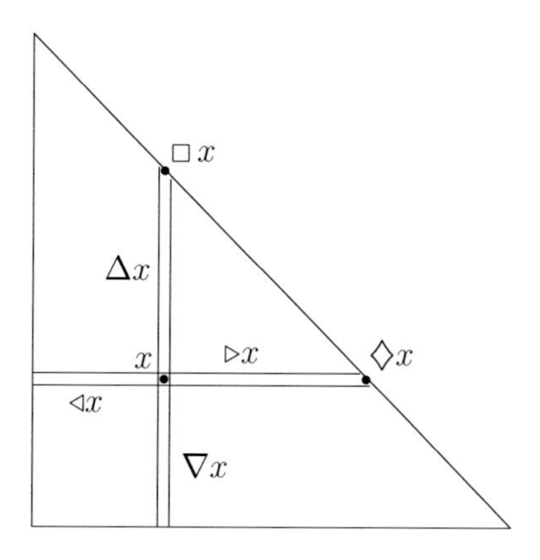


**Fig. 10** Geometrical interpretation of intuitionistic fuzzy operators of set-theoretical modal type

## 4 Modifications of the Four Intuitionistic Fuzzy Operators

Two modifications of the above operators are introduced here for the first time. Let $A$ be an arbitrary IFS and let, for $x \in E$, $\langle x, \mu_A(x), \nu_A(x)\rangle \in A$. Then, we define the following operators:

$$\overline{\Box}x = \{\langle y, \mu_A(y), \nu_A(y)\rangle | y \in E \,\&\, \langle \mu_A(x), \nu_A(x)\rangle \geq \langle \mu_A(y), \nu_A(y)\rangle\},$$
$$\overline{\Diamond}x = \{\langle y, \mu_A(y), \nu_A(y)\rangle | y \in E \,\&\, \langle \mu_A(x), \nu_A(x)\rangle \leq \langle \mu_A(y), \nu_A(y)\rangle\},$$

where for the two pairs $\langle a, b\rangle$ and $\langle c, d\rangle$,

$$\langle a, b\rangle \leq \langle c, d\rangle \quad \text{iff} \quad a \leq c \,\&\, b \geq d$$

and

$$\langle a, b\rangle \geq \langle c, d\rangle \quad \text{iff} \quad a \geq c \,\&\, b \leq d.$$

Obviously, for every $x \in E$, there hold:

$$\Delta x \subset^{st} \overline{\Box}x,$$
$$\lhd x \subset^{st} \overline{\Box}x,$$
$$\nabla x \subset^{st} \overline{\Diamond}x,$$
$$\rhd x \subset^{st} \overline{\Diamond}x,$$

where $\subset^{st}$ is the standard set-theoretical relation of "inclusion".

The geometrical interpretations of the new operators are shown in Fig. 11.

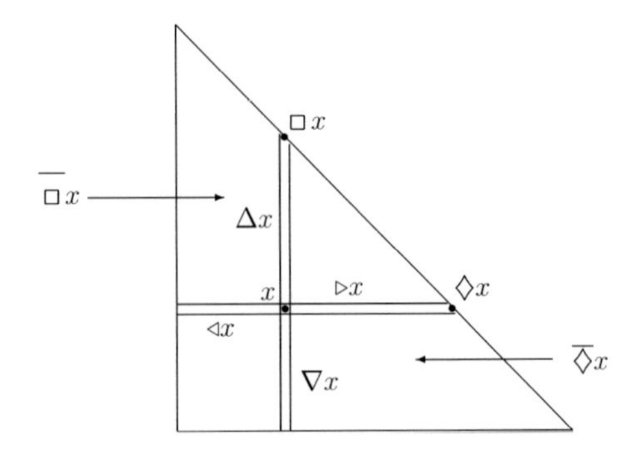


**Fig. 11** Geometrical interpretation of the modified intuitionistic fuzzy operators of set-theoretical modal type

Now, let $x \in E$ be fixed and let $y \in E$. Let

$$\langle y, \mu_A(y), \nu_A(y) \rangle \in \overline{\Box} x.$$

Then we have:

**Property 1**

$$\overline{\Box} y \subset^{st} \overline{\Box} x.$$

*Proof* Let, for some $z \in$ E,

$$\langle z, \mu_A(z), \nu_A(z) \rangle \in \overline{\Box} y.$$

Therefore, $\mu_A(z) \leq \mu_A(y) \leq \mu_A(x)$ and $\nu_A(z) \geq \nu_A(y) \geq \nu_A(x)$. Hence,

$$\langle z, \mu_A(z), \nu_A(z) \rangle \in \overline{\Box} x,$$

i.e., $\overline{\Box} y \subset^{st} \overline{\Box} x$.

Let

$$\langle y, \mu_A(y), \nu_A(y) \rangle \in \overline{\Diamond} x.$$

Then, we have

**Property 2**

$$\overline{\Diamond} y \subset^{st} \overline{\Diamond} x.$$

*Let now, for $x \in E$, $\neg x$ be its negation. In [3], there are definitions of about 40 different operations of "negation". For every $x \in E$ with degrees $\langle \mu_A(x), \nu_A(x) \rangle$, the simplest one is*

$$\neg\langle x, \mu_A(x), \nu_A(x)\rangle = \langle \neg x, \mu_A(\neg x), \nu_A(\neg x)\rangle = \langle \neg x, \nu_A(x), \mu_A(x)\rangle.$$

Clearly, this classic negation does not change the intuitionistic character of the IFSs. Here we use it but in future the discussion below will also be extended to the cases of the remaining types of negations.

Let for $x \in E$ and for an arbitrary negation $\neg$, $\neg A$ be the negation of the IFS $A$. The simplest negation defined for the IFS $A$ is

$$\neg A = \{\langle x, \nu_A(x), \mu_A(x)\rangle \mid x \in E\}.$$

Therefore, there is:

**Property 3** *For every $x \in E$,*

$$\neg\overline{\Box}\neg x = \overline{\Diamond} x,$$
$$\neg\overline{\Diamond}\neg x = \overline{\Box} x.$$

# 5 Conclusion

The new operators introduced in this paper imply a lot of new problems, and their solution will certainly contribute to a further development of the theory of intuitionistic fuzzy sets (IFSs). For example, as we mentioned above, it will be interesting to study Property 3 for different negations, defined over the IFSs. It is also very interesting to consider a possibility for defining operators that are extensions of the new modal operators in the direction towards the operators $F_{\alpha,\beta}$, $G_{\alpha,\beta}$, $H_{\alpha,\beta}$, $H^*_{\alpha,\beta}$, $J_{\alpha,\beta}$, $J^*_{\alpha,\beta}$, $X_{a,b,c,d,e,f}$.

# References


1. Atanassov, K.: Intuitionistic fuzzy sets, VII ITKR's Session, Sofia, June 1983 (Deposed in Central Sci. - Techn. Library of Bulg. Acad. of Sci., 1697/84) (in Bulg.)
2. Atanassov, K.: Intuitionistic Fuzzy Sets. Springer, Heidelberg (1999)
3. Atanassov, K.: On Intuitionistic Fuzzy Sets Theory. Springer, Berlin (2012)
4. Atanassov, K.: A short remark on operator $X_{a,b,c,d,e,f}$. Notes Intuitionistic Fuzzy Sets **19**(1) (in press) (2013)
5. Atanassov, K., Gargov, G.: Intuitionistic fuzzy logic operators of a set theoretical type. In: Lakov, D. (eds.) Proceedings of the First Workshop on Fuzzy Based Expert Systems, Sofia, Sept. 28–30, pp. 39–42 (1994)
6. Blackburn, P., de Rijke, M., Venema, Y.: Modal Logic. Cambridge University Press, Cambridge (2001)
7. Blackburn, P., van Bentham, J., Wolter, F.: Handbook of Modal Logic. Elsevier, Amsterdam (2007)
8. Carnap, R.: Meaning and Necessity. University of Chicago Press, Chicago (1947)

9. Carnielli, W., Pizzi, C.: Modalities and Multimodalities. Springer, Heidelberg (2008)
10. Chagrov, A., Zakharyaschev, M.: Modal Logic. Oxford University Press, Oxford (1997)
11. Feys, R.: Modal Logics. Gauthier, Paris (1965)
12. Goldblatt, R.: Mathematics of Modality. In: CSLI Lecture Notes No. 43. University of Chicago Press, Chicago (1993)
13. Zeman, J.: Modal Logic, The Lewis-Modal Systems. Oxford University Press, Oxford (1973)
14. Zadeh, L.: Fuzzy sets. Inf. Control **8**, 338–353 (1965)

# Uncertain Switched Fuzzy Systems: A Robust Output Feedback Control Design

**Vesna Ojleska, Tatjana Kolemishevska-Gugulovska and Imre J. Rudas**

**Abstract** The problem of robust output feedback control for a class of uncertain switched fuzzy time-delay systems via common Lyapunov function and multiple Lyapunov function methods is solved. Based on employing Parallel Distributed Compensation strategy, the fuzzy output feedback controllers are designed such that the corresponding closed-loop system possesses stability and robustness for all admissible uncertainties. An illustrative example and the respective simulation results are given to demonstrate the effectiveness and feasible control performance of the proposed design synthesis.

## 1 Introduction

In real world physical and other systems essential phenomena such as nonlinearity, uncertainty, and time-delay often co-exist simultaneously [1]. It is therefore that the issue on how to control a nonlinear time-delay system with uncertainties is a challenging, and important issue. Moreover, it is equally important in theory, in applications, and even more in practical implementations.

V. Ojleska (✉) · T. Kolemishevska-Gugulovska
Faculty of Electrical Engineering and Information Technologies,
Institute of Automation and Systems Engineering,
Ss. Cyril and Methodius University in Skopje, Rugjer Boshkovik 18,
1000 Skopje, Republic of Macedonia
e-mail: vojleska@feit.ukim.edu.mk

T. Kolemishevska-Gugulovska
e-mail: tanjakg@feit.ukim.edu.mk

I.J. Rudas
School of Information Science & Robotics, Obuda Univeristy,
H-1034 Budapest, Hungary
e-mail: rudas@uni-obuda.hu



V. Sgurev et al. (eds.), *Innovative Issues in Intelligent Systems*,
Studies in Computational Intelligence 623, DOI 10.1007/978-3-319-27267-2_10

Recently, switching systems, an important class of hybrid systems, which have wide background and applications, have been also one of the main research focuses in the control society. In turn, considerable number of fruitful results in analysis and design of switching systems have been derived too (see for example [3, 9, 10, 14, 15, 22, 35–38], and references therein).

On the other hand, the research activities on fuzzy systems based control, as an important intelligent control approach, combined with some of the math-analytical control theories has attracted great attention. In particular, the class of Takagi-Sugeno (T-S) fuzzy models has been found to be most effective for system modelling in various fuzzy systems based methods. Based on the T-S fuzzy model representations and the feedback control strategy, stability and robust analysis and design as well as handling parameter uncertainties for fuzzy systems have acquired considerable number of fruitful results (see for example [8, 13, 16, 17, 24–28], and references therein).

As a result of the positive notions in using switching systems and fuzzy systems strategies, alone, since the first paper [21], where these two types of strategies are combined, investigations of the synergy of fuzzy and switched systems in the sense of their control synthesis, appeared a logical and natural development. Conceptually, a switched fuzzy system is a type of switching systems in which all of the respective subsystems are fuzzy systems. Many nonlinear systems with switching features can be modelled as switched fuzzy systems or nonlinear systems with fuzzy switching control. However, the results for switched fuzzy systems and switching fuzzy control in the literature seem to be rather limited [23, 32], largely because the first relevant stability results for general nonlinear switched systems have been putted forward fairly recently in [38].

Kazuo et al. [7] and Hiroshi et al. [5], to the best of our knowledge, were the first reports on switching fuzzy control for nonlinear systems. Subsequently, based on the idea of switching Lyapunov function, Hiroshi et al. have finalized their research endeavours in [6]. Yang et al. [31] have contributed the stability solution of a class of uncertain systems based on fuzzy control switching. More recently, in [11], and [12] a new solution type to the robust output feedback control for a class of uncertain switched fuzzy systems was presented. At the same time, authors in [32] has thoroughly elaborated on representation modelling, stability analysis and control design for switched fuzzy systems for both continuous-time and discrete-time cases. Yang et al. [33] contributed a solution to the $H_\infty$ state feedback control for switched fuzzy systems. Authors in [18, 19] have explored and highlighted the influence of the state space partitioning when designing switched fuzzy controllers.

This paper is largely based on the work given in [20]. Inspired by works in [12] and [35], the problem of robust output feedback control for a class of uncertain switched fuzzy time-delay systems whose states are not measurable, hence not available, is further explored and solution given. Sufficient conditions and switching law are derived and formulated in the form of linear matrix inequalities (LMI) based on T-S fuzzy model. These are derived via both common Lyapunov function and multiple Lyapunov function approaches. The fuzzy output feedback controllers are designed by employing the Parallel Distributed Compensation-PDC—strategy. An illustrative example and the respective simulation results are given to demonstrate the effectiveness of the proposed control method and the closed-loop performance it can guarantee.

The next Sect. 2 is dedicated to the presentation of the output fuzzy control design. In Sect. 3, the stability analysis and switching law design are developed. In Sect. 4 the derived results are applied to an illustrative example. Concluding remarks and references follow thereafter.

# 2 Representation Modelling Preliminaries and Output Feedback Controller Design

## 2.1 Plant Fuzzy Rule Model and Switching Sequence Classes

In this study, the following class of switched fuzzy time-delay systems with uncertainty is considered:

$$\begin{aligned} &R_{\sigma}^{i} : IF\, z_1(\text{t}) \text{ is } M_{\sigma 1}^{i} \ldots \text{and } z_n(\text{t}) \text{ is } M_{\sigma n}^{i}, \text{ THEN} \\ &\dot{x}(t) = (A_{\sigma i} + \Delta A_{\sigma i})x(t) + (A_{h\sigma i} + \Delta A_{h\sigma i})x(t-h) \\ &\qquad + (B_{\sigma i} + \Delta B_{\sigma i})\, u_{\sigma}(t) \\ &y(t) = C_{\sigma i}x(t) \\ &x(t) = \Psi(t), t \in [-h, 0], \quad i = 1, 2, \ldots N_{\sigma}. \end{aligned} \tag{1}$$

The symbols used in (1) denote: $M_{\sigma j}^{i}$ represent fuzzy subsets defined by appropriate membership functions; $z(t) = [z_1(t), z_2(t), \ldots, z_n(t)]^T$ is a vector of the premise variables representing only the measurable system variables and not the entire state vector of the plant process; the sequence $\sigma \in M = \{1, 2, \ldots l\}$ is a piecewise constant function representing the switching signal; $x(t) \in R^n$ is the plant state vector; $u_\sigma(t) \in R^m$ is the input control vector; $y(t) \in R^{\vartheta}$ is the plant output vector. The plant systems matrices $A_{\sigma i} \in R^{n\times n}$, $A_{h\,\sigma i} \in R^{n\times n}$, $B_{\sigma\, i} \in R^{n\times m}$, $C_{\sigma\, i} \in R^{\vartheta\times n}$; $\Delta A_{\sigma\, i}$, $\Delta A_{h\,\sigma\, i}$, $\Delta B_{\sigma\, i}$ are time-varying matrices of appropriate dimensions that model system uncertainties. Quantity $h$ denotes the constant delay factor present in the plant, and $\Psi(t)$ is initial value of the state vector $x(t)$.

In the real-world plants often the system states are not all measurable. Hence, we consider introducing the switching law of the form $\sigma = \sigma(\hat{x}(t))$, where $\hat{x}(t)$ are observer generated estimates of system states, to generated the switching signal. That is, it is a sequence in time the piecewise constants of which comply with the estimated states. This way employing an output feedback control is enabled. Further, it is assumed a given partition of state space $R^n$ that is denoted as

$\{\tilde{\Omega}_1, \tilde{\Omega}_2, \ldots \tilde{\Omega}_l\}$, that is $\bigcup_{i=1}^{l} \tilde{\Omega}_i = R^n \backslash \{0\}$ and $\tilde{\Omega}_i \bigcap \tilde{\Omega}_j = \Phi, i \neq j$. The switching signal is represented as $\sigma = \sigma(\hat{x}(t)) = r$ if $\hat{x}(t) \in \tilde{\Omega}_r$. The switching signal is subject to the rule:

$$v_r(\hat{x}(t)) = \begin{cases} 1 & \hat{x}(t) \in \tilde{\Omega}_r \\ 0 & \hat{x}(t) \notin \tilde{\Omega}_r \end{cases}, \quad r \in M.$$

That is, $v_r(\hat{x}(t)) = 1$ if and only if $\sigma = \sigma(\hat{x}(t)) = r$. The partition $\{\tilde{\Omega}_1, \tilde{\Omega}_2, \ldots \tilde{\Omega}_l\}$ and the switching law $\sigma$ will be designed later.

## 2.2 *Employed Output Feedback Controller*

For a given $v_r(\hat{x}(t))$ and based on fuzzy-rule inference [34], the considered system (1) can be represented by means of:

$$\begin{aligned} \dot{x}(t) &= \sum_{r=1}^{l} \sum_{i=1}^{N_r} v_r(\hat{x}(t)) \mu_{ri}(z(t)) [(A_{ri} + \Delta A_{ri}) x(t) \\ &\quad + (A_{hri} + \Delta A_{hri}) x(t-h) + (B_{ri} + \Delta B_{ri}) u_r(t)] \\ y(t) &= \sum_{r=1}^{l} \sum_{i=1}^{N_r} v_r(\hat{x}(t)) \mu_{ri}(z(t)) C_{ri} x(t), \end{aligned} \tag{2}$$

where

$$\mu_{ri}(z(t)) = \frac{\prod_{j=1}^{p} M_{rj}^i(z_j(t))}{\sum_{i=1}^{N_r} \prod_{j=1}^{p} M_{rj}^i(z_j(t))},$$

$$0 \leq \mu_{ri}(z(t)) \leq 1, \sum_{l=1}^{N_r} \mu_{ri}(z(t)) = 1,$$

and $M_{rj}^i(z_j(t))$ represents the membership function of $z_j(t)$ belonging to fuzzy subset $M_{rj}^i$. The following assumption is needed in the sequel for deriving the new result.

**Assumption 1** The parameter uncertainty matrices are norm bounded, that is

$$[\Delta A_{ri} \quad \Delta A_{hri} \quad \Delta B_{ri}] = D_{ri} F_{ri}(t) [E_{1ri} \quad E_{hri} \quad E_{2ri}],$$

where $D_{ri}$, $E_{1ri}$, $E_{hri}$ and $E_{2ri}$ are constant matrices of appropriate dimensions, $F_{ri}(t)$ are unknown time-varying matrices, satisfying $F_{ri}^T(t) F_{ri}(t) \leq I, \ i = 1, 2, \ldots N_r$.

According to Parallel Distributed Compensation-PDC—design strategy [24, 27, 28], the fuzzy output feedback controllers and observers are designed via the following system architecture:

$$\begin{aligned} u_r(t) &= -\sum_{r=1}^{l}\sum_{i=1}^{N_r} v_r(\hat{x}(t))\mu_{ri}(z(t))K_{ri}\hat{x}(t) \\ \dot{\hat{x}}(t) &= \sum_{r=1}^{l}\sum_{i=1}^{N_r} v_r(\hat{x}(t))\mu_{ri}(z(t))\left\{\begin{array}{l} A_{ri}\hat{x}(t) + A_{hri}\hat{x}(t-h) \\ +B_{ri}u_r(t) + L_{ri}[y(t) - C_{ri}\hat{x}(t)] \end{array}\right\} \end{aligned} \tag{3}$$

In here, $\hat{x}(t) \in R^n$ is state vector of the fuzzy observer, and $L_{ri}$ represents observer gain matrix for the $i$th fuzzy rule of the $r$th switched plant subsystem. It is known [24] employing such controller designs in the system synthesis can guarantee the global asymptotic stability of the closed-loop system.

## 3 Stability Analysis and Main Results

In this section, sufficient conditions for global asymptotic stability of the uncertain switched fuzzy time-delay system (1) are given. For the observer equation, a common Lyapunov function is employed such that the observer error $e(t) = x(t) - \hat{x}(t)$ tends to zero under arbitrary switching law. By means of multiple Lyapunov function method, a switching rule is designed based on observed state $\hat{x}(t)$ such that the output-feedback control system is asymptotically stable.

Thus, in turn, the following system representation in closed loop is obtained:

$$\begin{aligned} \dot{x}(t) = \sum_{r=1}^{l}\sum_{i=1}^{N_r}\sum_{j=1}^{N_r} & v_r(\hat{x}(t))\mu_{ri}(z(t))\mu_{rj}(z(t))\{[A_{ri} + \Delta A_{ri} - (B_{ri} \\ & + \Delta B_{ri})K_{rj}]x(t) + (A_{hri} + \Delta A_{hri})x(t-h) \\ & + (B_{ri} + \Delta B_{ri})K_{rj}e(t)]\} \end{aligned} \tag{4}$$

$$\begin{aligned} \dot{\hat{x}}(t) = \sum_{r=1}^{l}\sum_{i=1}^{N_r}\sum_{j=1}^{N_r} & v_r(\hat{x}(t))\mu_{ri}(z(t))\mu_{rj}(z(t))[(A_{ri} - B_{ri}K_{rj})\hat{x}(t) \\ & + A_{hri}\hat{x}(t-h) + L_{ri}C_{rj}e(t)] \end{aligned} \tag{5}$$

$$\begin{aligned} \dot{e}(t) = \sum_{r=1}^{l}\sum_{i=1}^{N_r}\sum_{j=1}^{N_r} & v_r(\hat{x}(t))\mu_{ri}(z(t))\mu_{rj}(z(t))[(A_{ri} + \Delta A_{ri} \\ & - L_{ri}C_{rj})e(t) + (A_{hri} + \Delta A_{hri})e(t-h) \\ & + (\Delta A_{ri} - \Delta B_{ri}K_{rj})\hat{x}(t) + \Delta A_{hri}\hat{x}(t-h)] \end{aligned} \tag{6}$$

In what follows the next assumption and lemma are needed, which are given before proceeding to the proof of asymptotic stability for the proposed designs.

**Lemma 1** [29]. Given constant matrices D and E, and a symmetric constant matrix Y of appropriate dimension, the following inequality holds:

$$Y + DFE + E^T F^T D^T < 0,$$

*where F satisfies* $F^T F \leq R$*, if and only if for some* $\varepsilon > 0$,

$$Y + \varepsilon DD^T + \varepsilon^{-1} E^T E < 0.$$

**Lemma 2** [30]. *Given constant matrices X and Y, for exist appropriate dimension positive definite matrix S and arbitrary* $\varepsilon > 0$*, the following inequality*

$$X^T Y + Y^T X \leq \varepsilon X^T S X + \varepsilon^{-1} Y^T S^{-1} Y,$$

*holds.*

*A proof of the closed-loop of system (1) which guarantees global asymptotic stabilization is given as presented below.*

**Theorem 1** *If there exist the positive definite matrices* $P_r, S_r, P_e, Q$*, some matrices* $K_{ri}$ *and* $L_{ri}$*, switching law* $\sigma = \sigma(\hat{x}(t)) \in M = \{1, 2, \ldots l\}$*, some constants* $\beta_{r\lambda}(\ r = 1, 2, \ldots, l, \lambda = 1, 2, \ldots, N_r)$ *that are either all positive or all negative, and a group of positive constants* $\alpha_{rj}, \xi_{ri}, \varepsilon_{rij}(\ i, j = 1, 2, \ldots N_r)$*, such that the following LMI are satisfied*

$$\begin{bmatrix} \Pi_{rij} + \sum\limits_{\substack{\lambda = 1 \\ \lambda \neq r}}^{l} \beta_{r\lambda}(P_\lambda - P_r) & P_r A_{hri} \\ A_{hri}^T P_r & -S_r + E_{hri}^T E_{hri} \end{bmatrix} < 0, \tag{7}$$

$$\begin{bmatrix} \mathrm{T}_{rij} & P_e A_{hri} \\ A_{hri}^T P_e & -Q + E_{hri}^T E_{hri} \end{bmatrix} < 0, \tag{8}$$

where

$$\begin{aligned} \Pi_{rij} = {} & A_{ri}^T P_r + P_r A_{ri} - \alpha_{rj} P_r B_{rj} B_{ri}^T P_r - \alpha_{rj} P_r B_{ri} B_{rj}^T P_r + S_r \\ & + \varepsilon_{rij}^{-1} \xi_{ri}^2 P_r P_e^{-1} C_{ri}^T C_{rj} C_{rj}^T C_{ri} P_e^{-1} P_r + E_{1ri}^T E_{1ri} - \alpha_{rj} P_r B_{rj} E_{2ri}^T E_{1ri} \\ & - \alpha_{rj} E_{1ri}^T E_{2ri} B_{rj}^T P_r + \alpha_{rj}^2 P_r B_{rj} E_{2ri}^T E_{2ri} B_{rj}^T P_r \\ \mathrm{T}_{rij} = {} & A_{ri}^T P_e + P_e A_{ri} - \xi_{ri} C_{rj}^T C_{ri} - \xi_{ri} C_{ri}^T C_{rj} + Q \\ & + E_{1ri}^T E_{1ri} + 4 P_e D_{ri} D_{ri}^T P_e + \varepsilon, \end{aligned}$$

then under the output feedback controller (3), with

$$\left.\begin{array}{l} K_{ri} = \alpha_{ri} B_{ri}^T P_r \\ L_{ri} = \xi_{ri} P_e^{-1} C_{ri}^T \end{array}\right\}, r = 1, 2, \ldots l, i = 1, 2, \ldots N_r, \tag{9}$$

and switching law $\sigma = \sigma(\hat{x}(t)) \in M = \{1, 2, \ldots l\}$, the closed-loop system of system (1) is asymptotically stable for all admissible parameter uncertainty.

*Proof* It is known from (8) that

$$\begin{bmatrix} e(t) \\ e(t-h) \end{bmatrix}^T \begin{bmatrix} \mathrm{T}_{rij} & P_e A_{hri} \\ A_{hri}^T P_e & -Q + E_{hri}^T E_{hri} \end{bmatrix} \begin{bmatrix} e(t) \\ e(t-h) \end{bmatrix} < 0 \tag{10}$$

holds for any $e(t) \neq 0$. Then, under arbitrary switching law, the observer error satisfies $\lim_{t \to \infty} e(t) = 0$.

Without loss of generality, we assume that $\beta_{r\lambda} \geq 0$. It is apparent that for any $\hat{x}(t) \neq 0$ there exists at least one $r \in M$, such that $\hat{x}^T(t)(P_\lambda - P_r)\hat{x}(t) \geq 0, \forall \lambda \in M$. Applying inequality (7) yields

$$\begin{bmatrix} \hat{x}(t) \\ \hat{x}(t-h) \end{bmatrix}^T \begin{bmatrix} \Pi_{rij} & P_r A_{hri} \\ A_{hri}^T P_r & -S_r + E_{hri}^T E_{hri} \end{bmatrix} \begin{bmatrix} \hat{x}(t) \\ \hat{x}(t-h) \end{bmatrix} < 0. \tag{11}$$

For an arbitrary $r \in M$, let

$$\Omega_r = \{\hat{x}(t) \in R^n | \hat{x}^T(t)(P_\lambda - P_r)\hat{x}(t) \geq 0, \quad \forall \hat{x}(t) \neq 0\},$$

and thus $\bigcup_r \Omega_r = R^n \backslash \{0\}$. Thereafter, we construct the sets $\tilde{\Omega}_1 = \Omega_1, \ldots, \tilde{\Omega}_r = \Omega_r - \bigcup_{i=1}^{r-1} \tilde{\Omega}_i$. Obviously, we have

$$\bigcup_{i=1}^{l} \tilde{\Omega}_i = R^n \backslash \{0\}, \text{ and } \quad \tilde{\Omega}_i \bigcap \tilde{\Omega}_j = \Phi, i \neq j.$$

Next, we design a switching law as follows:

$$\sigma(\hat{x}(t)) = r \quad \text{when} \quad \hat{x}(t) \in \tilde{\Omega}_r, r \in M. \tag{12}$$

Consequently, we consider the following Lyapunov functional

$$\begin{aligned} V_r(t) &= \hat{x}^T(t) P_r \hat{x}(t) + \int_{t-h}^{t} \hat{x}^T(\tau) S_r \hat{x}(\tau) d\tau \\ &\quad + e^T(t) P_e e(t) + \int_{t-h}^{t} e^T(\tau) Q e(\tau) d\tau \end{aligned} \tag{13}$$

where $P_r$, $S_r$, $P_e$ and $Q$ are positive definite matrices, and let

$$V_{1r}(\hat{x}(t)) = \hat{x}^T(t)P_r\hat{x}(t) + \int_{t-h}^{t} \hat{x}^T(\tau)S_r\hat{x}(\tau)d\tau,$$

$$V_2(e(t)) = e^T(t)P_e e(t) + \int_{t-h}^{t} e^T(\tau)Qe(\tau)d\tau.$$

Then it follows:

$$\dot{V}_r(t) = \dot{V}_{1r}(\hat{x}(t)) + \dot{V}_2(e(t)). \tag{14}$$

Furthermore notice the following.

(1) The time derivative of $V_{1r}(\hat{x}(t))$ satisfies

$$\begin{aligned}
\dot{V}_{1r}(\hat{x}(t)) &= \dot{\hat{x}}^T(t)P_r\hat{x}(t) + \hat{x}^T(t)P_r\dot{\hat{x}}(t) \\
&\quad + \hat{x}^T(t)S_r\hat{x}(t) - \hat{x}^T(t-h)S_r\hat{x}(t-h) \\
&= \sum_{r=1}^{l}\sum_{i=1}^{N_r}\sum_{j=1}^{N_r} v_r(\hat{x}(t))\mu_{ri}(z(t))\mu_{rj}(z(t)) \\
&\quad \times \{\hat{x}^T(t)[(A_{ri} - B_{ri}K_{rj})^T P_r + P_r(A_{ri} - B_{ri}K_{rj})]\hat{x}(t) \\
&\quad + \hat{x}^T(t-h)A_{hri}^T P_r\hat{x}(t) + \hat{x}^T(t)P_r A_{hri}\hat{x}(t-h) \\
&\quad + e^T(t)(L_{ri}C_{rj})^T P_r\hat{x}(t) + \hat{x}^T(t)P_r L_{ri}C_{rj}e(t) \\
&\quad + \hat{x}^T(t)S_r\hat{x}(t) - \hat{x}^T(t-h)S_r\hat{x}(t-h)\}.
\end{aligned} \tag{15}$$

According to (9) and Lemma 2, it follows

$$\begin{aligned}
\dot{V}_{1r}(\hat{x}(t)) \le & \sum_{r=1}^{l}\sum_{i=1}^{N_r}\sum_{j=1}^{N_r} v_r(\hat{x}(t))\mu_{ri}(z(t))\mu_{rj}(z(t)) \\
& \times \{\hat{x}^T(t)[A_{ri}^T P_r + P_r A_{ri} - \alpha_{rj}P_r B_{rj}B_{ri}^T P_r - \alpha_{rj}P_r B_{ri}B_{rj}^T P_r \\
& + S_r + \varepsilon_{rij}^{-1}\zeta_{ri}^2 P_r P_e^{-1} C_{ri}^T C_{rj} C_{rj}^T C_{ri} P_e^{-1} P_r]\hat{x}(t) \\
& + \hat{x}^T(t-h)A_{hri}^T P_r\hat{x}(t) + \hat{x}^T(t)P_r A_{hri}\hat{x}(t-h) \\
& + \varepsilon_{rij}e^T(t)e(t) - \hat{x}^T(t-h)S_r\hat{x}(t-h)\}.
\end{aligned} \tag{16}$$

(2) The time derivative of $V_2(e(t))$ can be found as follows

$$\begin{aligned}\dot{V}_2(e(t)) &= \dot{e}^T(t)P_e e(t) + e^T(t)P_e\dot{e}(t)\\ &\quad + e^T(t)Qe(t) - e^T(t-h)Qe(t-h)\\ &= \sum_{r=1}^{l}\sum_{i=1}^{N_r}\sum_{j=1}^{N_r} v_r(\hat{x}(t))\mu_{ri}(z(t))\mu_{rj}(z(t))\\ &\quad \times \{e^T(t)[(A_{ri}+\Delta A_{ri}-L_{ri}C_{rj})^T P_e\\ &\quad + P_e(A_{ri}+\Delta A_{ri}-L_{ri}C_{rj})]e(t)\\ &\quad + e^T(t-h)(A_{hri}+\Delta A_{hri})^T P_e e(t)\\ &\quad + e^T(t)P_e(A_{hri}+\Delta A_{hri})e(t-h)\\ &\quad + \hat{x}^T(t)(\Delta A_{ri}-\Delta B_{ri}K_{rj})^T P_e e(t)\\ &\quad + e^T(t)P_e(\Delta A_{ri}-\Delta B_{ri}K_{rj})\hat{x}(t)\\ &\quad + \hat{x}^T(t-h)\Delta A_{hri}^T P_e e(t) + e^T(t)P_e\Delta A_{hri}\hat{x}(t-h)\\ &\quad + e^T(t)Qe(t) - e^T(t-h)Qe(t-h)\}.\end{aligned} \tag{17}$$

According to Lemma 1 and 2, and with regards to (9) and Assumption 1, one can derive

$$\begin{aligned}\dot{V}_2(e(t)) &\le \sum_{r=1}^{l}\sum_{i=1}^{N_r}\sum_{j=1}^{N_r} v_r(\hat{x}(t))\mu_{ri}(z(t))\mu_{rj}(z(t))\\ &\quad \times \{e^T(t)[A_{ri}^T P_e + P_e A_{ri} - \xi_{ri}C_{rj}^T C_{ri} - \xi_{ri}C_{ri}^T C_{rj} + Q\\ &\quad + E_{1ri}^T F_{ri}^T(t)D_{ri}^T P_e + P_e D_{ri}F_{ri}(t)E_{1ri}]e(t)\\ &\quad + e^T(t-h)A_{hri}^T P_e e(t) + e^T(t)P_e A_{hri}e(t-h)\\ &\quad + e^T(t-h)E_{hri}^T F_{ri}^T(t)D_{ri}^T P_e e(t)\\ &\quad + e^T(t)P_e D_{ri}F_{ri}(t)E_{hri}e(t-h)\\ &\quad + \hat{x}^T(t)(E_{1ri}^T - \alpha_{rj}P_r B_{rj}E_{2ri}^T)F_{ri}^T(t)D_{ri}^T P_e e(t)\\ &\quad + e^T(t)P_e D_{ri}F_{ri}(t)(E_{1ri} - \alpha_{rj}E_{2ri}B_{rj}^T P_r)\hat{x}(t)\\ &\quad + \hat{x}^T(t-h)E_{hri}^T F_{ri}^T(t)D_{ri}^T P_e e(t)\\ &\quad + e^T(t)P_e D_{ri}F_{ri}(t)E_{hri}\hat{x}(t-h) - e^T(t-h)Qe(t-h)\}\\ &\le \sum_{r=1}^{l}\sum_{i=1}^{N_r}\sum_{j=1}^{N_r} v_r(\hat{x}(t))\mu_{ri}(z(t))\mu_{rj}(z(t))\\ &\quad \times \{e^T(t)[A_{ri}^T P_e + P_e A_{ri} - \xi_{ri}C_{rj}^T C_{ri} - \xi_{ri}C_{ri}^T C_{rj} + Q\\ &\quad + E_{1ri}^T E_{1ri} + 4P_e D_{ri}D_{ri}^T P_e]e(t) + e^T(t-h)A_{hri}^T P_e e(t)\\ &\quad + e^T(t)P_e A_{hri}e(t-h) - e^T(t-h)(Q - E_{hri}^T E_{hri})e(t-h)\\ &\quad + \hat{x}^T(t)(E_{1ri}^T - \alpha_{rj}P_r B_{rj}E_{2ri}^T)(E_{1ri} - \alpha_{rj}E_{2ri}B_{rj}^T P_r)\hat{x}(t)\\ &\quad + \hat{x}^T(t-h)E_{hri}^T E_{hri}\hat{x}(t-h)\}\end{aligned} \tag{18}$$

Substituting (16) and (18) into (14) yields

$$\begin{aligned}
\dot{V}_r(t) \leq & \sum_{r=1}^{l}\sum_{i=1}^{N_r}\sum_{j=1}^{N_r} v_r(\hat{x}(t))\mu_{ri}(z(t))\mu_{rj}(z(t)) \\
& \times \{\hat{x}^T(t)[A_{ri}^T P_r + P_r A_{ri} - \alpha_{rj} P_r B_{rj} B_{ri}^T P_r - \alpha_{rj} P_r B_{ri} B_{rj}^T P_r \\
& + S_r + \varepsilon_{rij}^{-1} \xi_{ri}^2 P_r P_e^{-1} C_{ri}^T C_{rj} C_{rj}^T C_{ri} P_e^{-1} P_r + E_{1ri}^T E_{1ri} \\
& - \alpha_{rj} P_r B_{rj} E_{2ri}^T E_{1ri} - \alpha_{rj} E_{1ri}^T E_{2ri} B_{rj}^T P_r \\
& + \alpha_{rj}^2 P_r B_{rj} E_{2ri}^T E_{2ri} B_{rj}^T P_r]\hat{x}(t) + \hat{x}^T(t-h) A_{hri}^T P_r \hat{x}(t) \\
& + \hat{x}^T(t) P_r A_{hri} \hat{x}(t-h) - \hat{x}^T(t-h)(S_r - E_{hri}^T E_{hri})\hat{x}(t-h)\} \\
& + \sum_{r=1}^{l}\sum_{i=1}^{N_r}\sum_{j=1}^{N_r} v_r(\hat{x}(t))\mu_{ri}(z(t))\mu_{rj}(z(t)) \\
& \times \{e^T(t)[A_{ri}^T P_e + P_e A_{ri} - \xi_{ri} C_{rj}^T C_{ri} - \xi_{ri} C_{ri}^T C_{rj} + Q \\
& + E_{1ri}^T E_{1ri} + 4 P_e D_{ri} D_{ri}^T P_e + \varepsilon_{rij} I]e(t) + e^T(t-h) A_{hri}^T P_e e(t) \\
& + e^T(t) P_e A_{hri} e(t-h) - e^T(t-h)(Q - E_{hri}^T E_{hri}) e(t-h)\} \\
= & \sum_{r=1}^{l}\sum_{i=1}^{N_r}\sum_{j=1}^{N_r} v_r(\hat{x}(t))\mu_{ri}(z(t))\mu_{rj}(z(t)) \\
& \times \begin{bmatrix} \hat{x}(t) \\ \hat{x}(t-h) \end{bmatrix}^T \begin{bmatrix} \Pi_{rij} & P_r A_{hri} \\ A_{hri}^T P_r & -S_r + E_{hri}^T E_{hri} \end{bmatrix} \begin{bmatrix} \hat{x}(t) \\ \hat{x}(t-h) \end{bmatrix} \\
& + \sum_{r=1}^{l}\sum_{i=1}^{N_r}\sum_{j=1}^{N_r} v_r(\hat{x}(t))\mu_{ri}(z(t))\mu_{rj}(z(t)) \\
& \times \begin{bmatrix} e(t) \\ e(t-h) \end{bmatrix}^T \begin{bmatrix} \mathrm{T}_{rij} & P_e A_{hri} \\ A_{hri}^T P_e & -Q + E_{hri}^T E_{hri} \end{bmatrix} \begin{bmatrix} e(t) \\ e(t-h) \end{bmatrix}
\end{aligned} \tag{19}$$

From (10) and (11), we can infer that under the switching law (12), $\dot{V}(t) < 0$ holds for arbitrary $\hat{x}(t) \neq 0$ and $e(t) \neq 0$, i.e. $x(t) \neq 0$. Therefore, the closed-loop system of system (1) is asymptotically stable, and the observer error $e(t)$ asymptotically converges to zero. This concludes the proof. □

The stability conditions of Theorem 1 can be transformed into linear matrix inequalities (LMIs). In fact, in view of (8), and using Schur's complement [2], we obtain the following LMIs

$$\begin{bmatrix} \Theta_{rij} & P_e D_{ri} & P_e A_{hri} \\ D_{ri}^T P_e & -0.25I & 0 \\ A_{hri}^T P_e & 0 & -Q + E_{hri}^T E_{hri} \end{bmatrix} < 0, \tag{20}$$

where

$$\Theta_{rij} = A_{ri}^T P_e + P_e A_{ri} - \xi_{ri} C_{rj}^T C_{ri} - \xi_{ri} C_{ri}^T C_{rj} + Q + E_{1ri}^T E_{1ri} + \varepsilon_{rij} I$$

Upon substitution of the solutions $P_e$ of the LMI (20) into inequality (7) and multiplying both sides of inequality (7) by the matrix $diag[X_r \quad X_r]$ with $X_r = P_r^{-1}$, $W_r = X_r S_r X_r$, we obtain the following inequality result

$$\begin{bmatrix} \Psi_{rij} & A_{hri} X_r \\ X_r A_{hri}^T & -W_r + X_r E_{hri}^T E_{hri} X_r \end{bmatrix} < 0, \tag{21}$$

where

$$\begin{aligned} \Psi_{rij} = {} & X_r A_{ri}^T + A_{ri} X_r - \alpha_{rj} B_{rj} B_{ri}^T - \alpha_{rj} B_{ri} B_{rj}^T + W_r \\ & + \varepsilon_{rij}^{-1} \xi_{ri}^2 P_e^{-1} C_{ri}^T C_{rj} C_{rj}^T C_{ri} P_e^{-1} + X_r E_{1ri}^T E_{1ri} X_r \\ & - \alpha_{rj} B_{rj} E_{2ri}^T E_{1ri} X_r - \alpha_{rj} X_r E_{1ri}^T E_{2ri} B_{rj}^T + \alpha_{rj}^2 B_{rj} E_{2ri}^T E_{2ri} B_{rj}^T \\ & + \sum_{\substack{\lambda=1 \\ \lambda \neq r}}^{l} \beta_{r\lambda} (X_r P_\lambda X_r - X_r) \end{aligned}$$

Now with regard to (21), we have

$$\begin{aligned} & \begin{bmatrix} \Psi_{rij} & A_{hri} X_r \\ X_r A_{hri}^T & -W_r + X_r E_{hri}^T E_{hri} X_r \end{bmatrix} \\ & = \begin{bmatrix} \mathrm{N}_{rij} & A_{hri} X_r \\ X_r A_{hri}^T & -W_r \end{bmatrix} + \begin{bmatrix} X_r E_{1ri}^T E_{1ri} X_r & 0 \\ 0 & X_r E_{hri}^T E_{hri} X_r \end{bmatrix} \\ & = \begin{bmatrix} \mathrm{N}_{rij} & A_{hri} X_r \\ X_r A_{hri}^T & -W_r \end{bmatrix} + \begin{bmatrix} X_r E_{1ri}^T & 0 \\ 0 & X_r E_{hri}^T \end{bmatrix} \begin{bmatrix} E_{1ri} X_r & 0 \\ 0 & E_{hri} X_r \end{bmatrix} \\ & < 0, \end{aligned}$$

where

$$\begin{aligned} \mathrm{N}_{rij} = {} & X_r A_{ri}^T + A_{ri} X_r - \alpha_{rj} B_{rj} B_{ri}^T - \alpha_{rj} B_{ri} B_{rj}^T + W_r \\ & + \varepsilon_{rij}^{-1} \xi_{ri}^2 P_e^{-1} C_{ri}^T C_{rj} C_{rj}^T C_{ri} P_e^{-1} - \alpha_{rj} B_{rj} E_{2ri}^T E_{1ri} X_r \\ & - \alpha_{rj} X_r E_{1ri}^T E_{2ri} B_{rj}^T + \alpha_{rj}^2 B_{rj} E_{2ri}^T E_{2ri} B_{rj}^T \\ & + \sum_{\substack{\lambda=1 \\ \lambda \neq r}}^{l} \beta_{r\lambda} (X_r P_\lambda X_r - X_r). \end{aligned}$$

Next, via applying Schur's complement again, the LMI is obtained as follows

$$\begin{bmatrix} \Xi_{rij} & X_r & \cdots & X_r & A_{hri}X_r & X_rE_{1ri}^T & 0 \\ X_r & -\beta_{r1}^{-1}X_1 & \cdots & 0 & 0 & 0 & X_rE_{hri}^T \\ \vdots & \vdots & \ddots & \vdots & \vdots & \vdots & \vdots \\ X_r & 0 & \cdots & -\beta_{rl}^{-1}X_l & 0 & 0 & 0 \\ X_rA_{hri}^T & 0 & \cdots & 0 & -W_r & 0 & 0 \\ E_{1ri}X_r & 0 & \cdots & 0 & 0 & -I & 0 \\ 0 & E_{hri}X_r & \cdots & 0 & 0 & 0 & -I \end{bmatrix} < 0, \tag{22}$$

where

$$\begin{aligned} \Xi_{rij} = {} & X_rA_{ri}^T + A_{ri}X_r - \alpha_{rj}B_{rj}B_{ri}^T - \alpha_{rj}B_{ri}B_{rj}^T + W_r \\ & + \varepsilon_{rij}^{-1}\xi_{ri}^2P_e^{-1}C_{ri}^TC_{rj}C_{rj}^TC_{ri}P_e^{-1} - \alpha_{rj}B_{rj}E_{2ri}^TE_{1ri}X_r \\ & -\alpha_{rj}X_rE_{1ri}^TE_{2ri}B_{rj}^T + \alpha_{rj}^2B_{rj}E_{2ri}^TE_{2ri}B_{rj}^T - \sum_{\substack{\lambda=1 \\ \lambda\neq r}}^{l} \beta_{r\lambda}X_r. \end{aligned}$$

and $X_\lambda = P_\lambda^{-1}, \lambda = 1, 2, \ldots, l,\ \lambda \neq r$. Thus, the stability conditions of the uncertain switched fuzzy time-delay system are transformed into the LMI (20), and (22). These LMI are tractable by means of the LMI Toolbox of the MATLAB [4].

## 4 Illustrative Example and Simulation Results

The given uncertain switched fuzzy time-delay system is:

$$\begin{aligned} & R_1^1 : \text{if } x_1(\text{t}) \text{ is } M_{11}^1, \text{ then} \\ & \dot{x}(t) = (A_{11} + \Delta A_{11})x(t) + (A_{h11} + \Delta A_{h11})x(t-h) \\ & \qquad + (B_{11} + \Delta B_{11})u(t) \\ & y(t) = C_{11}x(t) \end{aligned} \tag{23a}$$

$$\begin{aligned} & R_1^2 : \text{if } x_1(\text{t}) \text{ is } M_{11}^2, \text{ then} \\ & \dot{x}(t) = (A_{12} + \Delta A_{12})x(t) + (A_{h12} + \Delta A_{h12})x(t-h) \\ & \qquad + (B_{12} + \Delta B_{12})u(t) \\ & y(t) = C_{12}x(t) \end{aligned} \tag{23b}$$

$$\begin{aligned} &R_2^1 : \text{if } x_1(\text{t}) \text{ is } M_{21}^1, \text{ then} \\ &\dot{x}(t) = (A_{21} + \Delta A_{21})x(t) + (A_{h21} + \Delta A_{h21})x(t-h) \\ &\qquad + (B_{21} + \Delta B_{21})u(t) \\ &y(t) = C_{21}x(t) \end{aligned} \tag{23c}$$

$$\begin{aligned} &R_2^2 : \text{if } x_1(\text{t}) \text{ is } M_{21}^2, \text{ then} \\ &\dot{x}(t) = (A_{22} + \Delta A_{22})x(t) + (A_{h22} + \Delta A_{h22})x(t-h) \\ &\qquad + (B_{22} + \Delta B_{22})u(t) \\ &y(t) = C_{22}x(t) \end{aligned} \tag{23d}$$

along with the system model matrices

$$A_{11} = \begin{bmatrix} 6.2 & -2.1 \\ 2.3 & -2.8 \end{bmatrix}; A_{12} = \begin{bmatrix} 6.5 & -2.8 \\ 2.7 & -3.0 \end{bmatrix}; \tag{24a}$$

$$A_{21} = \begin{bmatrix} -2.9 & 7.8 \\ -5.1 & 0.5 \end{bmatrix}; A_{22} = \begin{bmatrix} 2.2 & -7.6 \\ 3.5 & -4.6 \end{bmatrix}; \tag{24b}$$

$$A_{h11} = \begin{bmatrix} 1.1 & 0.8 \\ 0.1 & 0.5 \end{bmatrix}; A_{h12} = \begin{bmatrix} 0.6 & 0.5 \\ 1.1 & 0.2 \end{bmatrix}; \tag{25a}$$

$$A_{h21} = \begin{bmatrix} 0.1 & 1.6 \\ 1.3 & 0.6 \end{bmatrix}; A_{h22} = \begin{bmatrix} 1.1 & 1.4 \\ 0.4 & 0.2 \end{bmatrix}; \tag{25b}$$

$$B_{11} = \begin{bmatrix} 4 & 0.2 \\ 0.5 & 2 \end{bmatrix}; B_{12} = \begin{bmatrix} 4 & 0.1 \\ 0.2 & 2 \end{bmatrix}; \tag{26a}$$

$$B_{21} = \begin{bmatrix} 4 & 0.2 \\ 0.1 & 2 \end{bmatrix}; B_{22} = \begin{bmatrix} 4 & 0.1 \\ 0.4 & 2 \end{bmatrix}; \tag{26b}$$

$$C_{11} = [2 \quad 0]; C_{12} = [2 \quad 0]; C_{21} = [2 \quad 0]; C_{22} = [2 \quad 0]; \tag{27}$$

$$D_{11} = D_{12} = \begin{bmatrix} 0 & 0.2 \\ 0.2 & 0 \end{bmatrix}; D_{21} = D_{22} = \begin{bmatrix} 0 & 0.2 \\ 0.1 & 0 \end{bmatrix}; \tag{28}$$

$$E_{111} = E_{112} = \begin{bmatrix} 0 & 0.2 \\ 0.2 & 0 \end{bmatrix}; E_{121} = E_{122} = \begin{bmatrix} 0 & 0.1 \\ 0.2 & 0 \end{bmatrix}; \tag{29a}$$

$$E_{211} = E_{212} = \begin{bmatrix} 0 & 0.1 \\ 0.2 & 0 \end{bmatrix}; E_{221} = E_{222} = \begin{bmatrix} 0 & 0.2 \\ 0.1 & 0 \end{bmatrix}; \tag{29b}$$

$$E_{h11} = E_{h12} = \begin{bmatrix} 0 & 0.1 \\ 0.1 & 0 \end{bmatrix}; E_{h21} = E_{h22} = \begin{bmatrix} 0 & 0.6 \\ 0.1 & 0 \end{bmatrix}; \tag{30}$$

$$F_{11}(t) = F_{12}(t) = F_{21}(t) = F_{22}(t) = \begin{bmatrix} \sin t & 0 \\ 0 & \cos t \end{bmatrix}. \tag{31}$$

The membership functions of fuzzy subsets are chosen as follows:

$$\mu_{11}(x_1(\mathrm{t})) = \mu_{21}(x_1(\mathrm{t})) = 1 - 1/(1 + e^{-4x_1(\mathrm{t})});$$

$$\mu_{12}(x_1(\mathrm{t})) = \mu_{22}(x_1(\mathrm{t})) = 1/(1 + e^{-4x_1(\mathrm{t})}).$$

Next, let it be chosen $\xi_{11} = \xi_{12} = 1$, $\xi_{21} = \xi_{22} = 0.8$. Solving LMI (20) yields the positive definite matrix

$$P_e = \begin{bmatrix} 0.4397 & -0.3416 \\ -0.3416 & 0.9283 \end{bmatrix}.$$

The fuzzy state observer is designed as

$$\begin{aligned} \dot{\hat{x}}(t) = \sum_{r=1}^{2}\sum_{i=1}^{2} v_r(\hat{x}(t))\mu_{ri}(x_1(t))\{A_{ri}\hat{x}(t) + A_{hri}\hat{x}(t-h) \\ + B_{ri}u_r(t) + L_{ri}[y(t) - C_{ri}\hat{x}(t)]\}. \end{aligned}$$

With regard to (9), the obtained observer gain matrices are:

$$L_{11} = \begin{bmatrix} 6.3692 \\ 2.3440 \end{bmatrix}; L_{12} = \begin{bmatrix} 6.3692 \\ 2.3440 \end{bmatrix};$$
$$L_{21} = \begin{bmatrix} 5.0954 \\ 1.8752 \end{bmatrix}; L_{22} = \begin{bmatrix} 5.0954 \\ 1.8752 \end{bmatrix}.$$

By substituting $P_e$ into LMI (7), upon choosing $\alpha_{11} = \alpha_{12} = 6$, $\alpha_{21} = \alpha_{22} = 8$, $\varepsilon_{1ij} = \varepsilon_{2ij} = 2(i,j = 1,2)$, $\beta_{12} = \beta_{21} = 10$, the following positive definite matrices are obtained:

$$P_1 = \begin{bmatrix} 0.6191 & 0.1043 \\ 0.1043 & 1.2678 \end{bmatrix}; P_2 = \begin{bmatrix} 0.6103 & 0.0753 \\ 0.0753 & 2.9155 \end{bmatrix}.$$

Following Sect. 3, the partition is adopted as:

$$\Omega_1 = \{\hat{x}(t) \in R^2 | \hat{x}^T(t)(P_2 - P_1)\hat{x}(t) \geq 0, \hat{x}(t) \neq 0\},$$
$$\Omega_2 = \{\hat{x}(t) \in R^2 | \hat{x}^T(t)(P_1 - P_2)\hat{x}(t) \geq 0, \hat{x}(t) \neq 0\},$$

hence $\Omega_1 \bigcup \Omega_2 = R^2 \backslash \{0\}$.

Thus, we design a switching law as follows:

$$\sigma(\hat{x}(t)) = \begin{cases} 1, \hat{x}(t) \in \Omega_1; \\ 2, \hat{x}(t) \in \Omega_2 \backslash \Omega_1. \end{cases}$$

Also, the output feedback is designed as follows:

$$u_r(t) = -\sum_{r=1}^{2} \sum_{i=1}^{2} v_r(\hat{x}(t)) \mu_{ri}(x_1(t)) K_{ri} \hat{x}(t).$$

With regard to (9), the following output feedback controller gains are obtained:

$$K_{11} = \begin{bmatrix} 15.1709 & 6.3978 \\ 1.9950 & 15.6993 \end{bmatrix}; K_{12} = \begin{bmatrix} 14.9831 & 4.0616 \\ 1.6236 & 15.6367 \end{bmatrix};$$
$$K_{21} = \begin{bmatrix} 19.5894 & 4.7416 \\ 2.1811 & 46.7692 \end{bmatrix}; K_{22} = \begin{bmatrix} 19.7701 & 11.7389 \\ 1.6928 & 46.7090 \end{bmatrix}.$$

Computer simulation investigation was carried out using the initial condition state vector $x(0) = [5, \ -1]^T$ and the value $h = 0.6$.

The obtained computer simulation results are depicted on Figs. 1, 2, 3, 4, and 5. Figures 1 and 2 show the evolution of system state and the observer state trajectories with regard to time, respectively. Figure 3 depicts the time evolution of the control signals, whereas Fig. 4 gives an insight on how and when the switching signals changes its values between 1 and 2, selecting respectively between regions $\Omega_1$ and $\Omega_2$. According the result given on Fig. 5, it is evident that the time evolutions of the state observer errors, converge asymptotically to zero.

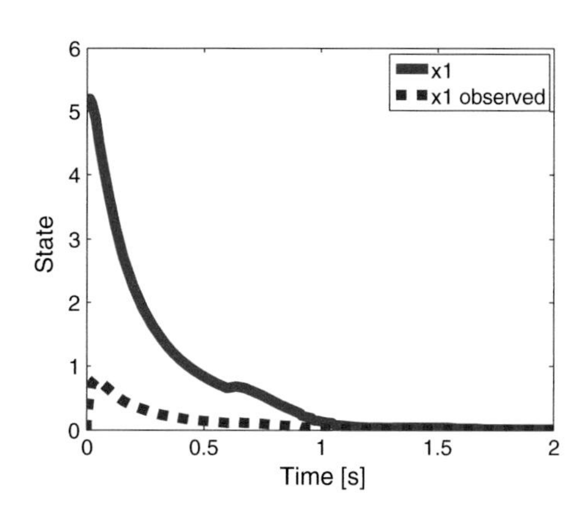


**Fig. 1** Time evolutions of state $x_1(t)$ and observed state $\hat{x}_1(t)$

**Fig. 2** Time evolutions of state $x_2(t)$ and observed state $\hat{x}_2(t)$

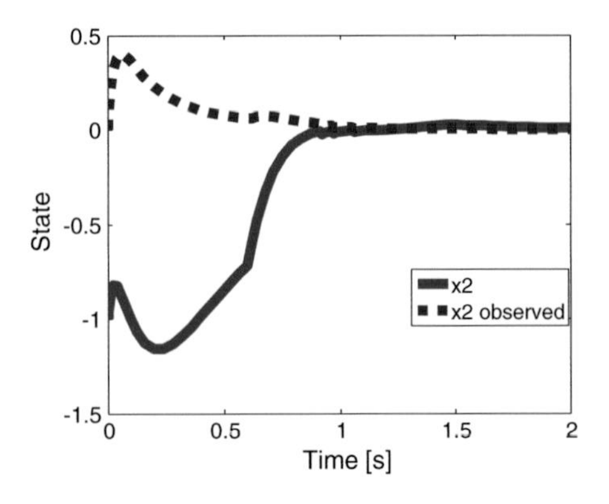


**Fig. 3** Time evolutions of the control inputs $u_1(t)$, and $u_2(t)$

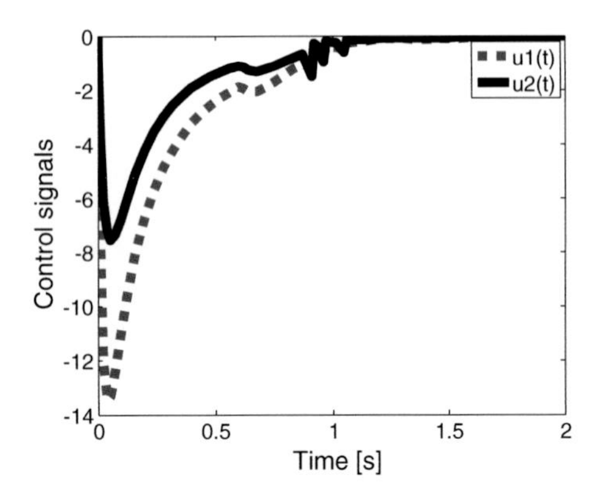


**Fig. 4** Time evolution of the switching signal, changing its value between 1 and 2, as it selects between region $\Omega_1$ and $\Omega_2$, respectfully

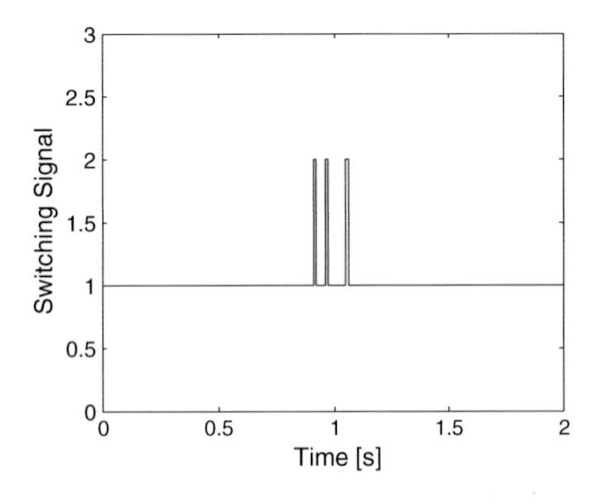


These simulation results demonstrate that the uncertain switching fuzzy time-delay system (23a–d) is asymptotically stabilized under the proposed design of output feedback controller and the appropriate switching law. In this way we showed the effectiveness of the proposed concept.

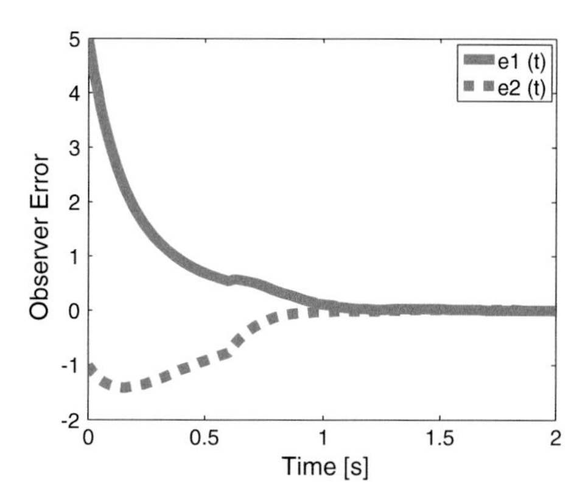


**Fig. 5** Time evolutions of the observer errors $e_1(t)$, $e_2(t)$

## 5 Concluding Remarks

On the grounds of fuzzy T-S model the problem of robust output feedback control for a class of uncertain switched fuzzy time-delay systems whose states are not measurable, hence not available, has been further explored and solved. Sufficient stability conditions and switching law are derived and reformulated as linear matrix inequalities. These are derived by using common Lyapunov function and multiple Lyapunov function approaches. The fuzzy output feedback controllers are designed by employing the strategy of Parallel Distributed Compensation. An illustrative example along with the respective simulation results demonstrates the effectiveness of the proposed control synthesis and the system performance in closed-loop achieved.

**Acknowledgments** The authors gratefully acknowledge the crucial contribution by Professor Georgi M. Dimirovski in proving the theoretical results reported in this article. Also, they are grateful to Dr. Yi Lin, Dr. Hong Yang, and Prof. Jun Zhao for their contributions during the early stage of this research endeavors as well as they thank to the respective university institutions in Shenyang, P.R. China, and in Skopje, R. Macedonia, for supporting their research on switched fuzzy systems.

## References

1. Aström, K.J., Albertos, P., Blanke, M., Isidori, A., Schaueflberger, W., Sanz, R. (eds.): Control of Complex Systems. Springer, London Heidelberg (2001)
2. Boyd, S., Ghaoui, E.L., Feron, E., Balakrishnan, V.: Linear Matrix Inequalities in System and Control Theory. The SIAM, Philadelphia (1994)
3. Branicky, M.S.: Multiple Lyapunov functions and other analysis tools for switched and hybrid systems. IEEE Trans. Autom. Control **43**(4), 475–482 (1998)
4. Gahinet, P., Nemirovski, A., Laub, A.J., Chilali, M.: LMI Control Toolbox. The Mathworks Inc, Natick

5. Hiroshi, O., Kazuo, T., Wang, H.O.: Switching fuzzy control for nonlinear system. In: Proceedings of the IEEE International Symposium on Intelligent Control, Houston, TX. Piscataway, NJ, pp. 281–286 (2003)
6. Hiroshi, O., Kazuo, T., Wang, H.O.: Switching fuzzy controller design based on switching Lyapunov function for a class of nonlinear systems. IEEE Trans. Syst. Man Cybern. B Cybern. **36**(1), 13–23 (2006)
7. Kazuo, T., Iwasaki, M., Wang, H.O.: Stability and smoothness conditions for switching fuzzy systems. In: Proceedings of the 19th American Control Conference, Chicago, IL, pp. 2474–2478 (2000)
8. Lee, H.J., Park, J.B., Chen, G.R.: Robust fuzzy control of nonlinear systems with parametric uncertainties. IEEE Trans. Fuzzy Syst. **9**(2), 369–379 (2001)
9. Liberzon, D., Morse, A.S.: Basic problems in stability and design of switched systems. IEEE Control Syst. Mag. **19**, 59–70 (1999)
10. Liberzon, D.: Switching in Systems and Control. Birkhauser, Boston Basel Berlin (2003)
11. Liu, Y., Dimirovski, G.M., Zhao, J.: Robust output feedback control for a class of uncertain switching fuzzy systems. In: Proceedings of the 17th IFAC World Congress, Seoul, KO, pp. 4773–4778 (2008)
12. Liu, Y., Zhao, J.: Robust Output Feedback Control of a Class of Uncertain Switching Fuzzy Systems", Report, Key Laboratory for Integrated Automation of Process Industry. Northeastern University, Shenyang (2008)
13. Lo, J.C., Lin, M.L.: Robust H∞ nonlinear modeling and control via uncertain fuzzy systems. Fuzzy Sets Syst. **143**(2), 189–209 (2004)
14. Momeni, A., Aghdam, A.G.: Switching control for time-delay systems. In: Proceedings of the 25th American Control Conference, Minneapolis, MN, pp. 5432–5434 (2006)
15. Nie, H., Zhao, J.: Hybrid state feedback H∞robust control for a class of time-delay systems with nonlinear uncertainties. Control Theory Appl. **22**(4), 567–572 (2005)
16. Ohtake, H., Tanaka, K., Wang, H.O.: Derivation of LMI design conditions in switching fuzzy control. In: Proceedings of the 43rd IEEE Conference on Decision and Control, Atlantis, Paradise Island, Bahamas, pp. 5100–5105 (2004)
17. Ohtake, H., Tanaka, K., Wang, H.O.: Switching fuzzy controller design based on switching Lyapunov function for a class of nonlinear systems. IEEE Trans. Syst. Man Cybern. Part B, Cybern. **36**(1), 13–23 (2006)
18. Ojleska, V.M., Kolemishevska-Gugulovska, T.D., Dimirovski, G.M.: Switched fuzzy control systems: exploring the performance in applications. In: Proceedings of the 4th European Symposium on Computer Modelling and Simulation, Pisa, Italy, pp. 37–42 (2010)
19. Ojleska, V.M., Kolemisevska-Gugulovska, T.D., Dimirovski, G.M.: Influence of the state space partitioning into regions when designing switched fuzzy controllers. Facta Universitatis Ser. Autom. Control Robot. **9**(1), 103–112 (2010)
20. Ojleska, V., Kolemishevska-Gugulovska, T., Rudas, I.J.: A robust output feedback control design for uncertain switched fuzzy systems. In: Proceedings of the IEEE 6$^{th}$ International Conference on Intelligent Systems - IS 2012, Sofia, Bulgaria, 2012, pp. 264–271
21. Palm, R., Driankov, D.: Fuzzy switched hybrid systems-modeling and identification. In: Proceedings of the IEEE International ISIC/ CIRA/ ISAS Joint Conference, Gaithersburg, MD, pp. 130–135 (1998)
22. Sun, Z.D., Ge, S.S.: Switched Linear Systems: Control and Design. Springer, New York (2004)
23. Tanaka, K., Iwasaki, M., Wang, H.O.: Switching control of an R/C hovercraft: Stabilization and smooth switching. IEEE Trans. Syst. Man Cybern. B Cybern. **31**(6), 853–863 (2001)
24. Tanaka, K., Wang, H.O.: Fuzzy Control System Design and Analysis: A Linear Matrix Inequality Approach. Wiley, Canada (2001)
25. Tong, S.C., Wang, T., Li, H.X.: Fuzzy robust tracking control design for uncertain nonlinear dynamic systems. Int. J. Approx. Reason. **30**, 73–90 (2002)
26. Tong, S.C., Li, H.X.: Observer-based robust fuzzy control of nonlinear systems with parametric uncertainties. Fuzzy Sets Syst. **131**, 154–165 (2002)

27. Wang, H.O., Tanaka, K., Griffin, M.: Parallel distributed compensation of nonlinear systems by Takagi and Sugeno's Fuzzy Model. In: Proceedings of the 4th IEEE International Conference on Fuzzy Systems - FUZZ-IEEE, 1995, pp. 531–538
28. Wang, H.O., Tanaka, K., Griffin, M.: An approach to fuzzy control of nonlinear systems: Stability and design issues. IEEE Trans. Fuzzy Syst. **4**(1), 14–23 (1996)
29. Xie, L.: Output feedback H∞ control of systems with parameter uncertainty. Int. J. Control **63** (4), 741–750 (1996)
30. Xie, L., De Souza, C.: Robust H∞ control for linear systems with norm-bounded time-varying uncertainty. IEEE Trans. Autom. Control **37**(8), 1188–1191 (1992)
31. Yang, H., Dimirovski, G.M., Zhao, J.: Stability of a class of uncertain fuzzy systems based on fuzzy control switching. In: Proceedings of the 25th American Control Conference, Minneapolis, MN, pp. 4067–4071 (2006)
32. Yang, H., Dimirovski, G.M., Zhao, J.: Switched fuzzy systems: Representation modeling, stability analysis and control design. In: Studies in Computational Intelligence 109, pp. 155–168. Springer, Berlin Heidelberg, DE (2008)
33. Yang, H., Dimirovski, G.M., Zhao, J.: A state feedback H∞ control design for switched fuzzy systems. In: Proceedings of the 4th IEEE International Conference on Intelligent Systems, Varna, BG, pp. 4.2–4.7 (2008)
34. Zadeh, L.A.: Inference in fuzzy logic. In: The IEEE Proceedings, vol. 68, pp. 124–131 (1980)
35. Zahirazami, S., Karuei, I., Aghdam, A.G.: Multi-layer switching structure with periodic feedback control. In: Proceedings of the 25th American Control Conference, Minneapolis, MN, 2006, pp. 5425–5431
36. Zhao, J., Spong, M.W.: Hybrid control for global stabilization of the cart-pendulum system. Automatica **37**(12), 1941–1951 (2001)
37. Zhao, J., Nie, H.: Sufficient conditions for input-to-state stability of switched systems. Acta Automatica Sinica **29**(2), 252–257 (2003)
38. Zhao, J., Dimirovski, G.M.: Quadratic stability of a class of switched nonlinear systems. IEEE Trans. Autom. Control **49**(4), 574–578 (2004)

# Multistep Modeling for Approximation and Classification by Use of RBF Network Models

**Gancho Vachkov**

**Abstract** In this chapter two multistep learning algorithms for creating Growing and Incremental Radial Basis Function Network (RBFN) Models are presented and analyzed. The first one is the algorithm for creating Growing RBFN models. It starts with a simple RBFN model that has only one RBF and then gradually increases the number of the RBF units until the predefined model accuracy is satisfied. A modified constraint version of the particle swarm optimization (PSO) algorithm with inertia weight is developed and used for tuning the parameters of the Growing RBFN model. It allows obtaining *conditional optimum* solutions within the user predefined boundaries that has real physical meaning. The second multistep learning algorithm creates Incremental RBFN models. It is in the form of a composite structure that consists of one initial linear sub-model and a number of incremental sub-models. Each of these sub-models gradually decreases the overall approximation error of the Incremental model, until a desired accuracy is achieved. The performance of both proposed RBFN models is analyzed and evaluated for nonlinear approximation of a synthetic test example. A real wine quality data set is also used for performance evaluation of the proposed Growing and Incremental RBFN models when used for solving nonlinear classification problems. A brief comparison of both models with the classical single RBFN model that has large number of parameters is conducted. It shows the merits of the Growing and Incremental RBFN models in terms of efficiency and accuracy.

G. Vachkov (✉)
School of Engineering and Physics, Faculty of Science,
Technology and Environment, The University of the South Pacific (USP),
Laucala Campus, Suva, Fiji Islands
e-mail: gancho.vachkov@gmail.com

 
V. Sgurev et al. (eds.), *Innovative Issues in Intelligent Systems*,
Studies in Computational Intelligence 623, DOI 10.1007/978-3-319-27267-2_11

# 1 Introduction

Radial Basis Function (RBF) Networks have been widely used in the last decade as a power tool in modeling and simulation, because they are proven to be universal approximators of nonlinear input-output relationships with any complexity [1–3]. The classical RBF Network (RBFN) is a typical case of a composite multi-input, single output model that consists of a predetermined number of $N$ RBFs, each of them performing the role of a local model [3, 4]. Then the aggregation of all the local models as a weighted sum of their local outputs produces the nonlinear output of the RBFN.

There are some important features that make the RBF networks different from the classical feed forward networks, such as the well known multilayer perceptron (MLP), often called back-propagation neural network [1, 2, 5]. The most important difference is that the RBFN models are not homogeneous in parameters. As it is discussed in the sequel, they have three different groups of parameters that need to be appropriately tuned by using different learning (oprtimization) algorithms. This makes the learning process of the RBFN more complex, because usually it is done in a sequence of two or three learning phases, each of them using different optimization strategy. Therefore the proper choice of the optimization strategy directly affects the accuracy and approximation ability of the produced model.

Another well known problem of the classical RBFN models is that they have usually a predefined structure, i.e. a preliminary fixed number of neurons (RBFs) has to be given by the user prior to the learning process. Obviously it is not straightforward to find or guess in advance the optimal number of the RBFs. Therefore a random selection of this number may lead to creating a RBFN model with insufficient accuracy (the case of insufficient number of RBFs) or to an overparameterized RBFN model with redundant complexity (the case of larger than needed number of RBFs).

In this chapter we propose two novel multistep learning procedures for creating two respective types of RBFN models, named *Growing* RBFN models and *Incremental* RBFN models. Despite the differences in their internal structure, both of them have the same significant property of being able to gradually improve the approximation accuracy with each subsequent learning step.

The multistep learning process of the Growing RBFN model starts with a RBFN model that has only one RBF unit and at each step of the algorithm a new RBF unit is inserted to the current RBFN model. This learning process leads to gradual increase of the model complexity (the number of RBFs), but at the same time the model accuracy is gradually improved.

The multistep learning processs for the Incremental RBFN model starts with creating a simple initial *linear sub-model*. Then at each subsequent learning step a new additional *incremental sub-model* is created in the form of a nonlinear RBFN model with small number of RBFs. When added to the layer structure of the current incremental model, the overall model accuracy is improved.

The rest of this chapter is organized as follows. In Sect. 2 we analyse the internal structure and the properties of the classical RBFN model with a pre-determined number of RBF units. Section 3 gives a brief explanation of the Particle Swarm Oprimization (PSO) algorithm with constraints [6, 7] that is further on used for tuning the parameters of the proposed Growing and Incremental RBFN models. The optimization criteria for creating of these multistep models, when used as nonlinear approximators and as nonlinear classifiers are discussed in Sect. 4, where two different quantization procedures are proposed.

Section 5 describes the proposed learning algorithm for creating the Growing RBFN model. Performance evaluation of this type of multistep model is done in Sect. 6 for the case when gthe the model is used as nonlinear approximator. Section 7 analyses the performance of the Growing RBFN model, when used as nonlinear approximator on a real data set for red wine quality.

Section 8 describes the proposed learning algorithm for creating the Incremental RBFN model, while Sects. 9 and 10 analyse the perormance of this type of model when used as nonlinear appriximator and nonlinear classifier, respectively. Finally Sect. 11 concludes the chapter with overall analysis and future work directions.

## 2 Basics of the Classical RBF Network Model

Our objective is to create a model of a real process (system) with $K$ inputs and one output based on a preliminary available collection of $M$ experiments (*input-output* pairs) in the form:

$$\{(\mathbf{X}_1, y_1), \ldots, (\mathbf{X}_i, y_i), \ldots, (\mathbf{X}_M, y_M)\} \tag{1}$$

Here $\mathbf{X} = [x_1, x_2, \ldots, x_K]$ is the vector consisting of all $K$ inputs for the system (process) under investigation. The measured output from the process is denoted by $y$. Then the predicted output, modeled by the RBF network is given, as follows:

$$y_m = f(\mathbf{X}, \mathbf{P}) \tag{2}$$

Vector $\mathbf{P} = [p_1, p_2, \ldots, p_L]$ represents the list of all $L$ parameters participating in the RBFN model. They will be explained and analyzed later in the text.

The classical RBFN model is described as a *three layer* structure, namely *input*, *hidden* and *output* layer as shown in Fig. 1.

The modeled output of the RBF network with preliminary given number of $N$ Radial Basis Functions is calculated as a weighted sum of the outputs from all $N$ RBFs:

$$y_m = w_0 + \sum_{i=1}^{N} w_i u_i \tag{3}$$

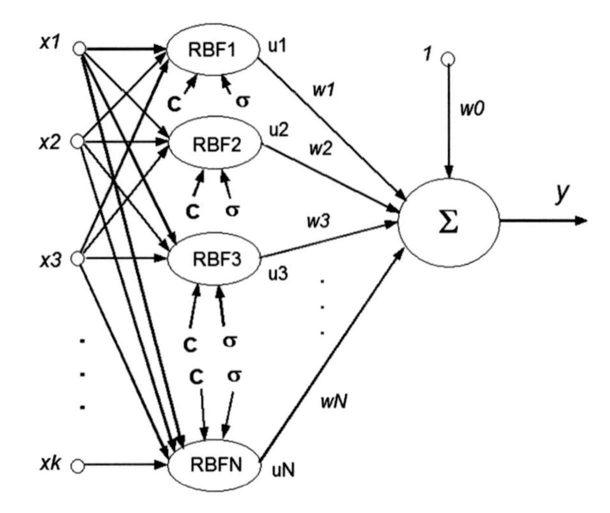


**Fig. 1** Structure of the classical radial basis function network with $K$ inputs and $N$ RBFs

Here $u_i, i = 1, 2, \ldots, N$ are the outputs of the RBFs with $K$ inputs $x_1, x_2, \ldots, x_K$ and $w_i, i = 0, 1, 2, \ldots, N$ are the *weights* associated with the outputs of the RBFs, including one additional *offset* weight $w_0$ as seen in the figure.

The output $u$ of each RBF is calculated by the following *Gaussian* function in the $K$-dimensional input space:

$$u = \exp\left(-\sum_{j=1}^{K} (x_j - c_j)^2 / (2\sigma^2)\right) \in [0, 1] \tag{4}$$

Is easy to realize that each RBF is determined by $K + 1$ parameters, namely the $K$-dimensional vector of the *Centers* (locations) $\mathbf{C} = [c_1, c_2, \ldots, c_K]$ of the RBFs in the $K$-dimensional space and the scalar *Width (spread)* $\sigma$. An example of a RBF with assumed scalar width of $\sigma = 0.15$ and a center $\mathbf{C} = [0.4, 0.6]$ in the two-dimensional space ($K = 2$) is shown in Fig. 2.

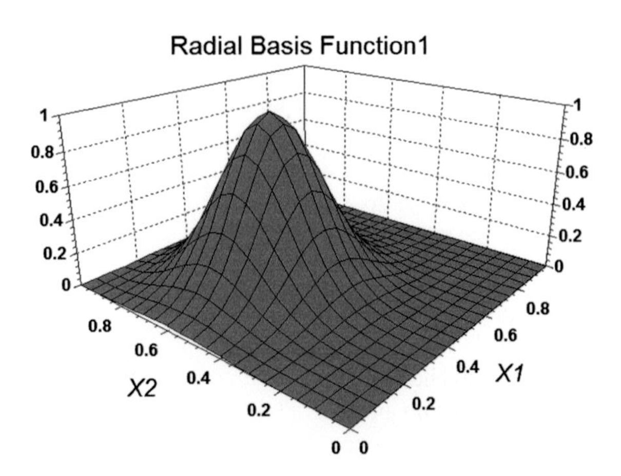


**Fig. 2** Example of a RBF with $K = 2$ inputs, center at [0.4, 0.6] and a single width $\sigma = 0.15$

It is straightforward to calculate the number $L$ of all parameters that characterize the classical RBF Network from Fig. 1 with $K$ inputs and $N$ RBFs, as follows:

$$L = N \times K + N + (N + 1) = N \times K + 2 \times N + 1 \tag{5}$$

All these parameters constitute the vector of *Parameters* **P** from (2). It is worth to note here that these parameters are *heterogeneous* in nature, unlike the parameters of the classical back-propagation neural network (the *multilayer perceptron*) where all parameters are homogeneous and represent one category—the connection weights between the units (neurons). In the classical RBF Network all $L$ parameters form 3 different sets (categories): the set **C** of *Centers* with $N \times K$ parameters; the set $\boldsymbol{\sigma}$ of all $N$ *Widths* (spreads) and the set **W** of $N + 1$ *Weights*, as follows:

$$\mathbf{P} = [p_1, p_2, \ldots, p_L] = \mathbf{C} \cup \boldsymbol{\sigma} \cup \mathbf{W} \tag{6}$$

It is seen from (5) that the number $L$ of all parameters in the RBFN model grows rapidly with increasing the size of the model, namely the number $N$ of RBFs and the number $K$ of the inputs. This creates the challenging task to alleviate the problem of computational complexity, which arises especially in RBFN models with large size. The multidimensional optimization problem becomes even more complex when the ultimate goal is to guarantee achieving the *global minimum* of the approximation error.

## 3 Tuning the Parameters of the RBF Network Model by Use of the Particle Swarm Optimization Algorithm with Constraints

The problem of tuning the parameters of the RBF network has been investigated by many authors for a long time by using different learning algorithms that include separate or simultaneous learning of the 3 groups of parameters [3–5] by applying different optimization approaches. In this chapter we use a nonlinear optimization strategy for *simultaneous* learning of all three groups of parameters in (6) and for this purpose we use a modified version of the Particle Swarm Optimization (PSO) algorithm with constraints, as described in the sequel.

The PSO algorithm [6–8] belongs to the group of the *multi-agent* optimization algorithms. It uses a heuristics that mimics the behavior of flocks of flying birds (particles) in their collective search for a food. The general concept here is that a single bird has not enough power to find the best solution, but in cooperation and exchanging information with other birds in the neighborhood it is likely to find the best (global) solution. The swarm consists of a predetermined number $n$ of particles (birds) that exchange information during their cooperative behavior. The details of the PSO algorithm are omitted here, but a very good overview of the PSO algorithm with its some of its popular modifications can be found in [7].

The classical version of the PSO algorithm in [6] does not impose any constraints (boundaries) on the search of the "birds" in the $K$-dimensional input space, so from a theoretical point of view it is an *unconstrained* optimization. The reason for such unconstrained assumption is that birds should be free and allowed to explore the whole unlimited space until eventually finding the global optimum.

However, many practical engineering problems require certain real constraints (limits) to the parameters $[x_1, x_2, \ldots, x_K]$ in the input space, in order to produce a physically meaningful optimal solution. This is in fact a *constrained optimization* procedure that produces the so called *conditional* optimum, which can be practically implemented.

In order to achieve such practical solutions by the PSO algorithm, we have made here a slight modification in the original version of the *PSO algorithm with inertia weight* [7] that uses predefined *minimum* and *maximum* constraints on all input parameters, as follows:

$$x_{i\min} \leq x_i \leq x_{i\max}, i = 1, 2, \ldots, K \tag{7}$$

The proposed modification of the PSO algorithm with constraints is very simple one. At each iteration of the algorithm, the new steps (velocities) for all particles (birds) in the input space are calculated, according to the original algorithm, but are not yet applied. Then these new *candidate* positions of the particles are checked against a possible *violation* of the given boundaries (7) of the input space. In case of no violation for a certain particle, it is allowed to be located to the new input space position. In the case of violation of (7), the particle's new position is put at the boundary (*minimum* or *maximum*) for this parameter, as follows:

$$\begin{aligned} &\mathit{if}\,(x_i < x_{i\min})\ \mathit{then}\ x_i = x_{i\min};\ i = 1, 2, \ldots K \\ &\mathit{if}\,(x_i > x_{i\max})\ \mathit{then}\ x_i = x_{i\max};\ i = 1, 2, \ldots K \end{aligned} \tag{8}$$

From a theoretical point of view, such strategy is actually a deviation from the original idea of the PSO for *free exploration* of the input space and instead keeps them to explore only the given restricted sub-area of the input space. In such way the particle will find eventually a plausible solution within this constraint sub-space and the benefit is that a solution with physical meaning and practicability will be found.

It is worth noting that sometimes not one, but a few iterations (trials) could be necessary for a certain particle to move away from the boundary and find a better location (if such exists) within the constrained sub-space. Then eventually this particle will find the *global optimum* within the constrained sub-space. In other cases, when the global optimum is outside the boundary, the particle will bounce back and forth repeatedly at the boundary, which means that it has found a *conditional optimum*.

It is worth saying here that the proposed PSO algorithm with constraints is by no means the only optimization tool that could be used for learning the parameters of the RBFN model. This is just our preferred choice in this chapter. Many other *meta-heuristic* methods and algorithms, such as those described in [9, 10] have their own merits and strengths in applications and therefore could be adapted for use in this case. Our main focus in this chapter is on developing concepts of different multistep algorithms for learning RBFN models subject to the use of proper and efficient optimization algorithm.

# 4 Optimization Criteria for Learning the RBFN Models

Depending on the purpose of using the RBFN model, there could be different optimization criteria. We describe below some of the most popular criteria used for nonlinear approximation and classification.

## *4.1 Optimization Criterion for Learning the RBFN Model When Used for Nonlinear Approximation*

When the RBFN model is used for performing a *nonlinear approximation* (also known as a *regression problem*), the most popular optimization criterion is to *minimize* the approximation error in the form of *rooted mean squared error* RMSE. This error takes into account the mean of the sum of all $M$ squared deviations (discrepancies) between the real measured output and the predicted output by the RBFN model, as follows:

$$RMSE = \sqrt{\frac{1}{M}\sum_{i=1}^{M}(y_i - y_{im})^2} \rightarrow min \tag{9}$$

Here the problem of learning the RBFN model is to properly optimize all $L$ parameters of the model from (6) so that to achieve the *global minimum* for the RMSE. This would produce the best approximation of the real experimental data.

## *4.2 Optimization Criteria for Learning the RBFN Model When Used for Classification*

For solving any kind of classification problem, a modification of the *statement-of-the problem* for learning the RBFN model should be done. The reason is that in the case of classification, the outputs of all $M$ samples (experiments) from the available training set are usually available as *integer numbers* that represent $T$ *ordered classes*, labeled as consequent numbers: $t = 1, 2, \ldots, T$.

Normally, the RBFN model produces real valued outputs, as given in (3). Therefore for classification purposes, this model should be transformed into a respective *Classifier* that is able to predict the class (an integer output), based on the given input vector with $K$ inputs (*features*). Therefore the Classifier can be viewed as a $K$-dimensional *Decision Making Unit*, which takes into account all $K$ real inputs (*features*) in order to make decision for the most proper integer output (the predicted *class*).

It is obvious that a kind of *quantization procedure* should be added to the output (3) of the RBFN model, so that its real valued outputs will be transformed into respective integer *predicted classes*.

There could be different ways to construct the quantization procedure, depending on the preliminary information and knowledge about the boundaries between the neighboring classes. Here we propose the following two quantization procedures:

*Quantization Procedure* 1:

*Initialization Step*. Define the *ranges* of the real values $\mathbf{R}_t = [B_{t\min}, B_{t\max}], t = 1, 2, \ldots, T$ that correspond to the respective classes, as follows:

$$B_{t\min} = (2t - 1)/2, \quad t = 2, 3, \ldots, T \tag{10}$$

$$B_{t\max} = (2t + 1)/2, \quad t = 1, 2, \ldots, T - 1 \tag{11}$$

Here different assumptions (heuristics) can be used for setting the *left* boundary of class 1 and the *right* boundary for class $T$, for example:

$$B_{1\min} = 0.0; \quad B_{T\max} = T + 1; \tag{12}$$

Then it is clear that the whole range $\mathbf{R}$ of the real output values from the RBFN model that correspond to all $T$ classes will be:

$$\mathbf{R} = \mathbf{R}_1 \cup \mathbf{R}_2 \cup \ldots \cup \mathbf{R}_T = [B_{1\min}, B_{T\max}] \tag{13}$$

The following two steps are executed for all $M$ samples: $i = 1, 2, \ldots, M$ from the training data set.

Step 1: Calculate the *predicted output* $y_i$ from the RBFN model (used as Classifier) for the given $i$-th sample;

Step 2: Calculate the integer *predicted class* $t_i^p$ based on the real output value $y_i$, produced by the RBFN model (the classifier), as follows:

$$if\ (B_{t\min} \le y_i < B_{t\max})\ then\ t_i^p = t, \quad t = 1, 2, \ldots, T \tag{14}$$

Step 3: Calculate the number $q$ of the *misclassified samples* for all $M$ available samples (experiments). This is a simple procedure of incrementing the number $q$ of misclassifications under the following condition:

$$if\ (t_i^p \neq t_i)\ then : q = q + 1; \quad i = 1, 2, \ldots, M \tag{15}$$

Here $t_i$ represents the *true* (given) class for the $i$-th sample in the training set, $i = 1, 2, \ldots, M$. Note that before the start of the calculations in Step 3, the number $q$ of misclassifications is set to *zero*.

The main characteristics of the above *quantization procedure* 1 is that here the boundaries for the real output values that are classified to a certain (integer) class are in the midpoint between the integer numbers of the neighboring classes, e.g. 1.5 and 2.5 for class 2; 3.5 and 4.5 for class 4 and so on.

A slightly different quantization procedure is proposed and given below. In fact the idea here is the same as in the quantization procedure 1, with the only difference that the left boundary for class 1 and the right boundary for class $T$ are open, but not constrained, as in (12).

*Quantization Procedure* 2:

*Initialization Step*: The predicted classes $t_i^p$ for all samples $i = 1, 2, \ldots, M$ are set to zero: $t_i^p = 0, i = 1, 2, \ldots, M$. which means that they are not yet defined. The number $q$ of all misclassified samples is also set to zero;

Step 1: Calculate the *predicted output* $y_i$ from the RBFN model (used as Classifier) for the given $i$-th sample;

Step 2: Calculate the *predicted class* $t_i^p$ for the $i$-th sample. This is done by comparison of the predicted real valued output $y_i$ with the respective available true class $t_i$ for this sample. The following sequence of calculations is performed here:

- Calculate the difference between the true class $t_i$ from the training set and the predicted real value $y_i$ from the RBFN model for the $i$-th sample: $\Delta_i = t_i - y_i; \quad i = 1, 2, \ldots, M$
- For classes $t = 2, 3, \ldots, T\text{-}1$ check the following:
  $if\ (\Delta_i > -0.5 \wedge \Delta_i \leq 0.5)\ then$ : the predicted class for the $i$-th sample is $t_i^p = t$;
- For class $t = 1$ check:
  $if\ (\Delta_i \leq 0.5)\ then$ : the predicted class for the $i$-th sample is $t_i^p = 1$;
- For *the last class* $t = T$ check:
  $if\ (\Delta_i > -0.5)\ then : the\ predicted\ class\ for\ the\ i\text{-}th\ sample\ is\ t_i^p = T;$

It is easy to notice that for each sample $i$, $i = 1, 2, \ldots, M$, only one of the above three conditions will be met, which means that the predicted class will be a unique integer number between 1 and $T$.

Step 3: Calculate the number $q$ of the *misclassified samples* for all $M$ available samples (experiments). This step is identical with *Step* 3 in the previous *quantization procedure* 1.

It is seen that both quantization procedures reach the step for calculating the number $q$ of the misclassified samples. This number is constrained between 0 (the ideal case of *perfect* classification) and $M$ (the worst case of *no correct* classifications).

The number $q$ of all misclassification for a given training set with $M$ samples is further used for calculating the *classification error Cerr*, as follows:

$$Cerr = q/M;\ 0 \leq q \leq M;\ 0.0 \leq Cerr \leq 1.0 \tag{16}$$

Obviously *Cerr* is a *real value* bounded between 0.0 and 1.0, which represents the fraction of all misclassified samples from the complete list of $M$ training samples. This error is used as p*erformance index* (criterion) that has to be minimized for learning the RBFN model as classifier.

In the sequel of the chapter two classification examples are given, namely a synthetic data set and a real data set. They are used for learning the proposed multistep RBFN models as classifiers, where the objective is to minimize the classification error *Cerr*.

## 5 The Concept of the Growing RBFN Model and Its Learning Algorithm

The simultaneous learning of all $L$ parameters of the classical RBFN model with predetermined number of $N$ RBFs has the serious disadvantage that there is no way to predict in advance the correct number of RBFs that will produce a model with the desired RMSE (9). Therefore it is a normal case to run several times the whole algorithm for tuning the RBFN model with different number *N of RBFs*, until achieving the right model with the desired accuracy (RMSE). This is a tedious process that leads to *redundant* and computationally expensive optimizations.

There is another serious demerit of the simultaneous learning of all $L$ parameters of the classical RBFN model. It is that the increase in the number $N$ of RBFs will linearly increase the number $L$ of the learning parameters, according to (5). As a result, the computation time for running the larger optimization task will also increase (even faster than linearly), because a larger number of iterations will be needed to get closer to the global optimum in the higher $L$-dimensional space.

Early works were reported in [11, 12] where the problem of simultaneous learning of a network with large number of parameters is replaced with the problem of learning a *growing network* that adds gradually a new computational unit at each growing step. In the sequel we describe our algorithm for creating a *Growing* RBF Network Model, which was firstly explained in our recent work [13]. It is experimentally shown there that this algorithm creates an economical model, in terms of computation time and reduced model complexity.

The main concept of the proposed Growing RBFN model is to avoid the simultaneous optimization of all parameters in the whole model. Here the original large size learning problem is decomposed into a sequence of smaller size optimizations that learn the parameters of the newly added RBF unit only. Thus the original large size optimization task is replaced by a sequence of smaller optimization sub-tasks.

The learning algorithm of the Growing RBFN model is a *multistep* procedure that starts with the simplest possible RBFN model consisting of only one ($N = 1$) RBF. It has a small number of parameters: $L = K + 2 + 1 = K + 3$ according to (5), namely: *K centers*, one *width*, one *weigh*t and the *offset weight*. These parameters are optimized by the PSO algorithm.

Then, if the desired accuracy *RMSEmin* is not yet achieved, the learning proceeds with another *step* of the algorithm by adding a *new* RBF to the current RBFN model. Thus $N$ becomes 2 and the number of all parameters will be: $L = 2 \times K + 2 \times 2 + 1$. This step is called *growing step* of the algorithm. It is clear that the new larger RBFN model has a higher complexity, however the PSO algorithm is run for only the parameters of the newly added RBF unit $L' = K + 2$, namely: *K centers*, one *width* and one *weight.* The parameters of the previous RBF unit (for $N = 1$) are kept at their optimized values in the previous step.

Again, if the desired approximation error is not yet achieved, another *growing step* is performed by adding one more RBF unit ($N = 3$) and the PSO algorithm is run for the new added $L' = K + 2$ parameters only. The process stops until the desired error is satisfied, that is: *RMSE* <= *RMSEmin.*

As a result, the learning algorithm automatically creates a Growing RBFN model with the desired accuracy and *minimal* number of RBFs, thus avoiding the problem of over-parameterizing.

## 6 Performance of the Growing RBFN Model for Nonlinear Approximation

A comparison of the standard learning algorithm for the classical RBFN model with the learning algorithm for the growing RBFN model was given in [13]. Here we summarize some of the results from comparing the two learning methods, based on the test nonlinear example shown in Fig. 3. This is a synthetic example with $K = 2$ inputs and one output, specially generated for testing purposes.

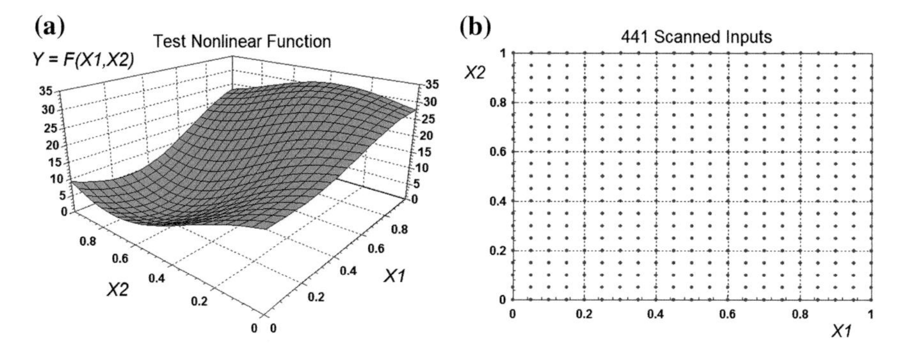


**Fig. 3** Test nonlinear example with two $K = 2$ inputs and the training set of evenly distributed $M = 441$ experimental data used for simulations

In order to partly eliminate the influence of the random nature of the PSO algorithm, both learning algorithms were run 10 times each, by using the same chosen tuning parameters of the PSO for both of them.

The standard RBFN model was learned with a fixed number of $N = 10$ RBF units, while the growing RBFN was learned with increasing number of the RBFs from $N = 1$ to $N = 10$ for each run.

The learning results from all 10 runs for the classical RBFN model are displayed in Fig. 4. In a similar way the results from learning of the growing RBFN model with up to $N = 10$ RBFs for all 10 runs are shown in Fig. 5. From this figure it is easy to notice the gradual and stable trend of decreasing the training error RMSE with increasing the number of participating RBFs in the growing RBFN model.

A comparison of the results from Figs. 4 and 5 also reveals that the growing RBFN model with a final number of 10 RBFs produces results with smaller deviation of the RMSE within all 10 runs, compared with the deviation in the results, produced by the classical RBFN model with the same number of 10 RBFs.

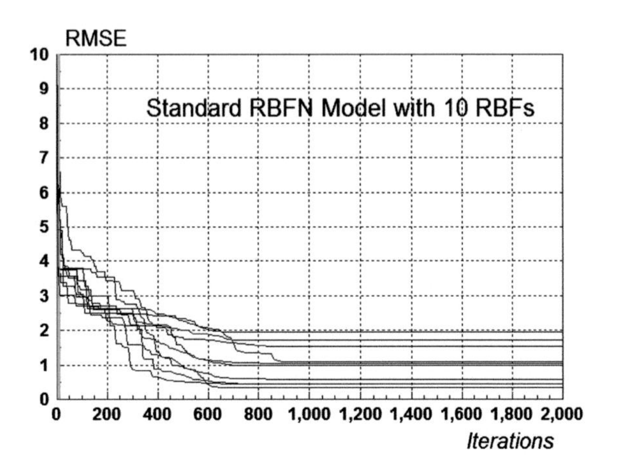


**Fig. 4** Learning results of the classical RBFN model after 2000 iteration and 10 runs

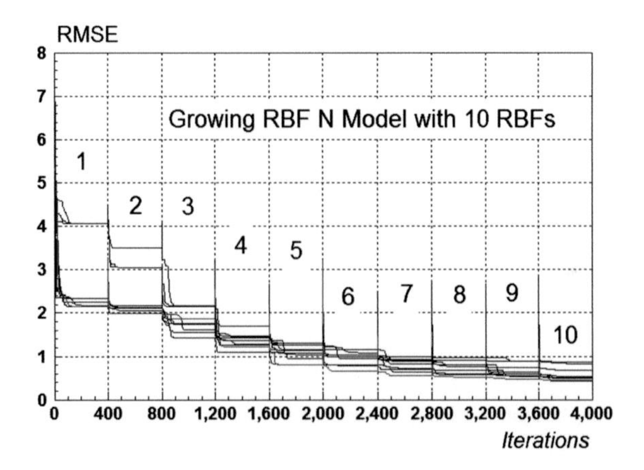


**Fig. 5** Learning results of the growing RBFN Model with 400 iterations for all 10 steps $N = 1, 2, \ldots, 10$ and for 10 runs with the same parameters of the PSO algorithm

In fact, the minimal obtained error RMSE from all 10 runs is similar for both models, but the mean RMSE for the growing RBFN model is $RMSE$ = 0.5785, which is much smaller than the mean RMSE for the growing RBFN model, which is $RMSE$ = 1.0196. In addition, the standard deviation of the RMSE for the classical RBFN model is 0.5641, while for the Growing model it is only 0.1556.

The above experimental results serve as a kind of an empirical proof that the learning process of the growing RBFN model has a stable trend of decreasing the RMSE. The obtained results from multiple runs of the Growing RBFN model have a smaller deviation of the RMSE, compared with the respective results from multiple runs of the Classical RBFN model.

In terms of computation cost both learning methods look similar, as follows. The average CPU time for the simultaneous learning of all $L = 10 \times 2 + 2 \times 10 + 1 = 41$ parameters of the classical RBFN model was about 51 s for 2000 Iterations with 1 GHz Intel CPU. In the case of learning the growing RBFN model, the total number of iterations was: $400 \times 10 = 4000$ and the CPU time was 52 s. However, this balance in the computation cost between the two learning algorithms may change rapidly in either direction, by changing the maximal number of iterations of the PSO for each step of the growing RBFN model.

# 7 Performance of the Growing RBFN Model in Solving Classification Problems

In this Section we analyze the experimental results from using the proposed growing RBFN model for solving classification problems, according to the explained *Quantization* Procedure 1 in Sect. 4, (10)–(15) and the criterion *Cerr* for calculating the *classification error* given in (16).

## *7.1 Solving the Classification Problem on a Test Nonlinear Example*

Here the same nonlinear test example as shown in Fig. 3 was used for classification purpose and $T = 3$ classes were artificially created by assuming the following levels of quantization for the output:

$$\begin{aligned} &\textit{if}\ (0 \le y < 10) \quad \textit{then Class} = 1; \\ &\textit{if}\ (10 \le y < 20) \quad \textit{then Class} = 2; \\ &\textit{if}\ (20 \le y < 35) \quad \textit{then Class} = 3; \end{aligned} \tag{17}$$

The surface of all these 3 classes in the 2-dimesional feature (input) space is shown in Fig. 6, while the boundaries between them is displayed in Fig. 7.

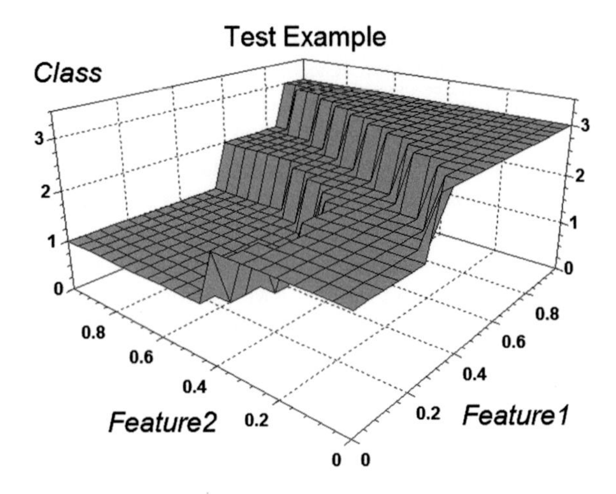


**Fig. 6** The 3-dimensional surface that shows the areas of the generated 3 classes

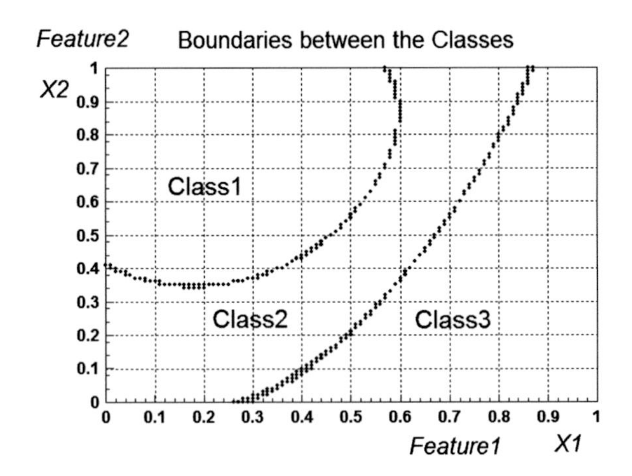


**Fig. 7** Boundaries between the three classes in the two-dimensional input space

The learning algorithm for the growing RBFN model was performed on a training set with 400 randomly generated samples in the 2-dimensional input space, as shown in Fig. 8. As a *Classifier*, the optimal Growing RBFN model with 8 RBFs was used, as shown from the training results in Fig. 9. Here the minimal classification error (16) was $Cerr = 0.0025$, which means *one* misclassification only.

The response surface of the optimal Classifier is shown in Fig. 10. The results from classification of two *test* data sets, consisting of 200 and 300 random samples are shown in the next Figs. 11 and 12 respectively.

It is worth noting that the popular *k*-nearest neighbors (kNN) algorithm has produced in average worse results for the same classification task with $k = 3$, 5 and 7, as follows:

Test Data Set with 200 *Samples*: 14, 14 and 11 misclassifications; Test Data set with 300 *samples*: 15, 10 and 14 misclassifications respectively.

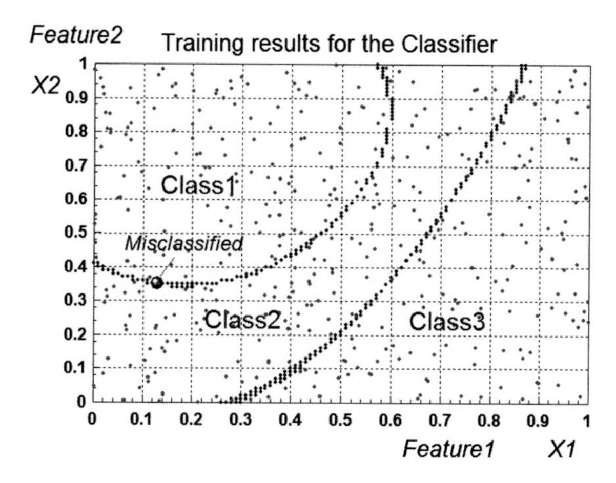


**Fig. 8** Training results of the classifier (growing RBFN model with 8 RBFs), based on 400 random samples

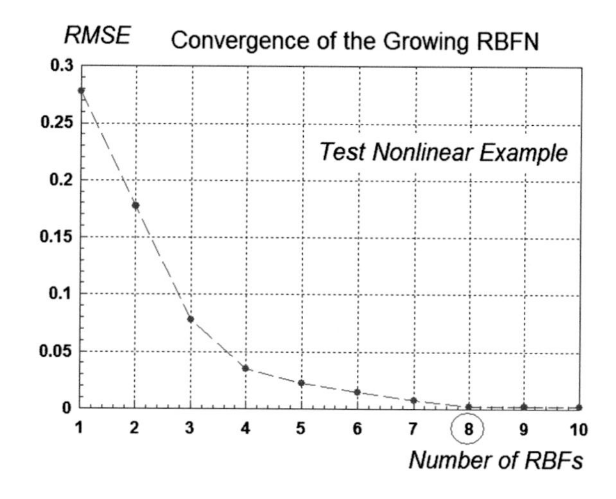


**Fig. 9** Monotonous decrease of the training error with increasing the number of the RBFs in the growing RBFN model

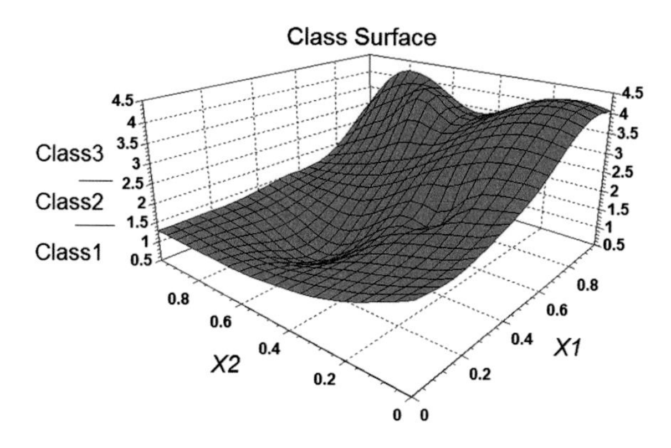


**Fig. 10** The response surface of the optimal classifier (the RBFN model with 8 RBFs)

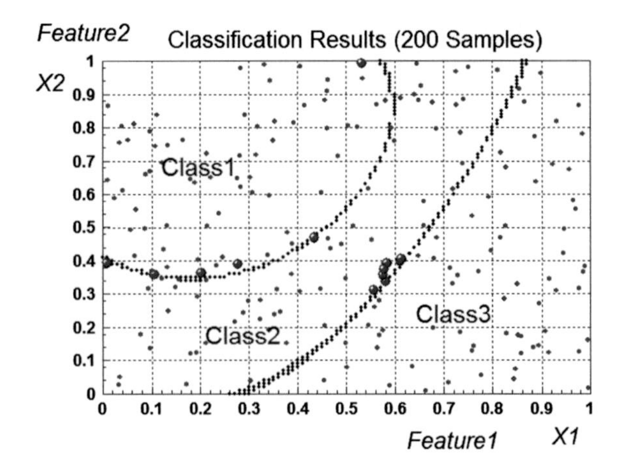


**Fig. 11** Classification results from a test set with 200 random samples with 12 misclassifications

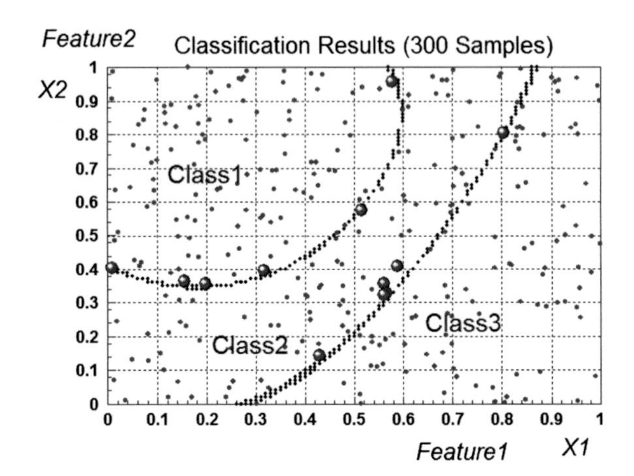


**Fig. 12** Classification results from the *test set* with 300 random samples, where 11 misclassifications were observed

## 7.2 *Solving the Classification Problem on the Example of Wine Quality Data Set*

In this sub-section we analyze the performance of the proposed incremental RBFN model for solving classification problems. In [14] two real datasets have been used for testing three different data mining methods for classification, namely the Multiple Regression (MR), Back-propagation Neural Networks (NN) and Support Vector Machines (SVM). These two datasets are related to the *red* and *white* variants of the Portuguese "*Vinho Verde*" wine. They comprise of 1599 samples for the red wine and 4898 samples for the white wine. The datasets are publicly available in *UCI Machine Learning Repository* [15]. The authors in [14] conclude that the SVM achieves the best classification performance with the least classification error.

In these data sets there are 11 input attributes that can be used as possible features for classification. They represent different real variables based on physicochemical laboratory tests. Some of the attributes are correlated, which encourages researchers to apply some techniques for feature selection before the actual classification.

The output variable (based on sensory data) is the *quality* of the wine, expressed as an integer value between 0 and 10. The worst quality is marked as 0 and the best wine quality is denoted by 10. The classes are ordered from 0 to 10, but are not balanced, i.e. there are much more normal wines than excellent or poor ones.

The histogram that shows the distribution of all classes for the complete data set for the red wine with 1599 samples is depicted in Fig. 13. It is seen from the figure that some classes are not represented by any samples. Therefore they have been omitted in our simulations and only the real existing classes: *3*, *4*, *5*, *6*, *7* and *8* have been taken into consideration. Then all the "active classes" *3*, *4*, *5*, *6* and *7* were renumbered as "new classes" with respective new labels: **1, 2, 3, 4, 5** and **6.** Their distribution can be also seen in Fig. 13.

Before training the proposed incremental RBFN model as a *nonlinear classifier* for this problem, we have conducted a simple *feature selection* procedure, as follows. From the whole list of 11 input attributes, the input 8 (wine density) was excluded, since it has very small variation throughout the whole range of samples [0.9901, 1.0037]. The remaining 10 inputs have been normalized in the range of [0.0, 1.0] and then used for trainning of the *nonlinear classifier* (the incremental RBFN model) and for classification of the test data into given number of 6 classes.

As for the inputs, we have omitted input 8 (wine density), since it has very small variation throughout the whole range of samples (between 0.9901 and 1.0037). Thus only the remaining 10 inputs were used as features for the classification problem with the new 6 classes.

The training curve for creating the Growing RBFN model is shown in Fig. 14. Here the first 1000 samples have been used as a *training set*. The minimal

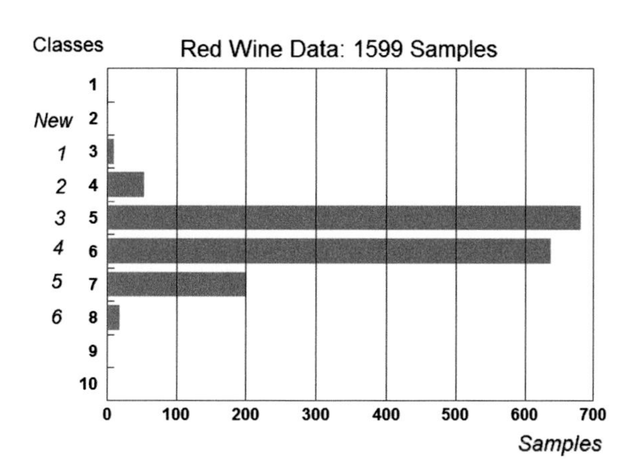


**Fig. 13** The histogram of all 6 new classes for the *Red Wine* data set with 1599 samples

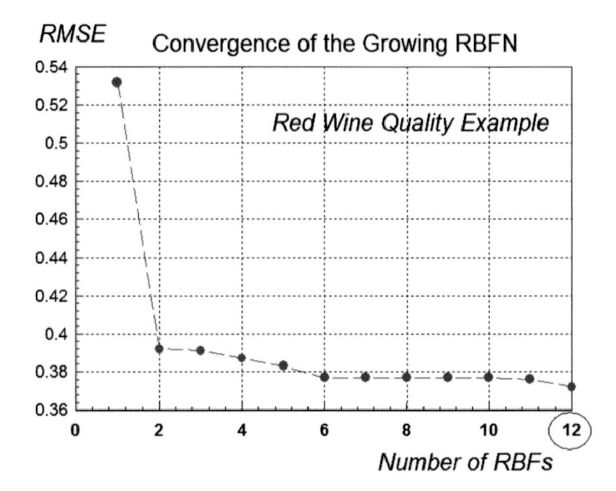


**Fig. 14** Gradual decrease of the training error RMSE during the learning process of the Growing RBFN model

classification error $Cerr = 0.3720$ was obtained with $N = 12$ RBFs. This translates into 372 misclassifications.

The above simulations show that the Data set with the red wine samples is much more "difficult" for classification than the test nonlinear example because it has a higher dimension and obviously some of the classes are overlapped. The performance of the Classifier, based on the Growing RBN model is slightly better than that one of the *kNN* algorithm for different assumed numbers of neighbors.

## 8 The Concept of the Incremental RBFN Model and Its Learning Algorithm

In this section we propose another multistep modeling concept for nonlinear approximation and classification based on the *Incremental* RBFN model. This is a kind of *composite model* with a *layer structure* that consists of several *sub-models*. The first sub-model is the *initial sub-model* and all others are called *incremental sub-models*.

The early ideas for creating of incremental type of models for the purpose of approximation have been explained in [11, 12]. Other ideas for creating incremental type of models are given in [16–18] with most applications for classification.

The difference between all those ideas and our proposed model here is that most authors are dealing with the problem of creating a single model that gradually improves its performance, when new examples are available, by appropriate change in its structure or parameteers. Others authors are interested in gradual (incremental) improvement of the model based on the same training data set, by gradual increase of the model complexity (inserting new units or changing the internal structure and parameters).

The proposed here incremental RBFN model is created by applying a sequential multistep identification procedure where each identification step defines one *sub-model*. The only available information before the start of the identification is the training data set (1) consisting of $M$ input-output samples.

The block diagram that visualizes the multistep identification procedure for creating the incremental RBFN model is shown in Fig. 15. It depicts the case of creating an Incremental RBFN model with *one initial* sub-model, denoted as $MOD0$ and *two incremental* sub-models, denoted as $MOD1$ and $MOD2$.

It is worth noting that the initial sub-model $MOD0$ is a compulsory first element in the structure of the composite incremental RBFN model, while the number of the subsequent incremental sub-models $MOD1$, $MOD2$, … is variable. This number can be predefined by the user or it can be found automatically during the multistep identification, based on the desired identification accuracy.

The general idea of creating the composite incremental model is that each new sub-model that is being added to the structure of the current incremental model will contribute to decreasing the approximation error RMSE (9) of the whole model with the current collection of already identified sub-models. The process of adding additional incremental sub-model continues until the desired value for the RMSE is obtained, or until the predefined number of incremental sub-models by the user is reached.

It is obvious that in this multistep procedure of subsequent identifications, the complexity of the whole incremental RBFN model and the number of all parameters in the model will gradually increase. However, the benefit here is that the computational cost is still low, because each identification step is performed on one

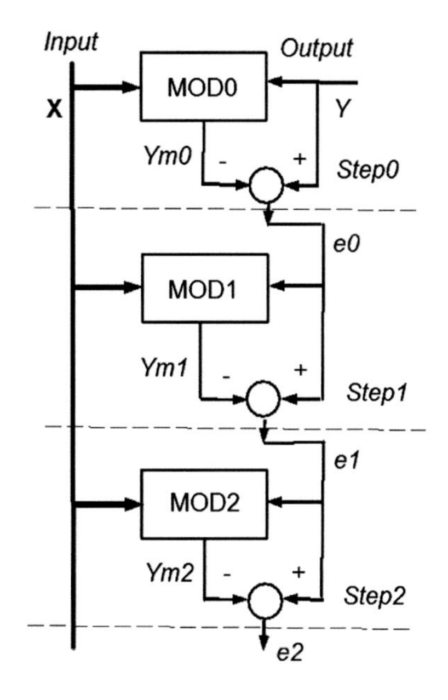


**Fig. 15** The multistep identification procedure for creating the incremental RBFN model

sub-model only that has a *small size* (small number of parameters). Since the identification process is actually an optimization procedure, this means that at each step, a small size optimization problem will be solved, compared to the need of solving *full scale* optimization problem that uses all model parameters, in the case of creating "one-step" full-size model.

## 8.1 The Initial Identification Step

At the initial identification step we create a relatively rough model that is aimed at capturing the most basic (general) input-output relationships. It is obvious that the best and yet simple model for such purpose would be the *linear sub-model*, denoted as *MOD*0 in Fig. 15. In case of a process with *K*-dimensional input vector **X**, this model has a total number of $K + 1$ parameters and is represented by the following equation:

$$y_{im0} = a_0 + \sum_{j=1}^{K} a_j x_{ij}, \; i = 1, 2, \ldots, M \tag{18}$$

According to the notations in Fig. 3 the model error after the *initial* identification step will be:

$$e_{i0} = y_{im0} - y_i, \quad i = 1, 2, \ldots, M \tag{19}$$

A graphical illustration of the result of the *initial* identification step is depicted in Fig. 16. The approximation error (9) at this step is denoted as *RMSE*0 and used as a measure of the model quality at this initial step.

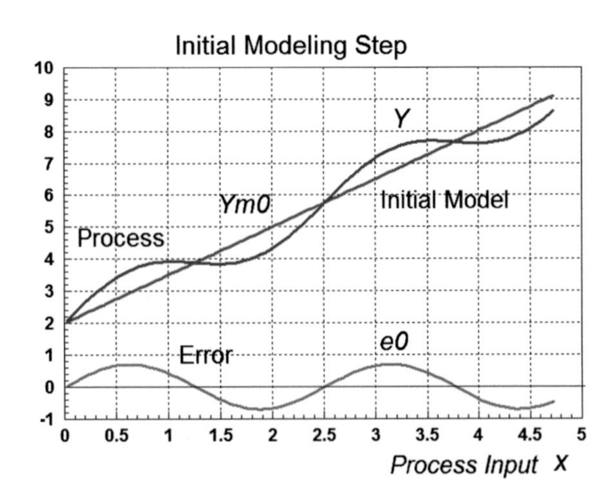


**Fig. 16** Illustration of the variables ivolved in the initial identification step

The errors (19) for all $M$ experiments create the so called error vector **E0**, which is saved for further use in the next incremental identification step.

## 8.2 The Subsequent Incremental Identification Steps

In the *first* incremental identification step one additional *sub-model MOD*1 is created, according to Fig. 15. This could be any kind of relatively simple model, but it is preferably to be a *non-linear* model that has a better approximation capability over the linear models. Further on we propose the use of RBFN models with a relatively small number of RBFs (e.g. $N = 3, 4, 5$) as *incremental sub-models*, because they are both relatively simple and nonlinear in nature.

In the *first* incremental identification step, the inputs are taken from the vector **X** of the original inputs, while the outputs needed for the identification are taken from the error vector **E0,** produced in the previous initial identification step.

As a result, the following new errors are obtained at the end of the first identification step:

$$e_{i1} = y_{im1} - e_{i0}, \quad i = 1, 2, \ldots, M \tag{20}$$

The new approximation error (9) obtained from the whole incremental model is denoted as *RMSE*1. The errors (20) for all $M$ experiments from this step create the error vector **E1** that is saved for further use in the next identification step.

In a similar way, at the *second* incremental identification step a new incremental *sub-model MOD*2, according to Fig. 15 is created with the following new errors:

$$e_{i2} = y_{im2} - e_{i1}, \quad i = 1, 2, \ldots, M \tag{21}$$

Here the new approximation error (9) obtained from the whole incremental model is denoted as *RMSE*2. The errors (21) for all $M$ experiments from this step create the error vector **E2** that is saved and used for the next identification step.

An illustration of the above explained two incremental identification steps is given in Figs. 17 and 18.

In a summary, if we like to create an incremental RBFN model with $r$ steps ($r > 0$), the result will be a composite model that consists of $r + 1$ sub-models: one linear sub-model *MOD*0 and $r$ non-linear RBF *sub-models MOD*1, *MOD*2, …, *MODr*. Then the apprroximation error *RMSE* will be monotonously decreasing function with each incremental step, as follows:

$$RMSE0 \geq RMSE1 \geq RMSE2 \geq \ldots \geq RMSEr \tag{22}$$

For all identification steps in this multistep procedure we use the PSO algorithm with constraints, as explained in Sect. 4.

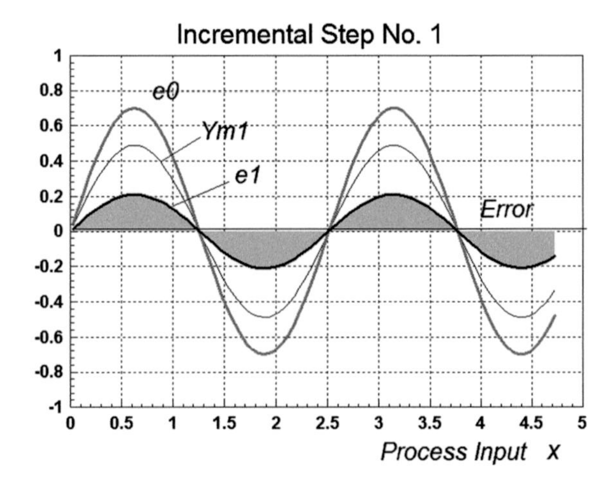


**Fig. 17** Illustration of the variables from the first incremental step

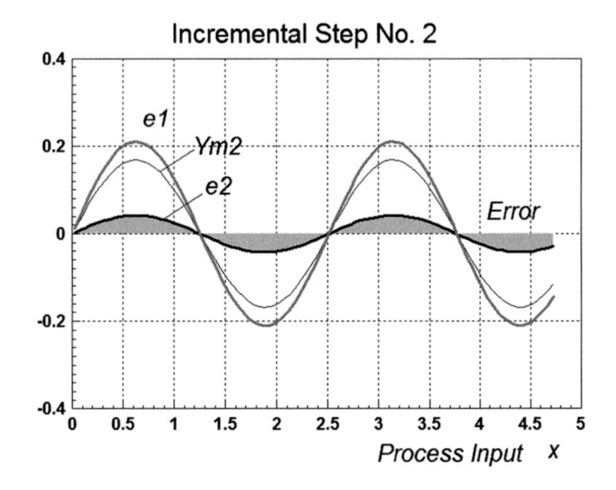


**Fig. 18** Illustration of the variables from the second incremental step

## 8.3 Calculation Procedure of the Incremental RBFN Model

From the above Eqs. (18)–(21) it is straightforward to conclude that for a given input **X**, the final output of the incremental RBFN model will be an *additive function* of the outputs of all $r + 1$ sub-models, included in its structure, i.e.

$$y_{m\,i} = \sum_{j=0}^{r} y_{i\,mj}, i = 1, 2, \ldots, M \tag{23}$$

The calculation structure of the incremental RBFN model is illustrated in Fig. 19. It is seen that in order to calculate the modeled output from the Incremental RBFN model, we need to keep separately all $r + 1$ sub-models and

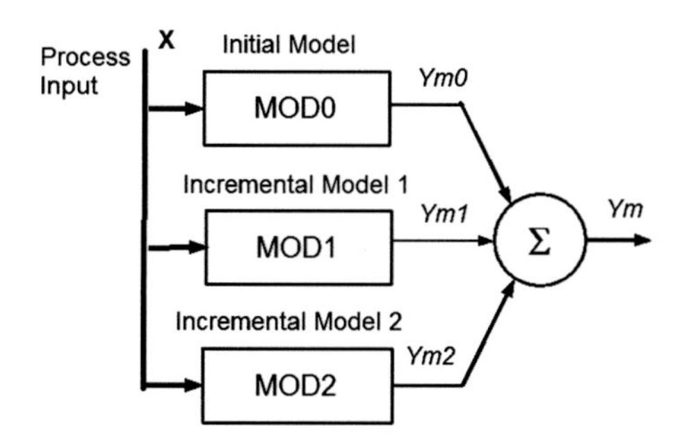


**Fig. 19** The procedure for calculating the composite incremental RBFN model

combine them according to (23). This multilayer structure gives certain flexibility of the whole incremental model, because it can be amended easily at any time by adding another preliminary identified sub-model to the current structure.

# 9 Performance of the Incremental RBFN Model for Nonlinear Approximation

The same synthetic nonlinear example from Sect. 6 and Fig. 3 has been used here for performance analysis of the Incremental RBFN model as nonlinear approximator.

A training set of $M = 400$ randomly generated inputs and the respective calculated outputs is shown in Fig. 20 and used for creating of the incremental RBFN model. Figure 21 shows the response surface of the identified *initial* linear *sub-model MOD*0 and Fig. 22 depicts the error surface after the initial identification step with approximation error *RMSE*0 = 3.4491.

Figure 23 is an illustration of the approximation capability of the incremental RBFN model consisting of the initial linear sub-model *MOD*0 and *one* only

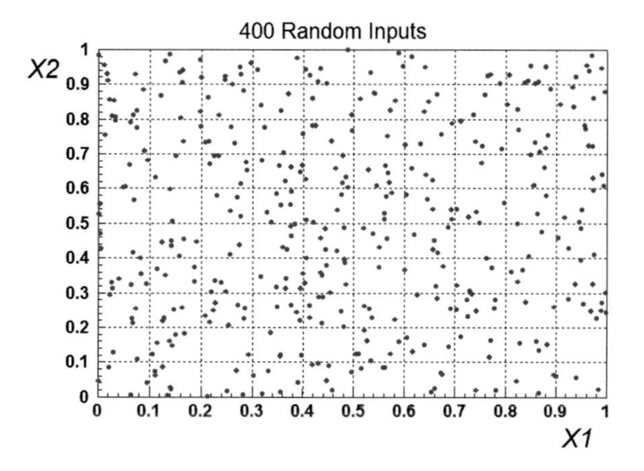


**Fig. 20** The training set of 400 random generated inputs for the test nonlinear example, used in the simlations

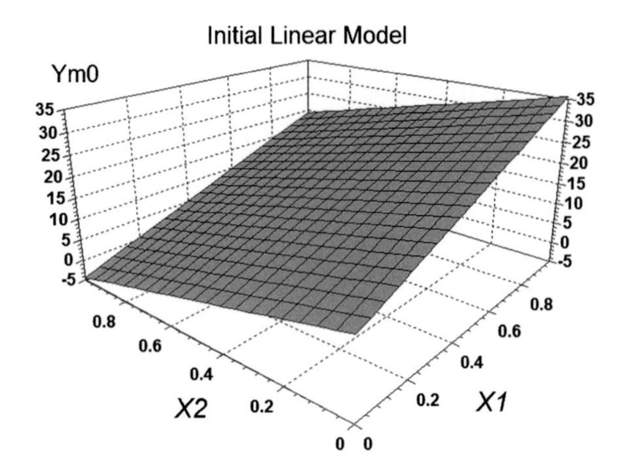


**Fig. 21** The response surface of the initial linear model *MOD*0. The approximation error at this initial step is: *RMSE*0 = 3.4491

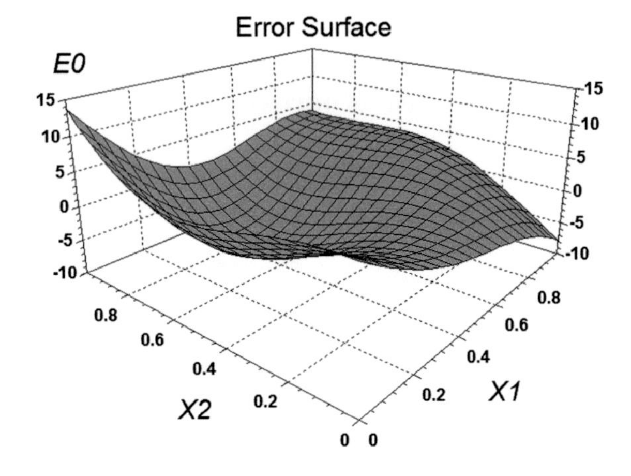


**Fig. 22** The response surface of the error vector **RMSE0** after the initial identification step

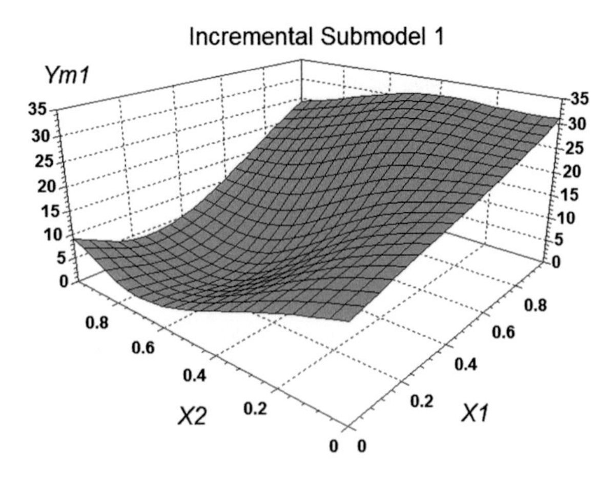


**Fig. 23** Response surface of the sub-model *MOD*1 created at *step* 1 of the incremental identification. The approximation error is: *RMSE*1 = 0.5267

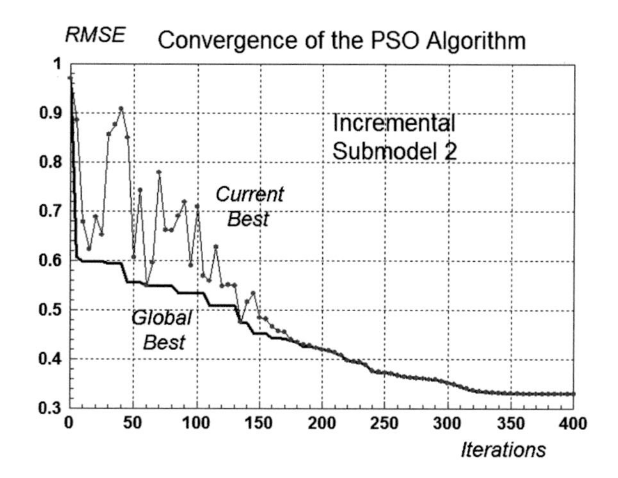


**Fig. 24** Convergence results for *RMSE*2 from the PSO algorithm used for training the incremental sub-model *MOD*2 at *step* 2

incremental sub-model, i.e. *MOD*1. Here the approximation error drops significantly from *RMSE*0 = 3.4491 to *RMSE*1 = 0.5267. The response surface obtained from the incremental RBFN model with one incremental sub-model is shown in Fig. 23. When comparing Fig. 23 with Fig. 3a that shows the original nonlinear example, it is easy to realize the high level of *similarity* between them.

Adding another incremental sub-model *MOD*2 in step 2 ($r = 2$) leads to further sharp decrease of the approximation error to *RMSE*2 = 0.3207. The fast and stable convergence of the PSO algorithm in this identification step is depicted in Fig. 24.

Finally, Fig. 25 show the simulation results from comparison of the performance of three different types of incremental RBFN models. They use the same initial linear sub-model and three different incremental sub-models with 3, 4 and 5 RFBs respectively.

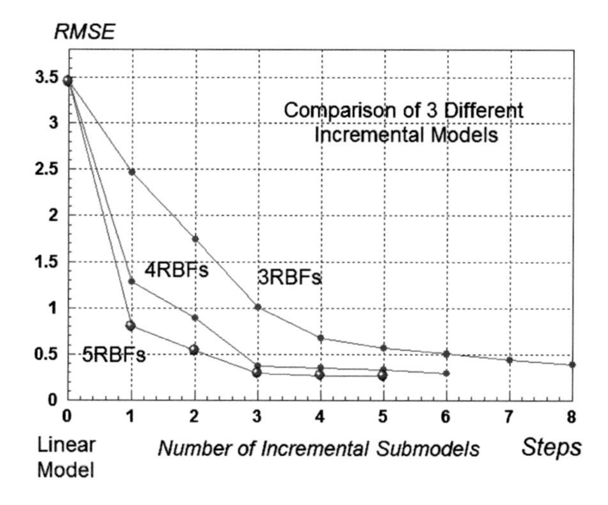


**Fig. 25** Comparison results of three incremental RBFN models that use different incremental sub-models with 3, 4 and 5 RBFs respectively

As expected, the more complex sub-models with 5 RBFs have produced results with better (smaller) approximation error, which resulted in a faster convergence of the RMSE, with using just a few ($r = 3, 4$) incremental steps.

For comparison, the simpler sub-models with 3 and 4 RBFs needed more identification steps (up to $r = 8$ shown here) to achieve a similar level of approximation error. This means that basically the same approximation error could be achieved in two different ways, namely by using smaller number, but more complex sub-models, or by using a larger number, but simpler sub-models in the incremental structure. Obviously a *trade-off* decision should be made, considering two additional factors: the CPU time and the total number of parameters of the model.

Another comparison of the approximation results of the incremental RBFN model with the results from a single large RBFN model with similar number of parameters was made. Here a single (classical) RBFN model with 20 RBFs was used with a total number of parameters: $20 \times 2 + 20 + 21 = 81$, according to (5). The respective number of parameters of the 3-step ($r = 3$) incremental RBFN model, consisting of one *linear* and 3 *nonlinear* sub-models with 5 RBFs is: $3 + [5 \times 2 + 5 + 6] * 3 = 3 + 21 * 3 = 63 + 3 = 66$.

The results from this comparison showed a clear advantage of the proposed incremental RBFN model, as follows: the approximation error *RMSE* of the single classical RBFN model was within the range [0.5767, 1.0370] for 6 different runs of the PSO algorithm with slight change in the tuning parameters at each run. The respective RMSE obtained from the incremental RBFN model with $r = 3$ sub-models was much smaller, namely: *RMSE* = 0.2665.

The above simulations serve as an experimental proof for the better average *accuracy* and *efficiency* (in term of a CPU time) of the proposed incremental RBFN model, compared with the classical single RBFN model. However we should note here that this is not a rigorous proof and may lead to slight variations in the results, because of the stochastic nature of the PSO algorithm and with different initial selection of the PSO parameters.

## 10 Performance of the Incremental RBFN Model in Solving Classification Problems

In this section we analyze the performance of the proposed incremental RBFN model for solving classification problems. The same Red Wine Data Set from [16], consisting of 1598 samples with 11 original input attributes, as explained in Sect. 7 is used to assess the classification ability of the proposed Incremental RBFN model.

The histogram that shows the distribution of all 6 *active* classes for the data set of red wine with 1599 samples has been already depicted in Fig. 15. All other preprocessing details are the same as used in Sect. 7.

For solving this classification problem, the *Quantization Procedure* 2 from Sect. 4 was used in conjunction with the Incremental RBFN model. The first 1200 samples from the entire data set of 1599 samples were used as a *training set* for creating the nonlinear classifier. The remaining 399 samples were used as a *test data set* to assess the classification results.

Here two Incremental RBFN models were generated and used for classification, as follows: a model consisting of one linear model and one incremental submodel ($r = 1$), as well as a model with one linear sub-model and two incremental sub-models ($r = 2$). The following classification results were obtained: the initial *linear sub-model* ($r = 0$) achiewed an overall classificationn error of $Cerr0 = 0.6307$ (757 misclassifications), according to the criterion (16). The incremented RBFN model with one additional sub-model ($r = 1$) achieved an improved classification error of $Cerr1 = 0.4017$ (482 misclassifications). Adding more identification steps ($r = 2, 3\ldots$) in the incremental RBFN model almost did not improve the clasification results, which suggests that we are around the limits of the classification accuracy for this "difficult" real data example.

The parameters of the PSO algorithm used for identifying the Incremental RBFN models with $r = 1$ and $r = 2$ were, as follows, according to the notations in [7]: Number of *particles*: $n = 20$; Initial *inertia weight*: $\omega_1 = 0.95$; Final *inertia weight*: $\omega_2 = 0.25$; *Acceleration* coefficients: $\psi_1 = 1.80$; $\psi_2 = 1.70$; Maximal size of the *velocity* (step): $\upsilon_{max} = 0.20$; Total number of *iterations*: $ITmax = 2200$;

When aplying the incremental RBFN model as a nonlienar classifier for the *test data set* (the remaining 399 samples) the classification error has slightly increased to $Cerr = 0.4812$ (192 misclassifications).

A straightforward comparison of gthe above classification results with those of the authors in [16] is difficult to be made, because we have excluded one of the input attributes (input 8) and also we have not used a *k*-fold cross-validation strategy for classification. Instead, our classification results are based on only one division of the whole data into a *training* data set with 1200 samples and a *test* data set with the remaining 399 samples.

The obtained classification results are promissing for the capability of the Incremental RBFN model, when used as a nonlinear classifier, because the classification error is very close to the *mean absolute deviation* $MAD = 0.46$, obtained in [16]. We would like to note here that the main purpose in our simulations was to show the applicability of the incremental RBFN model and to prove experimentally that its accuracy is gradually improved when adding more incremental sub-models to its structure.

## 11 Conclusions

There are obviously some common features and properties of the proposed two multistep modeling structures in this chapter. The first one is that the Growing and Incremental RBFN models use the same type of building blocks for their structure, namely the RBF units that serve as local sub-models.

The second common feature of the Growing and Incremental RBFN models is that both of them have a flexible structure that enables the number of the RBFs to grow during the learning process. As a result both models have the capability of gradually improving the model accuracy until the user predefined accuracy is achieved. This is in fact very useful feature in preventing the models from becoming over-parameterized, since the learning process automatically stops at the most appropriate model structure.

Another common feature of both multistep models is that they use the same optimization strategy for simultaneously tuning of all three groups of parameters: the *centers* and *spreads* of the RBFs and the *weights*. Here the essential detail is that the optimization is done separately for each modeling step over a small number of parameters that correspond to the currently added sub-model or RBF unit. Thus the overall complex problem of large scale optimization is decomposed into a series of smaller size optimization sub-problems.

There is also a distinct difference in the concept of creating the two multistep RBFN models, as follows. The Growing RBFN model is a single model with a network structure consisting of RBF units. This model can gradually grow in complexity with inserting of a new RBF unit at each step of the learning process. Once we stop the growing of the model after fixed number of steps, the growing model is completed and cannot be further amended.

In contrast, the Incremental RBFN model is a *composite multilayer* model, which means that it consists of layers of sub-models, starting with one *linear* sub-model and a fixed number of nonlinear *incremental* sub-models. The essential point here is that each of these sub-models represents a single layer, which is kept as a separate local sub-model. Then, if we would like to create a model with a better accuracy, we can do this easily by learning one new sub-model and adding it to the current multilayer structure. In other words, the incremental RBFN models possess more flexible structure.

The future directions of this research are aimed at solving some currently existing problems with the proposed multistep RNBFN models. First, the problem of automatic selection of the tuning parameters for the PSO algorithm should be solved in a satisfactory way, so that to decrease the variations in the optimization results between the different runs. This would also increase the probability of finding the real global optimum solutions. Second, a more detailed and elaborated method is needed for possible automatic selection of the boundaries between the different classes, when the multistep RBFN models are used as nonlinear classifiers. This could significantly improve the classification accuracy, especially when dealing with large and noisy data sets.

## References

1. Poggio, T., Girosi, F.: Networks for approximation and learning. Proc. IEEE **78**, 1481–1497 (1990)
2. Park, J., Sandberg, I.W.: Approximation and radial-basis-function networks. Neural Comput. **5**, 305–316 (1993)

3. Musavi, M., Ahmed, W., Chan, K., Faris, K., Hummels, D.: On the training of radial basis function classifiers. Neural Networks **5**, 595–603 (1992)
4. Yousef, R.: Training radial basis function networks using reduced sets as center points. Int. J. Inf. Technol. **2**, 21–27 (2005)
5. Peng, J.-X., Li, K., Irwin, G.W.: A novel continuous forward algorithm for RBF neural modeling. IEEE Trans. Autom. Control **52**(1), 117–122 (2007)
6. Eberhart, R.C., Kennedy, J.: Particle swarm optimization. In: Proceedings of the IEEE International Conference on Neural Networks, pp. 1942–1948. Perth, Australia (1995)
7. Poli, R., Kennedy, J., Blackwell, T.: Particle swarm intelligence. An overview. Swarm Intell. **1**, 33–57 (2007)
8. Zhang, J.-R., Zhang, J., Lok, T., Lyu, M.: A hybrid particle swarm optimization, back-propagation algorithm for feed forward neural network training. Appl. Math. Comput. **185**, 1026–1037 (2007)
9. Mashinchi, M.H., Orgun, M.A., Pedrycz, W.: Hybrid optimization with improved tabu search. Appl. Soft Comput. **11**(2), 1993–2006 (2011)
10. Crepinsek, M., Liu, Shih-Hsi, Mernik, M.: Exploration and exploitation in evolutionary algorithms: a survey. ACM Comput. Surv. **35**(3), 35 (2013)
11. Platt, J.C.: A resource-allocation network for function interpolation. Neural Comput. **3**(2), 213–225 (1991)
12. Fritzke, B.: Fast learning with incremental RBF networks. Neural Process. Lett. **1**(5), 1–5 (1994)
13. Vachkov, G., Sharma, A.: Growing radial basis function network models. In: Proceedings of the IEEE Asia-Pacific World Congress on Computer Science and Engineering, APWC on CSE, pp. 111-116. Plantation Island Resort, Fiji Islands, 4–5 Nov 2014
14. Cortez, P., Cerdeira, A., Almeida, F., Matos, T., Reis, J.: Modeling wine preferences by data mining from physicochemical properties. Decis. Support Syst. (Elsevier) **47**(4), 547–553 (2009) ISSN: 0167-9236
15. Asuncion, A., Newman, D.: UCI Machine Learning Repository. University of California, Irvine (2007). http://www.ics.uci.edu/_mlearn/MLRepository.htm
16. Yamauchi, K., Yamaguchi, N., Ishii, N.: Incremental learning methods with retrieving of interfered patterns. IEEE Trans. Neural Networks **10**(6), 1351–1365 (1999)
17. Okamoto, K., Ozawa, S., Abe, S.: A fast incremental learning algorithm of RBF networks with long-term memory. In: Proceedings of the International Joint Conference on Neural Networks, pp. 102–107. USA (2003)
18. Ozawa, S., Tabuchi, T., Nakasaka, S., Roy, A.: An autonomous incremental learning algorithm for radial basis function networks. J. Intell. Learn. Syst. Appl. **2**(4), 179–189 (2010)

Printed in the United States
By Bookmasters